QUALITY COSTS: IDEAS & APPLICATIONS, VOLUME 2

A Collection of Papers

QUALITY COSTS: IDEAS & APPLICATIONS, VOLUME 2

A Collection of Papers

ASQC Quality Costs Committee

Jack Campanella, Editor

Quality Press
Milwaukee

QUALITY COSTS: IDEAS & APPLICATIONS, VOLUME 2

A Collection of Papers

ASQC Quality Costs Committee

Jack Campanella, Editor

Library of Congress Cataloging-in-Publication Data
Quality costs : ideals and applications
    p. cm.
  Includes bibliographical references.
  ISBN 0-87389-047-7 (v. 2)
  1. Quality control — Costs. I. Campanella, Jack
TS156.Q3626 1989
658.5'62 — dc20

Copyright © 1989 by ASQC Quality Press

All rights reserved. No part of this book may be reproduced in any form or by any means, electronic, mechanical, photocopying, recording, or otherwise, without the prior permission of the publisher.

10 9 8 7 6 5 4 3 2 1

ISBN 0-87389-047-7

Acquisitions Editor: Jeanine L. Lau
Production Editor: Tammy Griffin
Cover design by Artistic License. Set in Times by DanTon Typographers.
Printed and bound by BookCrafters.

Printed in the United States of America

Quality Press, American Society for Quality Control
310 West Wisconsin Avenue, Milwaukee, Wisconsin 53203

In memory of William D. Goeller, chairman of the Quality Costs Committee, (July 1987 - August 1988); a friend and an inspiration.

Jack Campanella

TABLE OF CONTENTS

PREFACE xiii

ACKNOWLEDGMENTS xv

1983

ASQC ANNUAL QUALITY CONGRESS TRANSACTIONS

Cost of Quality System — A Management Tool *James Demetriou* 1

Managing Cost of Quality *Vyasaraj V. Murthy* 5

Quality Cost Breakthroughs in U.S. Production *August B. Mundel* ... 10

Quality Costs in a Non-Manufacturing Environment *Frank Scanlon* .. 18

Quality Costs in the Process Industries *Walter Siff* 25

Quality Costs: We Know Where We're Going! Do You?
James J. Wayne and Andrew F. Grimm 30

The Quality Manager's Job: Optimize Costs
William C. Noz, Jr., Bradley F. Redding, and Paul A. Ware 38

Quality + or − Quality Costs Equals Productivity
Charles A. Aubrey, II and Debra A. Zimbler 46

Reducing Failure Cost and Measuring Improvement
William O. Winchell 54

Using Quality Costs in Productivity Measurement
D. Scott Sink and John B. Keats 57

EUROPEAN ORGANIZATION FOR QUALITY CONTROL ANNUAL CONFERENCE PROCEEDINGS

Follow-Up and Analysis of the Costs of Rejections
Miguel Arenas Luega 67

The Significance of User-Consumer Quality Costs in the World of Unlimited Resources *Bronislaw Oyrzanowski and Tadeusz Wawak* 81

QUALITY PROGRESS

Consumer Product Quality Control Cost Revisited (April issue)
Harold L. Gilmore 87

Don't Be Defensive About the Cost of Quality *Philip B. Crosby* 95

Management Team Seeks Quality Improvement from Quality Costs
Joseph J. Tsiakals 98

Principles of Quality Costs *Jack Campanella and Frank J. Corcoran* . . 102

Quality Costs: Current Applications *Edward Sullivan* 116

Quality Costs: Current Ideas *Edward Sullivan* 122

COST AND MANAGEMENT

Measuring Quality Costs (July - August issue) *Wayne J. Morse* 125

MANAGEMENT ACCOUNTING

Let's Help Measure and Report Quality Costs (August issue)
Harold P. Roth and Wayne J. Morse 134

1984

ASQC ANNUAL QUALITY CONGRESS TRANSACTIONS

Application of Economic Principles to Quality
Charles W. Bradshaw, Jr. 142

Managing for Success Through the Quality System *Frank Caplan* . . . 152

Profit Improvement Through Scrap Reduction
J. P. "Jerry" Reames 167

ANNUAL INTERNATIONAL INDUSTRIAL ENGINEERING CONFERENCE
PROCEEDINGS

    Cost of Quality and Productivity Improvement
    Michael P. Quinn and Egbert F. Bhatty 176

EUROPEAN ORGANIZATION FOR QUALITY CONTROL ANNUAL
CONFERENCE PROCEEDINGS

    Costs Related to Quality Present Situation in the Federal Republic of Germany (FRG) and Future Aspects   *Rudiger K. Vocht* 191

    How to Institutionalize Quality Cost Improvement   *J. M. Groocock* .. 201

    How to Succeed in Actions to Reduce Quality Costs. Theory and Reality.
    K. Ullberg and M. Karnebjer 208

    Quality Costing in the UK
    Peter A. Daisley, J. J. Plunkett, and B. G. Dale 216

1985

ASQC ANNUAL QUALITY CONGRESS TRANSACTIONS

    Increased Profits Through Company-Wide Commitment
    William J. Ortwein 225

    On the Road to Quality Savings   *William D. Goeller* 235

    Quality Costs II: The Economics of Quality Improvement
    John T. Hagan 245

    Relationship of Financial Information and Quality Costs: A Tutorial
    Earl T. Szymanski 257

EUROPEAN ORGANIZATION FOR QUALITY CONTROL ANNUAL
CONFERENCE PROCEEDINGS

    Quality Costs — Failures and Potentials   *Frank M. Gryna* 268

FOOD TECHNOLOGY

    The Effects of Regulation on Quality Costs (September issue)
    William A. Golomski 279

QUALITY PROGRESS

Business Management and Quality Costs: The Japanese View
(May issue)   Hitoshi Kume 282

Quality Cost: Better Prevent than Cure (September issue)
Alain M. Chauvel and Yves A. Andre 292

1986

ASQC ANNUAL QUALITY CONGRESS TRANSACTIONS

Costing Quality for Sensitivity Analysis
Hugh E. Kroehling and Lawrence R. Petersek 299

Focusing Quality Costs Using the Basics   William O. Winchell 309

Predicting Quality Cost Changes Using Regression
K. S. Krishnamoorthi 315

Solving the Quality Cost Equation   David D. Kildahl 326

EUROPEAN ORGANIZATION FOR QUALITY CONTROL ANNUAL
CONFERENCE PROCEEDINGS

Quality and Economy More Emphasize the Role of Quality on Sales Rather than on Cost   Noriaki Kano 331

Quality Costs and Profits — Myth or Reality
Anthony R. Stephenson 346

INSTITUTE OF ELECTRICAL AND ELECTRONICS ENGINEERS
TRANSACTIONS ON ENGINEERING MANAGEMENT

Finding the Cost of Software Quality   Charles P. Hollocker 355

NORTHEAST QUALITY COSTS CONFERENCE PROCEEDINGS

Guide for Managing Vendor Quality Costs   William O. Winchell 368

QUALITY PROGRESS

Accounting for the Real Cost of Quality (January issue)
   Dean-Michael Lenane 374

An Engineering Organization's Cost of Quality Program
   (January issue)   *Lawrence J. Schrader* 384

Optimum Quality Costs and Zero Defects: Are They Contradictory Concepts?
   (November issue)   *Arthur M. Schneiderman* 394

CIM REVIEW

Accounting for Quality Costs — A Critical Component of CIM
   (Fall issue)   *Wayne J. Morse and Kay Poston* 400

1987

ASQC ANNUAL QUALITY CONGRESS TRANSACTIONS

Avoidance/Failure Costs Reversal: An Action Plan
   Andrew F. Grimm and James G. Fox 409

Do Controller Departments Measure Quality Costs?
   Thomas N. Tyson 420

A High Tech Sight for Cost Targets   *John M. Ryan* 426

Quality Costs in Staffs — Micro, Macro, or Both?
   William O. Winchell and Caroline J. Bolton 435

Quality Costs — New Concepts and Methods   *Edgar W. Dawes* 440

NORTHEAST QUALITY COSTS CONFERENCE PROCEEDINGS

Quality Costs: Prevention the Tool for Reduction
   Frank J. Corcoran 449

Spending Prevention Dollars Effectively   *Martin W. Wirt* 454

ROCKY MOUNTAIN CONFERENCE PROCEEDINGS

Qualty Costs: Principles and Implementation   *Jack Campanella* 460

QUALITY PROGRESS

Quality Cost Analysis: Extend the Benefits (September issue)
William O. Winchell and Caroline J. Bolton 474

THE QUALITY REVIEW

The Cost of Poor Quality (Spring issue)
Walter F. Raab and Edward P. Czapor 479

MANAGEMENT ACCOUNTING

Quality and Profitability: Have Controllers Made the Connection?
(November issue)   *Thomas N. Tyson* 483

Why Quality Costs Are Important (November issue)
Wayne J. Morse and Harold P. Roth 490

PREFACE

Quality Costs: Ideas and Applications, Volume 2 is a compilation of papers and articles from many available sources and contains today's thinking on quality costs, expanded to include application of quality costs in the service industries as well as manufacturing. Volume 1, first compiled in 1984, promised that since quality costs is an ever-developing field of interest, up-to-date information will be compiled and published in succeeding volumes over the coming years. To that end, Volume 2 begins where Volume 1 left off.

This book is intended to be a complete source of reference material as well as a chronological history of the development of quality costs. Therefore, no judgment has been made as to the content of the articles themselves and neither the editor nor the American Society for Quality Control takes responsibility for or endorses the philosophies contained within.

The editor hopes that this new edition will be helpful, easy to use, and has articles of interest to all. Suggestions for improvement of future volumes is encouraged and may be addressed to Quality Press, American Society for Quality Control, 310 West Wisconsin Avenue, Milwaukee, Wisconsin 53203.

Jack Campanella
Editor

ACKNOWLEDGMENTS

The editor would like to thank the Quality Costs Committee for its support, the authors of the compiled papers and articles who took the time and effort to share their expertise, and the various journals and organizations that gave permission to reprint the many articles appearing in this edition.

Special thanks is given to Quality Press and, in particular, Jeanine Lau, for her indispensable contributions to this project.

1983

ASQC ANNUAL QUALITY CONGRESS TRANSACTIONS

COST OF QUALITY SYSTEM — A MANAGEMENT TOOL

James Demetriou
Manager of Quality Administration
ITT Avionics Division
Clifton, New Jersey

Abstract

Quality Control professionals who consider themselves a part of management must also consider the profitability of their Company as part of their responsibility. Their contribution to profit is accomplished by maintaining the required level of quality while at the same time minimizing the cost of achieving that quality. To accomplish this, the cost of quality must be measured, analyzed and controlled. This paper will describe the fundamentals of selling, implementing and utilizing a cost of quality system which will result in increasing profit margins.

Introduction

Knowing the cost of doing business is an essential for those responsible for running the business, e.g., managers, the board of directors, the chairman of the board and the holders of an interest in the business such as stockholders, bondholders, banks. On this basic of business operations rests the profitability or non-profitability of a given business organization.

There are many components making up the cost of operation: material, labor, facilities, utilities, marketing, etc. Some of these are fixed, recurring costs. Others such as material and labor vary in proportion to the sales level at any given period. Within this material and labor cost segment of overall costs there is a cost element concerned with the quality of the product — the Quality Department. This department is essentially charged with the responsibility for assuring that only good products are delivered to the customer. Stated simplistically, the department performs an appraisal of the products before releasing them for shipment. The conventional term for this is inspection and in many companies the Quality Department is considered to be (and in some instances is) an inspection function. As such, the Quality function does not direct any of its efforts to prevent or reduce the levels of failure resulting from inadequate design or manufacturing operations. As a result, redesign, rework, repair and scrap costs are incurred. This is true to a greater or lesser degree for all manufacturing operations or office systems. These failure costs are not the result of any Quality Department actions but are nevertheless part of the total cost of quality for a company. Obviously, if the costs incurred due to failure can be reduced, the profitability of the company will improve. To achieve this, the Quality Department must increase its emphasis on prevention of failure. If this can be achieved, less of the Quality Department's efforts (and budgeted dollars) will be needed in the appraisal (inspection) activity, since statistical sampling of the inprocess product can be applied and there will be less delays in final acceptance operations which might otherwise result from borderline conformance of some products. More importantly, the costs incurred because of failures in the manufacturing operation will be significantly reduced.

Development of the System

The Quality Department has an excellent opportunity to more effectively meet its responsibilities as part of management by showing how a reduction in the cost of quality for the whole operation can result in improved profitability of manufacturing and reduced warranty costs, with improved customer relations as a bonus. How to do this? The first and most important step is to show top management the potential for reducing costs, by developing a presentation of present costs versus anticipated reduction in cost of quality. The data for the presentation will probably have to be extrapolated from the present accounting system, using all existing data plus some estimates of hours and dollars, depending on the type of accounting system in place.

Hours and dollars must be segregated into the following sub-elements of the three major categories of a cost of quality system, which have been discussed in the earlier part of this paper; prevention, appraisal, failure. These are defined as follows:

Prevention: Contract review; design and drawing review; design change board; component engineering; vendor support; quality system audits; preparation of Quality procedures and test specifications.
Appraisal: Inspection (receiving, in-process, final); test; Quality Control (vendor source inspection, in-process final acceptance, shipping); travel expense (source inspection); calibration costs; material evaluation costs.
Failure: Engineering changes; rework;repair; retest; scrap; warranty.

In presenting a cost of quality proposal to top management, a clear definition of goals, both tangible and intangible, must be made and must be presented in terms of dollars to be saved in cost of quality.

Having successfully presented the proposal and been given the go-ahead, the Quality Department must first determine whether the required expertise is available within the department or whether hiring of experienced personnel is necessary. The Accounting Department must now be approached.

Since all of the components of the cost of quality are present in the Accounting Department records, it is necessary to establish the mechanics within that department to categorize such costs. The Quality Department must assist Accounting in this exercise by providing them with the sub-elements listed above. The Accounting Department in turn will establish a chart of accounts, charge numbers, etc., for each of these sub-elements.

On a regular basis (usually monthly) this data will be provided to the Quality Department for inputting into a matrix to provide a display of what is happening with the cost of quality relative to sales, both monthly and on a cumulative year-to-date basis. As data is accumulated and trends become apparent, goals may be established both in dollars expended for each category and its sub-elements and as a cost of quality percent of sales.

An example of such a matrix is shown below:

COST OF QUALITY YEAR TO DATE AND CURRENT MONTH

CATEGORY	YEAR END ( ) % COQ	YEAR TO DATE			CURRENT MONTH		
		FORECAST ($000's)	ACTUAL ($000's)	% OF ACTUAL COQ	FORECAST ($000's)	ACTUAL ($000's)	% OF ACTUAL COQ
PREVENTION							
Quality Engineering	24.4	622.0	586.0	19.1	61.9	57.5	14.9
Quality Administration	6.0	129.9	129.7	4.2	12.7	13.3	3.4
Total Prevention	30.4	751.9	715.7	23.3	74.6	70.8	18.3
APPRAISAL							
Receiving Inspection	6.5	242.4	237.4	7.8	25.5	23.6	6.0
All Other Inspection	6.8	203.0	228.7	7.5	25.9	27.2	7.0
Test	6.2	420.9	183.6	6.0	91.6	28.8	7.4
Standards & Calibration	7.2	176.7	162.5	5.3	18.6	18.2	4.7
Vendor Quality	6.8	168.7	175.5	5.7	16.1	16.7	4.3
Travel Expense	1.6	76.7	59.6	1.9	9.7	9.3	2.4
Mat'ls. & Evaluation Lab.	5.0	107.0	101.2	3.3	11.6	9.2	2.4
Product Acceptance	13.9	355.8	314.0	10.3	35.2	36.4	9.4
Total Appraisal	54.0	1751.2	1462.5	47.8	234.2	169.4	43.6
FAILURE							
Rework Labor	8.7	240.8	271.3	8.9	22.4	47.1	12.2
ECR/ECN Labor	5.7	481.5	522.3	17.0	44.8	86.3	22.3
Scrap	0.4	43.3	85.0	2.8	4.2	11.9	3.1
Warranty	0.5	35.7	5.5	0.2	3.3	1.6	0.5
Total Failure	15.3	801.3	884.1	28.9	74.7	146.9	38.1
Total Quality Costs	2430.7	3304.7	3062.3		383.5	387.1	
Sales	71570.9	71654.0	71991.4		7222.0	8091.1	
COQ as a % of Sales	3.4	4.6	4.3		5.3	4.8	

Each business organization will necessarily have to develop their data collection and reporting system within their particular organizational structure. The foregoing text and matrix are intended only as guidelines.

A further refinement to the Cost of Quality Report will be a chart on which is projected a cumulative goal for a year, by month of the Cost of Quality as a percent of sales. Each month, the actual Cost of Quality/Sales ratio (percentage) will be plotted cumulatively and compared with the projection.

An example of such a chart is shown below:

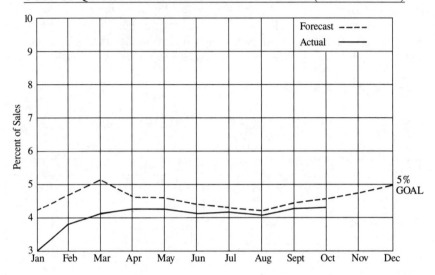

Summary

In conclusion, Quality professionals, as part of management, must extend their efforts beyond the traditional role of ensuring quality of product. They must be concerned with profitability rather than quality at any price. A system for measuring the cost of quality such as is described in this paper will provide the mechanism for obtaining such measurement and will result in reduced cost of quality and increased profits.

MANAGING COST OF QUALITY

Vyasaraj V. Murthy
Honeywell Information Systems, Inc.
Billerica, Massachusetts

Abstract

The paper deals with the most promising of all the currently emerging approaches to help increase the productivity of American Enterprise. American Business continues to be driven by short and medium range financial goals. The Cost of Quality is a significant part of the Product Cost and this paper investigates an approach to capture it. The paper stresses the point that for a Quality Cost program to be successful, it needs not only the blessings of top management but also their commitment and how to go about to secure the same. The line managers need help and this approach provides the much needed help for them to manage their part of the business. It opens their eyes to the Total Cost of Quality rather than the traditional emphasis on Scrap and Rework costs being held to an acceptable minimum.

Introduction

It is not an unfamiliar subject to this audience. It is not something new or even a new buzz word. It has been known to the quality professionals and management for a long time but still has not caught on despite the onslaught of the Japanese phenomenon. Gentlemen, I am talking about Cost of Quality.

I shall not attempt to talk to you about what Cost of Quality is. That subject has been covered adequately in the publications of the ASQC Quality Costs Committee. What I intend to do today is to give it a new perspective that will make it palatable to Management from a business perspective.

There are three basic ingredients to success in managing the Cost of Quality:

1. Acceptance by Top Management.
2. Acceptance by the Financial community which is concomitant with the first requirement.
3. Must be easily understood by lower levels of management.

All of the above seem fairly simple to state but very hard to implement. I can talk from my personal experience and I am sure many of you might attest to the same fact.

Fortunately for me, miracle of the computer age and working in a company like Honeywell it was a fairly easy task to develop the algorithm of distributing the departmental costs into the proper buckets of Quality Cost Categories of Prevention,

Appraisal, Internal and External Failures. There existed a sophisticated financial and labor cost reporting system which could be accessed to develop the rudimentary Cost of Quality reporting system. With the help of a friend in the university I even managed to get a software program in BASIC language which automated the total effort. The manager could have the Cost of Quality Report for his department within 10 days after the fiscal month had ended. Alas! even with all of this the Cost of Quality Report was not selling like a hot cake. There was a nod of appreciation, nay a left handed compliment, an opportunity to be a spokesman on the subject but no sale. A breakthrough occurred with a change of management. The new manager knew even less, by his own admission, than the previous one. He turned out to be my best ally in the Cost of Quality game. He realized the potential value of this management tool and put it in that perspective. He was not going to sell the Cost of Quality in the conventional sense. He would create an appetite by interesting the financial community and the top management as to what was costing us for quality including all the hidden costs which they as management should know but were never privy to that information directly. It was presented to them in such a fashion that they would not even recognize its impact.

A series of presentations were made to the top echelons of management with prime emphasis on product cost and how much of it was product quality cost along with all the associated overhead. At the same time, there was a push by Corporate Finance group who had heard about the Cost of Quality from a different source and set up a task force comprising of the controllers of the various divisions of the company. That was a shot in the arm for the whole project.

Cost of Quality and Its Relation to Product Cost

Cost of Quality based on a product line throws a new light on profit margins. The efforts of cost accountants are directed towards accumulating all of the identifiable costs that can be charged to the product. This traditionally includes inspection and test labor but treats other elements of the Cost of Quality as overhead. In a way Cost of Quality by product line gives them a new tool for assessing the true product cost and enables them to help the top management make a better business decision for marketing the product or profitability analyses.

Traditionally, the managers look at the financial reports and as long as their department's financial performance is within the norms that they are used to from past experience they tend not to pay much attention to their costs. In this age of keen competition and the constant pressure for improving productivity the traditional managers need help badly. The Cost of Quality reports give him that kind of help for making intelligent, conscious decisions. Cost of Quality associates the cost of internal failure such as rework and retest and rescheduling because the job was not done right the first time.

For every action he takes within the context of the Cost of Quality he is making a decision knowing full well that it has impact not only in the near future but in the years to come. This in no way takes away his fundamental right to make a decision. It only enhances the quality of his decision. He is still within his rights not to do anything but this time around it was not because of lack of time or other priorities but because he really feels that such a decision is warranted under the present conditions and the business climate.

In summary, what we needed was the top management commitment and not just lip service, a report that jived with the financial and accounting documents and finally a torch bearer who could make it seem the best thing since Mom's apple pie and created an appetite within the rank and file management. It helps to know that top management is looking at the report so that other echelons of management do not want to be left behind or worse still look ignorant in front of their managers.

Let me now recap the developments that took place at Honeywell and where we are today in using it.

The Cost of Quality Program efforts began in earnest in late 1980 with the limited scope of defining, designing, and developing a Cost of Quality Report for the manufacturing operations on a division wide basis. Early efforts dealt with the manual collection of cost data and manual collation and classification of Quality Cost Elements, namely, prevention, appraisal, internal failure and external failure.

To be able to properly categorize the cost elements it is essential to know the types of activities within each of the Product Assurance Engineering departments and the Manufacturing departments. This was accomplished by using a questionnaire specially designed for the purpose with listing of all the conceivable activities and requesting the respondents to answer the questionnaire. The data was then collated and an algorithm was developed for allocating the departmental costs into the proper Quality Cost categories. The expenses associated with the central P.A.E. Operations consisting of the director and his direct reports were proportionately allocated to the plants based on the plant headcounts.

The first report was generated in the 4th quarter of 1980 which highlighted the problem areas and significant opportunities for improvement. The manual collation and verification took two months and the 1st quarter, 1981 report was made available by May 1981. The cost data had to be compatible with the financial accounting data and had to be timely to coincide with the monthly financial reports. We therefore decided to use the financial data that was readily available for generating the Cost of Quality Reports.

Since the Financial data base was already available, it was decided in January 1981 to obtain a data strip for the relevant P.A.E. and Manufacturing departments using internal programming resources.

The Cost of Quality Program was automated using an application software package especially developed for us by a contract programmer tailored to our specific needs. The first output from the automated program came out in July 1981. The 2nd and 3rd quarter reports were generated at the operations level. It was then deemed necessary to have the capability of departmental level reporting. This was successfully incorporated in the 4th quarter 1981 and at present we have the capability of publishing the Manufacturing Cost of Quality Report Package consisting of the operation level, plant level, and department level reports within a week of publication of the Financial reports and since we use identical data base the credibility of the reports is unquestioned.

The reports are published on a quarterly basis because we felt that the changes in the Cost of Quality Parameters and their tracking is not meaningful for shorter duration.

The Cost of Quality is reported as a percentage of Proforma Revenue or the Net Sales Billed and since it is volume sensitive a scaling factor is employed using Direct Labor as the activity level indicator.

Now let us look at some of the Cost of Quality report formats and typical reports that are produced and used by management in their business environment. The numbers are deliberately altered because of the confidential nature of the data. Suffice it to say that we have let the chips fall where they were and this really opened the eyes of our management.

It is an enlightened management that sees the usefulness of such a report rather than bury their heads in sand or worse still have a tunnel vision. I am happy and proud to have been instrumental in starting up such a system and am thankful to my company and my management for giving me this opportunity.

Some Theoretical Considerations

By the way, more often than not, there is a clash between altruism and reality. Referring to the famous Quality Cost Curve in Dr. Juran's handbook, the optimum quality cost is when the total Cost of Quality is minimum which occurs at the intersection of Discretionary and Consequential cost curves. So often it is difficult or even impossible to obtain the true External Failure Costs. Does this mean that we stop our plans for a Cost of Quality Program? Of course not!! What we can do is to treat only the variable portion of the exernal and internal failure costs. The optimum point for minimizing the total Cost of Quality occurs where the two variable cost curves intersect. Fixed portion of either Discretionary or Consequential costs really does not alter the optimum point.

More basic to the above point is the fact that in reality, an absolute, optimum cost does not exist. There is always a time lag between corrective actions taken and the manifestation of its effects. Sometimes this takes 12 months and more. Every time a change is made the cost curve changes its slope resulting in a new intersecting point and when the effect has taken place fully that changes the cost curve again and this process keeps on going. Furthermore, the chances of getting different points on the cost curves imply that there are varying levels of quality during the manufacture of a product. Needless to say that there is some variation in quality levels which in itself is insufficient to complete the cost curves. I propose that it is only a theoretical exercise and that too much time is wasted to expound this theory. Let us instead get on with it by taking a picture of today's Quality Costs and strive to minimize the total Cost of Quality.

Initially it is not important to know the ratio between discretionary and consequential costs. Guidelines as to what is the magic combination between Appraisal, Prevention and Failure costs are non-existent and that these ratios have to be worked out independently for processes and plants and which vary not only from industry to industry but also within the same industry and between manufacturers of similar products. Recently, I had the opportunity to discuss this very point with Dr. Gryna who agreed with me that it is not possible to arrive at the optimum Cost of Quality point other than by hypothetical, theoretical solution of successive integration using varying Quality levels and cost inputs associated with it.

The theoretical concept is useful in bringing home the point the concept of optimum quality cost. For a novice who finds it difficult to comprehend, it provides a logical solution to this complex problem.

Finally to recapitulate the high points in managing Cost of Quality the program should be:

- Based on the existing Financial documents readily available. Management trusts the financial documents and so build upon that trust.
- Use the Questionnaire process. This will involve the people who have to make use of the reports eventually. The confidence they will have since they filled out the information and you did not arbitrarily make them up.
- Display and create acceptance of the absolute dollars of internal failure costs.
- Gently suggest that a redistribution from consequential costs to discretionary costs such as Prevention and Appraisal gives you a better return on investment.

References

1. *Quality Costs - What & How?*, 2nd ed., (1971), ASQC Quality Costs Technical Committee.
2. *Guide for Reducing Quality Costs*, (1977), Ibid.
3. *Quality Planning and Analysis*, Juran & Gryna, (1970), McGraw-Hill Book Company, New York.
4. *Quality Control Handbook*, 3rd ed., (1979), Juran, McGraw-Hill Book Company, New York.

QUALITY COST BREAKTHROUGHS IN U.S. PRODUCTION

August B. Mundel, P.E.
August B. Mundel Associates
White Plains, New York

Abstract

Quality control, though widely advertised by industry in its effort to impress customers, is assiduously avoided by many organizations in the belief that it is costly and will adversely affect profits.

This belief in the adverse cost impact has been confirmed by experience with inefficient quality systems, systems not properly tailored to the operation. In contrast, where appropriate quality control procedures have been installed, breakthroughs to new levels of quality and improved costs have resulted.

This paper will discuss the essentials of efficient systems, proper programs, and system evaluation methods.

Introduction

Some quality activities are associated with each of the life phases of a product. The operational procedures and activities at each phase may be different, but each has similar requirements: the need to efficiently collect and analyze reliable data and to use it advantageously to economically produce salable, safe, conforming product.

Differing techniques may be more effective at different stages of product life. Product life can be divided into five phases: the design phase, the manufacturing phase, the distribution phase, the use phase, and the end-of-life or discard phase. Several of these phases may exist simultaneously.

Traditionally the quality function has been a major concern in the manufacturing stage and in many industries has been given a major share of the blame for the customer complaints and problems referred to the manufacturer during the use and end-of-life stage. Restrictions surrounding and still encumbering the quality function in many U.S. industrial plants make some of these accusations unwarranted. In reality they are a reflection on improper organizations, management deficiencies, and an incorrect concept of how to operate.

A major improvement in the quality of product and the efficiency of the entire operation can readily be made when management recognizes that a quality system is a cooperative venture, and requires the effort of all departments. The quality manual is a company system, not a quality department constitution or directive. It should describe the responsibilities of each department in the development of a system intended to design, produce, and deliver a product that is safe, suitable, satisfactory, sufficient to meet the economic needs of the user, fit for use, salable, and profitable.

Another error frequently made is the use of only one standard sampling procedure, Z1.4 (MIL STD 105D) or Z1.9 (MIL STD 414). Other sampling plans and statistical procedures are better suited to aid in the control and manufacture of acceptable product.

Design Phase

The collection of manufacturing line data and its proper analysis provides useful data for the designer on tolerances, trouble areas, and smoothly running operations. These data are essential if a good design is to be developed. These data can be collected by inspectors and operators. Most quality system manuals now require that data from production and field experience be collected, complaints and failures analyzed, and the results reported to the manufacturing, design, and other departments on a regular basis and at design reviews.

These data are therefore available for discussion at concept and other design review meetings. (Design review meetings are required by U.S. manuals.) Corrections for untoward past experiences can be integrated into the new design. This in itself is a break with the traditional redesign-the-wheel philosophy, and can be a major step in product improvement.

Another advantage obtained through the presence of quality management at design review meetings is that the quality staff receives advance notice of what is about to occur, what products, parts, components, and materials are involved, and what must be measured, tested, and inspected.

The initial design review meeting held at the concept stage provides interchanges which help the design, prepares the factory for special machine tool and skill requirements, and provides advance notice of the tests that will be performed on the new product.

Manufacturing Phase

The surveying and choice of vendors, the inspection of incoming materials, the control of processes, and inspection and test of manufactured product are all portions of the manufacturing phase. The more competent producers often have joint vendor/designer meetings to obtain the most efficient form, fit, and function from the vendor.

Although many contracts call for the use of the standard sampling plans previously mentioned, it is during the procurement and manufacturing phase that large savings can be generated by proper selection of the sampling procedure, and by reductions in the fraction nonconforming.

The advantages enjoyed by a manufacturer with a competent quality organization are that the inspection, control, and sampling procedures are adequate, appropriate, economical, reliable, accurate, and timely. They are promptly reviewed and analyzed. It thereby becomes possible to

1. Identify changes when they occur.
2. Promptly facilitate corrective action.
3. Promptly identify the need for design, process, and material changes.
4. Reduce scrap, rework, and sorting.
5. Improve the product and reduce costs.
6. Perpetuate successful designs and manufacturing procedures.
7. Properly protect consumer and producer interests.

The use of control systems has major advantages. Most manufacturers, even those who perform one or more 100 percent inspections, wish to sample product as a final and independent check on their procedures. The better the product, the less inspection of this type that is necessary, particularly where there are stable products and/or efficient 100 percent inspections.

The choice of efficient inspection procedures and the selection of more efficient sampling plans contributes to productivity. Table I shows a few sampling plans taken from Z1.4 and Z1.9, the current matching versions of the attributes and variables sampling plans. The table lists the lot sizes, the sample size code using Level II inspection, the sample sizes for attributes inspection for 1.0 percent AQL, and the sample sizes for s, R, and σ known variables inspection plans.

The fraction of the lot inspected is also shown in the right-hand column. A review of this right-hand column shows quite clearly the well-known phenomenon that, as the lot size increases, the fraction of the product inspected decreases.

Table I also shows that the variables acceptance plan Z1.9 uses a sample size approximately one-half to one-eighth as large as the attributes plan Z1.4, under the conditions assumed. Larger lots usually provide a higher probability of acceptance at the AQL, and better protection against poor lots. The handling of larger lots with their attendant economy and better operating characteristic is particularly recommended where good control exists. Even greater economy can be achieved if reduced sampling is justified, which is common under conditions of good control. Variables plans using reduced sampling procedures yield further economies. In addition, a variables acceptance plan, where usable, provides information as to the value of the average and the variability of the characteristic, useful for the designer, the user, and the manufacturing department.

Even in cases where a customer calls for a 2.5 percent or 4 percent AQL, it may be more economical to operate at a 1 percent or fractional percent process average. The reasons are several:

1. The yield is better.
2. The returns will be decreased.
3. Sample sizes can be reduced.
4. Fewer lots will be in dispute or returned.
5. The customer will be much more satisfied.

The effort to accomplish this improvement may be far less than the benefit. The cost is usually a one-time cost yielding a perpetual gain. If variables sampling can be used and the variability (under control) is stable, the sample sizes are markedly reduced.

Table I Comparison of Sample Sizes for Attributes and Variables Sampling Plans, AQL 1.0 Percent

Lot Size	Code Letter	Sample Sizes				Fraction of Lot in Sample
		Z1.4	Z1.9 Method			
			s	R	σ	
91-150	F	13				0.14--0.087
			10			0.11--0.067
				10		0.11--0.067
					4	0.044-0.027
501-1200	J	80				0.16--0.067
			35			0.07--0.03
				40		0.08--0.033
					12	0.024-0.01
3201-10,000	L	200				0.062-0.02
			75			0.023-0.0075
				85		0.027-0.0085
					25	0.008-0.0025

The use of control charts facilitates such a procedure. Control charts have the further advantage of providing information at the process site so that corrective action can be initiated when the process has shifted from its most advantageous level. The production of nonconformities can be markedly reduced or eliminated, thereby increasing the yield.

The result of reduced sampling and reduced inspection combines herein to provide greater productivity and better quality. On-line inspection can produce a similar effect, particularly where a continuous inspection plan is adopted. The major factor is the rapid feedback of information which can effectively reduce the marginal and nonconforming product.

Marginal product, the units on the borderline of the specification, is costly and troublesome. If there are many units on the borderline, there are more that are beyond the specification and incontrovertibly nonconforming. The presence of the marginal units becomes an incentive to buy more accurate test devices. It also makes for more frequent rejection when lots are sampled, and in the final analysis leads to factory limits which are tighter than customer limits and which thereby increase the quantity of nonconforming product.

The elimination of marginal product is not always easy. It may require new production equipment, specifications, process and design changes, or sometimes the initiation of a control procedure. When accomplished, it is always a major breakthrough to a new level of quality and a new minimum in costs.

Variables data are of greater use to the designer and to the production engineer than attributes data, as they provide a realistic view of what tolerances can be maintained. Variables data make it possible to obtain information on the capability and the need for repair or replacement of equipment.

Automatic test equipment (ATE), operating on a go-not go basis, cannot always do this, but sample information obtained from ATE with a variable output can do this effectively.

One of the major problems facing purchasers and analysts is a mixed lot. The instructions for attributes sampling plans (Z1.4, *et al*) state that a lot shall, as far as is practicable, consist of products of a single type, grade, class, size, and composition, manufactured under essentially the same conditions, and at essentially the same time. (See 105D, ¶5.2.)

This rule may be violated whenever material of more than one type, class, size, or one product made on two or more machines, is assembled into a shipping lot. It is a common cause of problems in the marketplace. It can plague producers as well as customers.

Distribution and Use Phases

The quality manual usually calls for a procedure whereby the manufacturer maintains a record of complaints and returns, replacements, warranty repairs, subsequent repairs, and field failures. The data made available through this analysis can provide information which will eliminate major problems, improve the product, and enhance sales.

The real test of a product occurs when it is placed in service. The satisfactory product does what it was purchased to do. It does this satisfactorily, economically, and reliably. The unsatisfactory product may perform unsatisfactorily, uneconomically, or unreliably, or it may cause injury or damage. The manufacturer/seller, or his agents, will in this latter case hear from at least some of the users.

Another source of information during the use phase is the need for replacement parts and service, both within the warranty period and thereafter. Certainly the reduction of within warranty service and replacement costs is advantageous. Some may argue that replacement parts, costs, and service subsequent to warranty are profitable, but if these greatly exceed competition or the user's expectations, there is the risk of loss of customers.

End-of-Life Phase

When a product reaches the end of its life, it is usually replaced. If the service has been satisfactory, the customer may replace it with a unit from the same source. Some users will make major repairs. The factors which affect the decision to repair, replace with the same brand, or buy another brand are many. They include the question of whether the product performed satisfactorily, the availability of repair parts, repair and service, and whether the product was efficient and of the correct size. Products which have been troublesome will seldom be replaced with one from the same producer.

Brand loyalty and the demand for old product replacement parts are both indications of good product performance.

The Breakthrough

The breakthrough has been initiated. Companies are beginning to recognize that quality is important, that good quality management can help produce a better, safer, more salable product more economically, that better quality costs less. Competition is proving this.

Companies are beginning to judge whether they have a successful quality system, based on:

1. The absence of rework, rejects, and selective assembly, or a marked movement toward their elimination.
2. The use of sampling, inspection, or control charts at the source or process, so that immediate corrective action is feasible.
3. Cooperation among design, manufacturing, sales, procurement, and quality departments (this is an indication of management participation).
4. The use of systematic design review procedures.
5. The use of sampling methods that are appropriate.

The breakthrough includes wider recognition of quality engineering certification. The certified quality engineer should be able to design a control procedure that produces conforming product at minimum expense. He should be able to advise and work with the design and manufacturing engineers to eliminate designs and processes which lead to uncontrollable variation and the production of scrap and rework.

This is accomplished by the corporate adoption of a systematic method of recognizing the interrelation and responsibility of each of the corporate departments. Each has a contribution to make to the product, starting with the design review and continuing through the product life.

The breakthrough is also being hastened by the development of certifying agencies which are beginning to demand conformity of the organization to a quality system, as well as requiring that a product conform when measured in accord with a Standard Method of Measuring Performance (SMMP).

The American Society for Quality Control's development of the curriculum of the CQE and the certification of CQE has been a major factor. The requirement that CQEs regularly obtain a minimum number of CEUs to maintain proficiency and certification have made the CQE certification more valuable.

Finally, there has been the recognition that the most frequent cause of excessive costs is lack of quality, and specifically the failure to adopt a comprehensive quality system.

Conclusion

The breakthrough has occurred when better quality of product is produced without rework and scrap, the costs of control have been reduced by proper selection of methods, and the customer has become satisfied to the extent that returns, complaints, and in-warranty repairs have been reduced by an order of magnitude or more. With the breakthrough, the economics of corporate operation will have been markedly improved.

When the breakthrough occurs:

1. The engineers can work on new product, not redesign and try to improve the product in production.
2. The manufacturing group produces, rather than reworks, product.
3. The quality people control, do not simply sort.
4. The sales people can sell rather than try to placate dissatisfied customers.

Bibliography

1. ANSI/ASQC Z1.4-1981, *Sampling Procedures and Tables for Inspection by Attributes,* Milwaukee: American Society for Quality Control.

2. ANSI/ASQC Z1.9-1980, *Sampling Procedures and Tables for Inspection by Variables for Percent Nonconforming,* Milwaukee: American Society for Quality Control.

3. International Standard 3951 Sampling Procedures and Charts for Inspection by Variables for Percent Defective (graphical methods).

4. Mundel, August B. Quality and Productivity, October 4-7, 1982, QualTest I Conference, IQ82-370, Society of Manufacturing Engineers.

5. Mundel, August B. Achieving Productivity and Manufacturing Excellence by Appropriate Sampling and Control, October 4-7, 1982, QualTest I Conference, American Society for Quality Control.

QUALITY COSTS IN A NON-MANUFACTURING ENVIRONMENT

Frank Scanlon
Director of Education/Quality
The Hartford Insurance Group
Hartford, Connecticut

Abstract

This presentation will focus on the utilization of Quality Costs as part of a TOTAL QUALITY PROGRAM IN A NON-MANUFACTURING ENVIRONMENT. Included are service functions for service companies and administrative functions for service and manufacturing companies.

A logical, practical approach including the methodology necessary to develop Quality Costs in the non-manufacturing environment will be presented. The *proper* use of the Quality Cost concepts might be the ultimate financial justification for top management making the decisions to support Quality. Quality Cost is another part of your total Quality management program.

Introduction

Since 1976, many of the successes of manufacturing Quality programs have been integrated into the administrative and service functions at The Hartford and other ITT subsidiaries. Over the years, we have shared many of these successes and problems at the ASQC Annual Quality Congress. This year we will focus on Quality Costs as part of your overall Quality management program.

It is important to understand Quality Costs is not the answer to all of the non-conformance problems within American business. At best, a good Quality Cost Program will supplement, strengthen and support an already good Quality organization. If the proper environment has been set, the development of a practical Cost of Quality Program will allow your company to identify and measure your present cost of Quality and provide the vehicle to improve the accepted, agreed upon base dollar figure.

Sell Quality Concepts

The Quality professional must have a positive attitude in order to sell Quality concepts to management. Is Japanese Quality better than ours, or are their public relations better? If we want to sell Quality and Cost of Quality Programs to our General Managers, we will have to identify what's in it for them:

- Improve customer relations, market share, employee satisfaction, more profit
- Eliminate or reduce expenses, unproductive work, waste
- Keep our job.

There is no General Manager today who does not want to improve Quality. Some just do not know how to achieve Quality improvement. Cost of Quality just might give the General Manager the financial justification to pursue a sound Quality improvement program.

Once management at ITT accepted this concept, top management provided the commitment necessary to implement a Quality Cost Program including:

- Financial resources
- Human resources
- Visible leadership and support.

Once we have gained management's attention, then selling the Cost of Quality philosophy becomes very practical. Quality itself was always intended to be very simple.

Cost of Quality

There are three premises associated with the philosophy of the Cost of Quality:

- Failures are caused
- Prevention is cheaper
- Performance can be measured.

Failures are found through appraisal. The *cause* of failure is identified and eliminated through corrective action or prevention. The further along in the process the errors are found, the more expensive they are to fix. If failure cost is reduced, appraisal efforts can also be reduced in statistically sound increments. When the level of prevention is such as to maintain a minimum of appraisal and failure costs, the state of "Quality Control" will exist. If this is simple, common sense and practical, why is it a problem?

This system will *not* work without a basic Quality measurement system to determine and clearly identify the correctable elements of failure that comprise the initial essence of cost improvement.

Management that does not accept Cost of Quality is in the dark ages of Quality improvement.

Time does not permit us to go into all of the details associated with the implementation and detail of a Quality Cost Program. However, we do have the time to emphasize the highlights and some of the methodology used in a non-manufacturing environment.

Definitions

The definitions of prevention, appraisal and failure, although translated somewhat differently, are identical to those which have been published for years in the Quality Cost Technical Manuals.

Appraisal — The cost of the planned checking/inspection of products and/or processes to the standard.

Failure — The cost to correct defects or otherwise dispose of problems created by identified defects.

Escape — Although the Quality purist will identify this as a failure cost, we believe in all businesses it is large enough to identify separately. This represents the costs of those defects which get through the system undetected.

Prevention — The cost to:
 - Analyze defect information for the purposes of identifying defect *cause*
 - Develop and implement corrective action plans that address the *elimination* of defect causes
 - Review, simplify or clarify instructions at operator's level
 - Equipment or systems to prevent defects
 - Train and research to prevent defects.

Methodology in Developing Quality Cost

Appraisal — Cost to verify that the end product meets the published specifications. This could be determined by taking the total number of hours spent during this verification, times the standard average salary. Also included in this category would be the cost to maintain and update machine edits or other software programs, which had been developed for the sole purpose of defect identification.

Examples of Appraisal Cost:
100 hours × $8.32 (standard average salary) = $832 Appraisal Cost.
Cost of new edit package to detect defects = $10,000 Appraisal Cost.

Failure — Failure can be classified into three categories:
 - Processing Failure — That is, the defects found during the appraisal process.
 - Internal Failure — Those defects found by the subsequent unit or function within the company prior to being delivered to "the end customer."

- External Failure — Those defects which have been detected after they have been received by the customer.

Three separate Quality Cost calculations are necessary to determine failure costs, as it is more expensive to find and fix external defects than the processing defects.

Methodology for Failure Costs

Processing Failure — Cost would be the accumulation of the supervisor's review time necessary to determine if a defect actually occurred and if so, who caused the defect plus the actual time spent by the individual to correct the defect times a predetermined average Quality Cost.

Example of Processing Failure:

1/2 hour supervisor's time × $12 (supervisor's standard average salary-SAS)	= $ 6.00
+ 1/2 hour employee's time × $8.32 (standard average salary-SAS)	= +$ 4.16
= total processing failure cost	= $10.16

Internal Failure — The internal failure costs would be the total of the following:
- Time of the individual finding the defect in a subsequent unit
- Plus the supervisory time of the subsequent unit to identify and route the defect to the unit causing the defect
- Plus the time of the supervisor of the unit creating the error to review the defect, identify the cause of the defect and record to the individual responsible
- Plus the time it takes the employee to correct the defect.

Example of Internal Failure:

15 minutes to find error 1/4 × $8.32 (SAS)	= $ 2.08
+ 15 minutes of supervisor in subsequent unit 1/4 x $12.00 (SAS)	= $ 3.00
+ 15 minutes of supervisor in original unit 1/4 × $12.00 (SAS)	= $ 3.00
+ 30 minutes of employee time to fix 1/2 × $8.32 (SAS)	= $ 4.16
= total Internal Failure Cost	$12.24

External Failure — Quality Cost to the company would be the cost of the internal defect plus the cost to process the item through the mail room plus cost of the supervisor to pull the file for the investigative work necessary to determine who caused the defect. All this time would then have to be added together. The grand total would equal the external Quality Cost.

Example of External Failure:

15 minutes for mail room to handle second time 1/4 × $8.32 (SAS)	= $ 2.08
+ total of Internal Failure example above	= $12.24
= total External Failure Cost	$14.32

Escape — The most costly defects in the Administrative/Service area are defects that get through the system undetected. These are the defects which get in the hands of our customer and cause us to lose dollars or the customer themself. An example would be an undercharged bill which is never surfaced by our customer.

The methodology for costing this category consists of a pre-test and a post-test. Pre-test would be developed by sampling the company's historical files to determine the amount of defects accumulating the plus or minus dollar values calculating a percentage by dividing net dollars lost by the total dollars available for review. Apply this percentage to the total dollars available which will determine the pre-test cost of undetected defects; this obviously does not include business loss due to overcharging customers. To test if Quality improvement is taking place, a post-test utilizing the same criteria on current work should be done periodically to determine if there is improvement or slippage in this "escape" category. This will result in an estimate of lost revenue and dollar improvement if the Quality Improvement Program is successful. Incidentally, the dollars saved in this category go directly to the bottom line.

Example of Escape Costs — Pre-Test:
A. $ 100,000 of accounts receivable reviewed
B. $ 10,000 in unfavorable errors found
C. $1,000,000 total accounts receivable available
D. Ten percent (10%) error rate
E. $100,000 total receivable lost - $1,000,000 × 10%

Example of Escape Costs — Post-Test:
A. $ 100,000 of accounts receivable reviewed
B. $ 5,000 in unfavorable errors found
C. $1,250,000 in total accounts receivable available
D. Five percent (5%) error rate
E. $ 72,500 total receivable cost - $1,250,000 × 5%

Quality improvement is taking place.

Methodology for Preventing Costs

Actual dollar spent to specifically address the prevention of defects.

This is the total time spent by management and staff times an average standard salary rate participating in actions specifically devoted to the prevention of defects. This can be "one-on-one" or "classroom." For example:

1. If the major cause of defects is the lack of knowledge by five (5) individuals within a given function and the manager calls an hour meeting for these individuals, the Cost of Quality would be the manager's average standard hourly rate plus the total of the five (5) individuals' hourly rate. This would equal the total preventive Quality Cost for this defect.

2. If equipment is purchased or modified specifically for the purpose of preventing defects, this should be considered a Quality Cost. Fox example, implementing an automated system to eliminate a manual task.
3. Supervisor or manager's time to analyze the error statistics for the purposes of developing corrective action can also be considered prevention time. This would be calculated by taking the average manager/supervisor's standard cost times the hours spent.

Summary

Please understand, the foregoing is a "broad brush" overview of where we can and should be going. Our question is how do we get there?

1. By professionalism
   The analysis of data and the development of a corrective action plan resulting in the reduction or elimination of failure cost causes is not for amateur Quality inspectors/checkers. They will not get the job done. It requires a Quality professional to act as a technical catalyst to properly identify the "real" problem and the courage to coordinate a practical corrective action plan that will *eliminate* failure cause.

2. The corrective action system must be formal and all encompassing, including the following:
   - Documentation
   - Tracking
   - Evaluation of the process of planned actions leading to the *elimination* of the identified problems
   - Staying on track.

   A formal corrective action system does not allow major problems to be camouflaged in many activities and fall into the cracks without *eliminating* the cause.

3. Get the professional to work with you:
   - Marketing
   - Manufacturing
   - Management
   - Comptrollers
   - Whoever.

4. Note, all human defects are not limited to clerical staff; consider:
   - Design engineer
   - Process design engineer
   - Marketing
   - Management
   - Anyone directly or indirectly associated with the end product.

5. Prevention is the name of the game.

   Without analysis and corrective action, all the other efforts are a waste of resources and should not be done. Prevention activity initially will push the discovery of the defects back from the most expensive to the least expensive area:
   - Customer
   - Final test
   - Final inspection
   - End process inspection
   - Back to the source of defect and the people who can prevent the failure in the first place.

Finding a temporary solution while looking for a permanent one is OK. You can be flexible without compromising. With prevention you can increase cost in qualification testing, as failure detection is cheaper there than finding the failures in process or by your customer. Prevention activities are elements of a management system that are built in and scheduled in learning from our mistakes. Mistakes that allow the perpetuation or birth of defect causes are now eliminated. Building this system and enforcing it is the right way to close the loop of Quality measurement, and corrective action experience will cement your victory over defects. ITT continues its refinements in Quality Cost practices in manufacturing and non-manufacturing areas. The documented savings of hundreds of millions of dollars just might be a drop in the bucket, as it has covered less than 50% of all ITT's employees.

Reference: Cost of Quality One-Day Seminar

QUALITY COSTS IN THE PROCESS INDUSTRIES

Walter Siff
James River Graphics
South Hadley, Massachusetts

Abstract

Although the principles of quality cost analysis are valid for all kinds of businesses, applying them in a practical way to a specific situation requires a flexible approach to match the economics and practices of an individual business. Taken as a group, the process industries are characterized by a distinctive set of quality costs, different in detail from those of other industries. This paper discusses the special characteristics of the process industries and suggests useful ways to classify, collect, and analyze their quality costs.

Text

Compiling quality costs to measure the current economic state of quality in a business is hardly new. The ideas which shape our current views of quality costs go back more than twenty-five years. A number of people proposed various ways to classify and group quality costs to provide a much needed insight into the economic effect of quality on a business, to guide work on reducing quality costs, and to measure progress in that direction.

Masser (1) appears to have been the first to propose grouping quality costs into categories of Prevention Costs, Appraisal Costs, and Failure Costs. These broad categories are mentioned by Juran (2) and Seder (3), but not emphasized. Feigenbaum (4) adopted the terms and divided Failure Costs into separate classes of Internal Failures and External Failures. He suggested in some detail the cost elements which should be included in each of the four categories. With the publication of *Quality Costs — What and How* (5) in 1967, Internal Failure, External Failure, Appraisal, and Prevention became firmly established in the quality engineering lexicon as the names of the basic categories of quality costs.

The utility of these categories as the primary classification of quality costs has been amply demonstrated over the years in almost every type of business. Within these categories, however, the specific elements of quality costs appropriate for inclusion are not universally applicable across all businesses. In this paper, I would like to talk about some of the elements of quality costs which have been found useful in process industries.

To begin with, we should describe a process industry to distinguish it from a mechanical, electronic, service, or other kind of industry. The products of process industries are usually in some bulk form rather than in discrete units. These may be liquids (like many chemical or petroleum products) or solids in the form of powders (like some other chemicals, or cement), pellets (like some plastics), or rolls (like paper or some other plastics). Some businesses are hybrids with many of the characteristics of a process industry in the upstream operations, but with a final product in discrete units. For example, photographic film, flashlight batteries, glass bottles, certain prepared foods, and tires are in this class.

The primary characteristic of a process industry, however, is that it is materials intensive. That is, materials represent a large fraction of the total cost of the product, typically 50 to 80% or more. A major determinant of success in process industries is how effectively those materials are managed through the process, so an important criterion of plant performance is yield, measured as the ratio of pounds out/pounds in or square yards out/pounds in, etc.

This point of view has a profound impact on the perception of Internal Failure Costs in process industries. It is common for manufacturing, technical, and accounting people to treat all forms of yield loss with equal concern, without distinguishing between defective product and process wastes. Although it has been stated (6) that including "unavoidable" costs, such as process waste, in the quality costs will expose them to challenge, my experience has been that process industry people relate better to an enumeration of Internal Failure Costs which includes process waste. The reason is that both process waste and true defectives are perceived simply as different forms of yield loss. Purcell's description (7) of broke (scrap) in a paper mill as "planned," "expected," or "unexpected" illustrates this way of looking at yield loss.

When the various reasons for yield loss are analyzed by arranging them in order of their magnitudes in what is commonly (if erroneously) called a Pareto distribution, the process waste items are appropriately ranked among the defective product items. In process industries, where it is not unusual to find the quality engineering function combined with process engineering, product development, or other technical functions, work on process waste or product defects frequently involves the same people and the same investigative approaches. As long as the item is among the vital few, work to understand its causes and correct them has the same effect on yield improvement, whichever kind of item it is.

The foregoing argument implies that process wastes are not necessarily unavoidable and process industry experience has shown this to be so. An example is found in a plant making vinyl fabrics. The vinyl coating was formed in a calender and laminated to a textile substrate; in the same pass through the machine, a texture was embossed into the vinyl surface. It was well known that the adhesion between the vinyl and the textile was inadequate in the first roll of any order because the process required the frictional heat generated in steady state operation to reach a temperature high enough to ensure good adhesion. The first roll of each order was customarily rejected as startup waste and sent for reembossing to an off-line machine which was kept hot enough to secure adequate adhesion. When it turned up in the Pareto distribution as a significant item of Internal Failure Cost, a study of the problem was made. Adhesion was measured on samples taken at frequent intervals along the length of a number of first rolls. The investigation disclosed that the adhesion reached a satisfactory level after about 125 to 175 yards were coated. By changing the manufacturing procedure to begin each run with a 200 yard roll (which was automatically rejected and sent for reembossing) followed by full 1000 yard rolls of acceptable product, an 80% reduction in startup reembossing was achieved.

It is often difficult to distinguish clearly between a process waste and a defective product. An extruded plastic film provides an example. The plastic was extruded through a flat die and cooled on a chilled roll. As the film was wound into a finished roll, 4 inches was trimmed from each side of the web. This represented 17% of the material processed, so it quickly came to the top of a Pareto distribution of yield losses. An investigation showed that the reason for the wide trim ribbons was the inability of the process to maintain the thickness tolerance at the edges of the web. The question of whether, as side trim, the loss is "unavoidable" process waste, or whether as failure to meet the thickness specification, it is defective product never came up in this plant. Instead, the extrusion die was simply repaired to improve its thickness capability near the edges. This allowed the edge trim to be reduced to 1 inch per side, or just over 4% of the material throughput.

In many process industries, batches of intermediate or finished product which do not meet specifications, and which represent a relatively large Internal Failure Cost, can be recovered by blending with other good lots in some appropriate proportion. This produces a series of lots which are all inside the specification limits, leaving only a very small net loss. A better insight into the pattern of quality costs can be had if the Quality Cost Report shows the gross Failure Cost of the rejected batches. If it happens with any frequency, monitoring the bad batches may allow a product performance problem to be anticipated, suggest a potential problem with raw materials or process conditions, or signal an excess of batches requiring blending. When a nonconforming batch is recovered by blending, the recovered value should be shown on a separate line of the Quality Loss Report as an offsetting credit to the gross Failure Cost. The difference between the two should also be reported as the net Failure Cost.

Another kind of yield loss, which is not an obvious process waste and which is found in some process industries, is the variance from standard material usage not accounted for by reported scrap. Some examples of the sources of this loss are the variation from standard in the solids content of liquid chemicals (either as the seller's product or the buyer's raw material), or in the moisture content of paper. If the raw materials for a product are bought by weight, but the product is sold by area or volume, such things as variation from standard in the thickness of a plastic film or in the weight of a glass bottle fall into the same class. Also included are spillage and sweepings, and evaporation, both literal and figurative.

An element of Internal Failure Cost which I have occasionally found useful is Unexplained Inventory Adjustments. This is based on the theory that inventory adjustments for which no reason can be found are a hideout for unreported scrap. Although direct action can seldom be taken to reduce this cost, including it in the total quality cost removes an incentive to misreport. I have noticed that a campaign to reduce unexplained inventory adjustments can result in an offsetting increase in reported scrap. The benefit is that the reported scrap is identified by cause, making it possible to analyze the data and — where it is warranted by the magnitude of the loss — to investigate and correct the problems.

In one instance, it was possible to act directly on an unexplained inventory adjustment. In this case, a plant was experiencing chronic shortages of a liquid plasticizer, which was bought in bulk in tank cars. The engineer who investigated the problem found that the accounting department was paying the invoice and booking into inventory the invoiced quantity. The warehouseman, on the other hand, was pumping out the tank cars without using a flow meter. In addition, through careless placement of the pump suction line, the end of the shift, or other interruptions, he was not completely emptying tank cars.

The first action was to install a flow meter in the pump system to measure the actual quantity received. Inventories were then charged with only the measured amount and invoices paid on the same quantity. The supplier quickly complained about the short payments and added that he had noticed for some time that tank cars were being returned with some product left in them. To close the loop, the warehouseman was required to certify that he had emptied the tank car, and the supplier agreed to report to the plant any returned product. As a result, the chronic plasticizer shortage was eliminated.

Because of the nature of their products, process industries do not usually have warranty costs as an item of External Failure Costs. Adjustments, however, are made for defective product found by customers and quantities of unsatisfactory products are returned. External Failure Costs should include these amounts in total as a partial measure of customer dissatisfaction. Sometimes returns can be salvaged by resale to a customer with a less stringent requirement, by blending with good lots, or by recycling through the process. When this happens, the Quality Cost Report should show the offsetting credit for the value of the recovered material, and the net External Failure Cost.

Appraisal Costs in process industries have, in general, the same elements as in other industries, but the proportion of Appraisal Cost allocated to chemical, physical, or metallurgical testing tends to be much higher. In recent years, increasing trends towards on-line instrumentation to measure both product and process characteristics and towards computers to process the data and report it (or to adjust the process automatically) has modified the nature of Appraisal Costs. Process instrumentation should be included as an element of Appraisal Cost.

The elements of Prevention Costs listed in standard textbooks or in *Quality Costs — What and How* (8) apply equally well in the process industries, and no special items appear to be needed.

Summary

Process industries are materials intensive and generally deal with bulk products. Measuring quality costs in these industries is the same in principle as for other industries, and the usual categories of Appraisal Costs, Prevention Costs, Internal Failure Costs, and External Failure Costs apply. The major differences between standard practice and the recommended quality cost elements for process industries are in the Internal Failure Costs, where these elements are included:

- Process Wastes
- Excess Material Usage
- Unexplained Inventory Adjustments

References

(1) Masser, W. J., *The Quality Manager and Quality Costs,* Industrial Quality Control, vol 14, no 6, October 1957, pp 5-8

(2) Juran, J. M., ed., *Quality Control Handbook (2nd Ed),* McGraw-Hill, NY 1962, p 1-46 (Juran)

(3) Ibid, p 11-4 (Seder, L. A.)

(4) Feigenbaum, A. V., *Total Quality Control,* McGraw-Hill, NY 1961, pp 83-89

(5) Quality Costs Committee, *Quality Costs — What and How,* ASQC, Milwaukee 1967

(6) Juran, J. M., and Gryna, F. M., Jr., *Quality Planning and Analysis, (2nd Ed),* McGraw-Hill, NY 1980, pp 23-24

(7) Juran, J. M., ed., Op cit, p 30-6 (Purcell, W. R.)

(8) Quality Costs — Cost Effectiveness Technical Committee, *Quality Costs — What and How, (2nd Ed),* ASQC, Milwaukee 1971, pp 8-12

QUALITY COSTS: WE KNOW WHERE WE'RE GOING! DO YOU?

James J. Wayne
Manager, Accounting
Andrew F. Grimm
Manager, Quality Control
Harnischfeger Corporation
Construction Equipment Division
Escanaba, Michigan

Abstract

The Quality Cost System is relatively new to the Escanaba operations. This paper comments on the problems in developing the system, the accounts to include, and the review of the rough system outline with the Division General Manager. Once totally developed, the system was presented to the Division Management Staff. Historical data for the previous three years and the first quarterly report was scrutinized. In evaluating the data, we found a gap between External Failure and Prevention costs. The conclusion of the paper discusses the development of the plans for improving the condition and the outcome from these plans.

Text

In the process of establishing a Total Quality Control program for the Escanaba operations, it became apparent that a quality cost reporting system was needed to determine the allocation of resources, track quality cost trends and establish economic goals.

An evaluation of the economic significance of the elements in the TQC program had to be made to determine priorities needed for reaching rational decisions concerning resource allocation. Indeed, what better way is there to establish an optimal TQC program than from the financial viewpoint of the business. The ultimate goal of the TQC program is to improve operating profits. Increased profits are anticipated as a result of a reduction in total failure costs and corresponding improvement of product acceptance by customers.

In order to accomplish the task of creating the Quality Cost Report, the Manager, Accounting and the Manager, Quality Control met at the request of the General Manager to work out a plan for designing, developing and implementing a Quality Cost Reporting system. In terms of Managerial Accounting, the Quality Cost Report is a segment report.

Segment reports illuminate financial results of particular aspects of the business. Quality Costs are the financial results of one segment of the business, i.e., product quality.

The first step in developing the report format was for accounting personnel to learn the concepts and terminology of quality costs and the Quality Cost Reporting system. ASQC's publication, "Quality Costs, What and How," was selected as the primary reference and, served as the outline for extracting the components of an effective quality cost report from the existing financial reporting system. The first outline of the Escanaba Operation's Quality Cost Report listed accounts from the existing chart of accounts. After a final review of the the report's design, historical quality cost data was generated in order to establish a base line demonstrating the nature of past costs (see Figure 1).

The first Quality Cost Report, 4th quarter fiscal 1981, was prepared with actual amounts compiled from the financial reporting system. At this time, it was decided to issue the Quality Cost Report on a quarterly basis to avoid the "report card" syndrome. Instead, it was resolved that the Quality Cost report would become management's tool for analysis, planning and control of the business activities needed to assure the intended quality levels of products.

THREE YEAR QUALITY COST DATA BASE

		Avoidance Costs			Failure Costs		
Fiscal Year	Quarter	$ (000's)	Cost Per EPH*	% of Net Sales	$ (000's)	Cost Per EPH*	% of Net Sales
1979	1	395	1.46	3.2	1,050	3.89	8.6
	2	460	1.57	2.8	1,096	3.73	6.7
	3	452	1.65	2.2	843	3.08	4.1
	4	434	1.71	2.1	1,391	5.45	6.7
1980	1	244	1.43	2.0	796	3.31	4.5
	2	367	1.61	1.6	1,038	4.54	4.4
	3	333	1.95	1.7	906	5.32	4.6
	4	333	1.86	1.6	1,627	9.10	7.8
1981	1	375	2.24	2.0	913	5.46	4.9
	2	859	4.08	5.4	1,140	5.41	7.1
	3	481	2.45	2.2	1,049	5.34	4.9
	4	596	3.12	2.7	1,622	8.49	7.3

*Earned Productive Hours   Figure 1

The Manager, Accounting and the Manager, Quality Control met again with the General Manager to discuss the rough draft of the fourth quarter, fiscal 1981 report. The General Manager recommended changes that further tailored the report to represent the operating environment. It was also decided to present quality cost theory and concepts, and the structure of the Quality Cost Reporting system to the Division's management staff because these are the individuals responsible for the success of the TQC program.

The presentation to the staff consisted of two parts. The first part described the quality cost reporting system and the second part was a discussion of the first edition of the Quality Cost Report.

A handout was prepared for the staff managers. An overhead transparency series was used to highlight the components of the reporting system. Following are brief statements representing the actual text of the handout:

Purpose — The quality cost system provides financial information to management for use in *analyzing, planning* and *controlling* product quality costs.

Scope — The reporting instrument of the system is the Quality Cost Report. The report covers avoidance and failure costs. Avoidance Costs are composed of prevention and appraisal costs and both internal and external failure costs comprise the Failure Costs structure.

Basic Philosophy — The Quality Cost system is basically oriented to "Management by Objectives." The object is to find and plan for the achievement of the optimal quality cost point. By tracking avoidance costs vs failure costs, a break-even point can be established at the intersection of the two cost lines. As expenditures for avoidance costs are increased, corresponding reductions in failure costs should be expected. If allocated fixed costs are added to the sums of cost values at succeeding points in time, a total cost curve is developed. The vertex of the total cost curve (its first derivative) is the optimal point. The optimal point is usually located at the same time space as the break-even point. If no further changes are made in costs, then the optimal point represents the operative cost system (see Appendix A).

If management is dissatisfied with the optimal point in terms of the dollar cost level and the time space level, then, in reality, it is dissatisfied with the current quality cost structure. Current planning would have to be changed to produce a cost structure more to management's liking. The Quality Cost Report should then generate the questions:

> "What must we need to do so that we know what the operation of the cost structure looks like?"
> "Where is the break-even point located?"
> "What does the corresponding optimal point look like?"
> "What must we do to achieve the optimum break-even point?"

The answers to these questions lead to fulfillment of the requirements of the Managerial Accounting process, i.e., *Analysis, Planning* and *Control*.

Historical Overview — A three year review of data, by quarter, indicates that the avoidance and failure cost lines run parallel instead of converging (see Figure 3).

Also, avoidance costs are maintained at a significantly low level. Analysis of the prevention and appraisal cost components of avoidance costs indicate the majority of expenditures were for appraisal when the Prevention Cost element, Quality Planning Other Than QC is subtracted (see Figure 2).

Quarterly Average (Rolling Three Year)
As of 4th Quarter 1981

	Quality Costs $	Quality Costs/EPH	% to Net Sales
Prevention Costs			
Quality Engineering	1,629	.008	.01
Quality Planning Other Than QC	214,047	.959	1.11
Quality Training Programs			
Quality Administration	13,503	.061	.07
Subtotal	229,179	1.028	1.19
Appraisal Costs			
Vendor Quality Assurance	1,770	.008	.01
Laboratory Acceptance Tests	6,610	.030	.04
Plant Inspection and Tests	189,161	.847	.97
Checking Labor	21,560	.097	.12
Product Quality Audit			
Calibration Control			
Inspection Tools and Supplies	4,157	.019	.03
Subtotal	223,258	1.001	1.17
TOTAL AVOIDANCE COSTS	452,437	2.029	2.36
Internal Failure Costs			
Scrap - P&H Error	52,871	.237	.28
Rework - P&H Error	191,492	.857	.98
Scrap - Vendor Error	73,265	.329	.39
Rework - Vendor Error	8,636	.039	.05
Pickup Labor	106,745	.479	.56
Subtotal	433,009	1.941	2.26
External Failure Costs			
Warranty-Marketing Error	111,284	.499	.58
Warranty-Engineering Error	173,356	.778	.91
Warranty-Manufacturing Error	108,390	.486	.57
Warranty-Purchasing Error	39,246	.176	.21
Liability Insurance Program	257,482	1.154	1.33
Subtotal	689,758	3.093	3.60
TOTAL FAILURE COSTS	1,122,767	5.034	5.86

Figure 2

This analysis shows that expenditures have to be increased in the prevention category elements associated with Quality Control in order to reduce failure costs. The objective of planning is to direct the quality cost lines toward a break-even point.

Quality Cost Report Accounts Outline — The outline suggested for the Harnischfeger Quality Cost Report follows the outline contained in the booklet, "Quality Costs — What and How." A condensed version of the full Harnischfeger outline follows. The first element of each category is shown in detail. The remaining elements in the category are just listed.

Quality Cost Element	Accounts
A) Prevention Costs 　1) Quality Engineering	Compensation cost for Quality Engineers Account Numbers: 8575-202 - Salaries 　　　　etc.
2) Information Equipment Engineering 　3) Quality Planning, other than QC 　4) Quality Training 　5) Quality Administration	
B) Appraisal Costs 　1) Vendor Surveillance	Compensation costs for Assistant Quality Control Manager-VQA, Receiving Inspectors, Layout Inspector plus all other costs pertaining to Vendor Surveillance Account Numbers: 8575-202 - Salaries 　　　　etc.
2) Laboratory Acceptance Testing 　3) Inspection and Test 　4) Checking Labor 　5) Product Quality Audits 　6) Calibration Control System 　7) Test and Inspection Materials	
C) Internal Failure Costs 　1) Scrap - All Scrap Losses	Scrap costs divided as to P&H error or vendor error Account Numbers: 8502-558 - Scrap-Vendor Error 　　　　etc.
2) Rework and Repair 　3) Pickup Labor	
D) External Failure Costs 　1) Warranty Replacement Costs	Costs are divided as to domestic or export sales. Account Numbers: 4207-XXX - Warranty - Domestic Machines 　　　　etc.
2) Liability Insurance Premiums	

Construction of the Quarterly Quality Cost Report — There are three sections contained in the Quality Cost Report:

1) Financial Data
2) Trend Charts — Breakdown Analysis
3) Evaluation and Recommendations

Accounting and Quality Control Contributions — The Quality Cost Report is constructed by contributions from both the Accounting and Quality Control Departments. Accounting constructs the Financial Data section, Quality Control constructs the Trend Charts section, and, both departments jointly develop the Evaluation and Recommendations section.

The second part of the presentation covered the first Quality Cost Report issued for the fourth quarter, fiscal 1981. (Note: Harnischfeger Corporation's fiscal year runs from November 1st to October 31st.) The report's financial data section is divided into three periods in order to provide comparisons for determining progress or pointing up danger signals. The three parts are:

the actual results for the reporting quarter,
fiscal year to date results, and,
a three year quarterly rolling average

Within each time period, actual costs, and, costs related to Earned Productive Hours and to Net Sales, are listed for each quality cost element.

The first and most obvious success of the quality cost program was to give management its first look at the true picture of quality cost performance. The quality cost reports demonstrated where the operation had been and how management could expect quality costs to behave in the future with proper planning. Charted historical data showed that the Avoidance and Failure Cost lines ran almost parallel instead of intersecting at a break-even point. In addition, avoidance costs remained significantly lower than failure costs. Avoidance costs had been expended primarily in the Appraisal category, and specifically in Inspection costs. With the inspection function being emphasized, the best that could be hoped for would be the transfer of failure costs from external to internal. However, these hopes did not materialize. Even with the emphasis on inspection, significant external failure costs were continually being incurred. This was due, in part, to the relatively low expenditure for prevention (see Figure 3).

Consequently, the historical cost data generated by the quality cost system has found management to evaluate avoidance cost expenditures. More emphasis must be placed on Prevention Costs in order to address and correct the "causes" of unquality rather than operating on the principle of separating the good from the bad. This new emphasis on Prevention must be in the areas of Quality Engineering and Quality Training. It is these two areas that offer the greatest potential for causing a convergence of the avoidance and failure cost trend lines to a break-even point.

Historical cost data also showed that the cost element, Quality Planning Other Than QC, has little impact on total failure costs. Therefore, an evaluation of this area of Prevention Cost is required to determine if the continued expense level is justifiable.

Four fiscal quarters of quality cost data have been compiled from the time of issue of the first report to the time for constructing the Escanaba operation's five year plan for 1983 to 1987. The recently designed TQC program was incorporated into the five year plan for the first time.

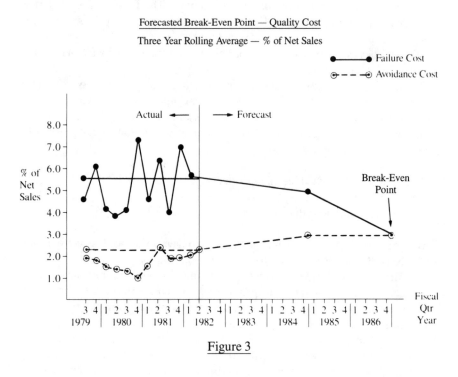

Figure 3

An important tool used for establishing priorities for TQC program activities implementation was the quality cost report and its historical data base. Since the Quality Cost Report has defined the need for added emphasis on Prevention activities and since the TQC program is Prevention oriented, the five year plan reflects the needs by planning for an increase in the Quality Engineering section and an increased emphasis in Quality Improvement programs and their training sessions. Figure 3 shows the estimate of the break-even point for the Escanaba operations. Note that the historical data is expected to continue until 1982 when the earlier initiated expenditures for prevention were commenced. The convergence of the quality cost trend lines at the break-even point is expected in 1986.

We know where we're going, do you?

Note: Data in this paper have been changed from actual. However, facsimile data are valid for evaluations and conclusions.

Appendix A

Theoretical Quality Costs Chart

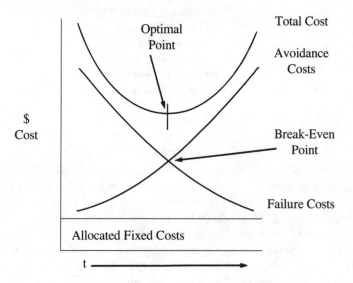

Bibliography

1. ASQC, "Guide for Reducing Quality Costs," ASQC Quality Costs Committee, 1977
2. ASQC, "Quality Costs — What and How," 2nd Edition, Quality Cost — Cost Effectiveness Committee, 1971

THE QUALITY MANAGER'S JOB: OPTIMIZE COSTS

William C. Noz, Jr.
Principal Engineer
Bradley F. Redding
Quality Manager
Paul A. Ware
Senior Engineer
Polaroid Corporation
Cambridge, Massachusetts

Abstract

The quality manager's fundamental job is to optimize the quality cost for the company's products. The accepted principles of quality costs and their category relationships provide the foundation for a new approach to identifying the opportunities for savings and the steps required to attain optimization. The paper examines a unique process for analyzing these opportunities on a product by product basis, and describes a powerful technique for establishing priorities for improvement projects. The methodology presented provides the quality manager with a strategic tool to achieve and maintain optimum quality costs.

Introduction

The importance of total quality cost data to the successful enterprise is well documented. Quality Assurance literature is replete with definitions and descriptions of how the quality manager can use this information to quantify objectives, set priorities, focus management attention, measure success, integrate it into strategic plans, and convert quality activities into the language of management.

Regardless of the known importance of quality costs, however, the fact remains that most quality managers find it difficult to make *effective* use of total quality cost data. This is most likely due to three major factors: the pressures of the working environment, the difficulty in getting the numbers, and the lack of simplicity in their application.

In this paper, we will develop the significance of the historical information pertaining to quality costs and their relevance to the overall strategy of any organization. We will present a new technique for using the quality cost facts on a product by product basis to analyze cost reduction opportunities, set priorities, and progress toward optimum total quality cost. Our aim is to re-emphasize the importance of this process and re-focus the manager's efforts toward this objective.

The Quality Manager's Job

If you look at any typical manufacturing organization today, you will find the Quality Manager virtually flooded with a variety of input about every aspect of the job as depicted in Figure 1.

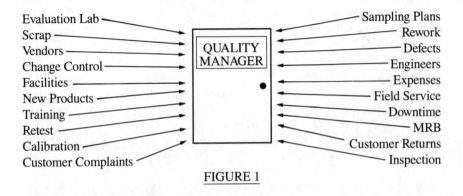

FIGURE 1

These problems are all real issues competing for the manager's attention, negotiation, and decision. The list of projects is endless; ideas for new approaches to the quality system abound. While all these projects and approaches have legitimate ends in themselves, their priorities are often not focused effectively toward an overall goal. The pressure to join with the manufacturing "team" for a quick solution of problems is pervasive because these solutions seem to be "vital" to the survival of the organization.

Since this environment is not likely to change in the short range, the manager needs a system to set priorities for the resources that *are* available. But how does the manager do this? What is the long run responsibility? What plan should be used for overall effective control of that responsibility?

We must begin with the reason for having a Quality Control organization and a definition of this manager's function:

> *The Quality Manager's job is to optimize the quality cost of the manufacturing organization.*

To accomplish this task the Quality Manager must focus on all the quality costs and know how to use them. Done correctly, an efficient program for driving the total cost toward optimum can be developed to the best advantage of the organization and the individuals involved. This program requires:

1. getting the total cost facts
2. knowing what they mean
3. understanding the optimum target
4. setting priorities by opportunity dollars
5. executing the plan and tracking progress

Quality Costs

The definition common to the literature is restated here for completeness:

1. *General Definition:* Quality Costs are *all* those costs due to *defective* product, the cost associated with making, finding, repairing and avoiding defects. The cost of making good products is not included.
2. *Categories:* Quality Costs are usually expressed in the following categories:
   External Failure: all activities resulting from failure to meet the product and service requirements of the customer to his satisfaction.
   Internal Failure: all activities before receipt of product by customer caused by failure of vendors or internal personnel or equipment to produce a satisfactory product.
   Appraisal: inspection and test (other than development) of the product, processes and all their constituent or replacement parts.
   Prevention: activities designed to keep defects or customer dissatisfaction from occurring in the first place.

The concept of optimum total quality cost is also well known. Figure 2 shows the general relationship and interaction of the four cost categories as quality of conformance ranges from one extreme to the other. The region of optimum total cost is easily identified. In Figure 3, five specific phases are identified within this region. Each phase represents the progression of activities in the development of the quality system and the drive toward optimum. The height of the bars is significant in that their relationship to one another represents the total quality cost which should be present in each phase if taken as a percent of phase A.

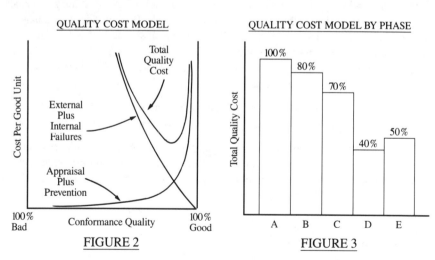

FIGURE 2   FIGURE 3

Table I describes the typical operational conditions of the quality system in each phase and the action items required to move the system into the next phase.

QUALITY COST PHASE

	A	B	C	D	E
OPERATIONAL CONDITIONS	Unfocused effort	Internal product sorting in place	Process controls installed	Improvement projects defined and active	Too much appraisal and planning
ACTION ITEMS	Establish final acceptance to sort product internally Develop data system Analyze problems	Establish process controls based on problem analysis	Establish improvement projects based on process control data	Maintain the status quo	Relax controls as appropriate

TABLE I

Figure 4 shows that certain relationships exist amongst the four quality cost categories *within* each phase. Figure 5 further expands these cost relationships as a percent of total cost to show clearly the reallocation of quality dollars to each category in each of the five phases.

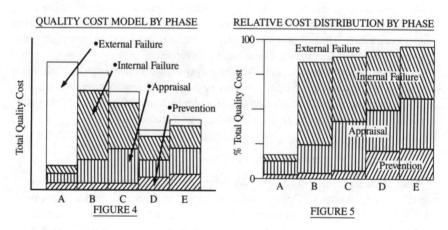

FIGURE 4   FIGURE 5

The significance of these phases and their definitions is that they provide both direction to an evolutionary system and the ability to set priorities for the effort to implement such a system. Table II shows the opportunity for real cost savings in moving from any given phase to improved phases.

	PERCENT OF TOTAL QUALITY COST SAVED BY MOVING TO PHASE		
CURRENT PHASE	B	C	D
A	20	30	60
B	——	12.5	50
C	——	——	43
E	——	——	20

TABLE II

The Optimization Process

The process of quality cost optimization can be traced in steps related to the defined phases in the development of the overall quality system. It is tempting to attack a move to Phase D immediately by simultaneously instituting sample plan changes, process controls and improvement projects.

This, however, is the same as attacking all problems at once in the current environment and leads to confusion and dilution of effort. The correct approach requires discipline and concentration. The action items of each phase must be implemented before progress can be made to the next phase. There are no shortcuts.

With these concepts in mind, the job of the quality manager is twofold: (1) to determine the phase of his quality system and (2) separate the total job into manageable elements and set priorities to handle each element. In order to identify the phase of the quality system *two* criteria must be analyzed with respect to each other:

- quality cost relationships
- statements of operational condition

Both of these criteria must agree on the phase identification before the phase is determined with certainty. One expected problem arises from incomplete cost analysis, where hidden costs have not been included, resulting in too optimistic a statement of phase. Conversely, too often the tools for an advanced phase are put in place but are shown to be ineffective when tested against the actual cost data.

The separation of the total job into manageable elements for priority setting is most easily and powerfully accomplished by breaking out the total quality costs *by product*. Each product will usually be found to be in its own phase, requiring actions different from other products. Figure 6 illustrates an example of the quality costs of five different products. Their total quality costs and proportions by category are shown, together with their current phase.

There are two keys to setting priorities for action. First, the current phase of each product must be converted into opportunity dollars to be saved in moving each product to the next phase. This is done in accordance with Table II, and for the situation of Figure 6 yields the following hierarchy of cost objectives, with Product IV showing the largest opportunity dollars to be saved:

PRODUCT	OPPORTUNITY DOLLARS
I	$ 80
II	20
III	25
IV	130
V	30

However, the *relative difficulty* in moving from one phase to another forms a second hierarchy which must be considered. This scale takes into consideration the fact that it becomes progressively more difficult to accomplish the action items of each successive phase as optimum is approached.

Together these two hierarchies can be combined into a single "priority matrix" as illustrated in Figure 7. When this is done for the example products, Product IV becomes third in priority for action as a result of applying the priority setting mechanism. The priorities for each of the nine cells are indicated in their lower left hand corner. As can be seen, the priorities are determined by moving diagonally from the upper right cell to the lower left, with the upper cell having top priority in each diagonal. The vertical axis shows the current phase of each product, with only the most common phases of A, B, and C shown here. Many schemes are possible for determining the cell size on the horizontal axis, depending upon the desires of the manager.

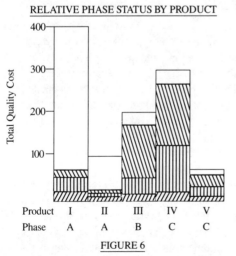

FIGURE 6

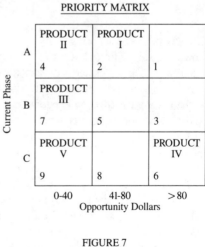

FIGURE 7

Getting Started

Although the literature describes what the quality cost numbers are and how to get them, getting these facts may prove difficult for a number of reasons. The manager is wise to approach this task carefully because it requires the joint effort of several members of the management team, including leadership by the quality and financial managers, as well as agreement from the line managers. In this activity, good sense must prevail over excessive detail.

We recommend that the first step in this program be a preliminary overview or a "rough cut" by the quality manager requiring a small amount of time. The objective here is to solidify the concepts and to draw attention to the magnitude of the dollars at stake. Table III illustrates this preliminary overview of costs.

PRELIMINARY COST/PHASE ESTIMATE

	EF	IF	A	P	TOTAL
MANPOWER - QUALITY - PRODUCTION					
MATERIAL - QUALITY - PRODUCTION					
EXPENSES (Extraordinary)					
OCCUPANCY (Extraordinary)					
TOTAL					
% OF TOTAL					

TABLE III

Here, the cost categories are listed at the top and various types of costs are arranged down the side. Through a series of only 18 questions, the costs, *by category,* are determined as a percentage of the total quality cost. Using these percentages and Figure 5, an estimate is made of the current phase. However, these costs must now be reconciled with the statements of condition for the identified phase. As stated earlier, *both* the costs and the condition statements should be in agreement in order to identify an overall phase.

The second step in this process is a detailed study *by product.* Here, the rough numbers of step one are developed more thoroughly and the real costs emerge for application in the priority system described above.

This technique then provides the Quality Manager with an overall strategy for developing a long-range optimization policy for the quality system while working within the short-range, task-oriented environment. It allows him to focus on the job of optimizing quality cost in a systematic, evolutionary manner. The paybacks for Quality are several: the manager can focus on his *real* job while he develops and controls his own resources; the individual is assigned to a more fulfilling job; the organization gets the biggest payback in the fastest manner and understands its real problems.

Summary

The methodology presented here incorporates into the fundamental quality cost thought process a unique way of determining priorities for optimizing this cost. We have described a simplified concept which is one part of a complex issue, but a concept which we believe is very useful to analyze the quality cost position, set priorities for improvement projects, and strive for optimization. From this base, other techniques can be employed to carry out the optimization effort in more depth by identifying the specific project areas to attack. There are also many techniques to report and monitor progress toward optimization, many of which could be used with equal success.

Our objective is to emphasize the importance of quality cost optimization as a primary goal. We believe that the cost phase model and the priority matrix are essential tools for this process. These techniques should become part of the quality control "toolbox" and should be used regularly to assess and improve the quality cost process.

Bibliography

American Society for Quality Control, *Guide for Reducing Quality Costs,* ASQC, Milwaukee.

American Society for Quality Control, *Quality Costs — What and How,* ASQC, Milwaukee.

Juran, Joseph M.(ed.), *Quality Control Handbook,* 3rd Edition, McGraw-Hill Co., New York, Chapter 5.

Juran, Joseph M. and Frank M. Gryna, *Quality Planning and Analysis,* 5th Edition, McGraw-Hill Co., New York.

Veen, B., Quality Costs, *Quality,* No. 2, Journal of the EOQC, 1974.

QUALITY + OR − QUALITY COSTS EQUALS PRODUCTIVITY

Charles A. Aubrey, II
Manager, Quality Assurance
Debra A. Zimbler
Quality Control Analyst
Continental Illinois National Bank
and Trust Company of Chicago
Chicago, Illinois

Abstract

This paper will explain quality costs in a bank or clerical environment, how quality costs in bank operations relate to quality in manufacturing operations and how an awareness and preventive attitude can be developed within a bank organization. The paper will also demonstrate how quality costs can be isolated, measured, analyzed, and reduced or shifted and how quality can be defined and measured in various ways.

Introduction

At Continental Bank, we define quality as the degree to which a product or service conforms to predetermined standards. The standards are established based on the necessary requirements as perceived by the customer and the performance for which the product was designed. In the banking industry, this definition is translated into providing accurate and timely information and transaction services to our customers while minimizing costs and maintaining competitive prices.

We measure our quality in two general ways: 1) adherence to standards; and 2) optimizing quality costs. The basic philosophy of establishing quality standards and then using them as a determinant of quality is the same for a manufacturing concern as for a service industry; however, the application differs. The quality of the manufacturing concern's output is evaluated by more tangible criteria than is a service industry's.

Whereas a manufacturing concern initially inspects tangible items based on physical and functional conformance, and then accepts or rejects the product, a service industry establishes standards for a product or service which are much less tangible. A manufacturing concern sets internal standards which are often based on external requirements, and then tests the production process to insure functional uniformity, product reliability and product safety. The service industry also sets standards, but because if offers a service rather than a manufactured item, the standards are based on fewer tangible, safety-related criteria. A service industry's standards reflect its concern for reliability. Timeliness and accuracy are the two critical components that determine a bank's quality, both internally and externally; however, timeliness and accuracy are less concrete than whether or not a metal fastener will withhold 2,000 pounds of pressure.

The use of quality standards is one method of determining a service industry's quality. In an effort to be more concrete and quantitative when evaluating quality levels, a service industry, and particularly a bank, can examine costs.

The costs of producing or offering a product or service are evaluated to determine the efficiency of the operation. If the cost of an operation is high due to rework or scrap, then two assumptions can be made. First, the operation is not as efficient as it could be, and second, the employees are not working as productively as possible. Either of these assumptions, or both, may apply and can be tested. Development of quality costs can help to quantify specific quality levels and ultimately improve productivity. If the quality costs reflect high failure costs, a quality improvement project can be undertaken to reduce costs and, in turn, increase productivity. However, before developing quality costs, the definition of each cost category should be clearly understood.

Quality Costs

For every industry, there are costs associated with providing and insuring a high-quality product and/or service. These costs are designated as quality costs and apply to both manufacturing and service industries. The four quality cost categories are prevention, appraisal, internal failure, and external failure. We have defined them as they relate to the banking industry.

1. *Prevention:* Prevention costs are those costs associated with operations or activities that keep failure from happening, and keep appraisal costs to a minimum. Examples of prevention activities are new product reviews, quality planning, Quality Circle meetings, training programs, written procedures, analysis of quality information and quality improvement projects.
2. *Appraisal:* Appraisal costs are those costs incurred to ascertain the condition of a product or service, in order to determine its degree of conformance to quality standards. Examples of appraisal activities are inspection of incoming work, supplies and materials, periodic inspection of work in process, checking, balancing, verifying, final inspection, and collecting quality data.
3. *Internal Failure:* Internal failure costs are those costs that are incurred as a result of correcting non-conforming service or products produced, prior to delivery to the customer. Examples of internal failure are machine downtime, scrap due to improperly processed forms and reports, and rework of incorrectly processed work.
4. *External Failure:* External failure costs are those costs that are incurred as a result of correcting non-conforming service or products produced, *after* delivery to the customer; or correcting a product or service that the customer perceives does not

conform to *his* specified standards. Examples of external failure costs are investigation time, payment of interest penalties, reprocessing of an item, scrap due to improperly processed or incorrect forms reports, or lost or never acquired business due to providing poor service or having a poor quality reputation.

Failure costs should always be considered the costs that would disappear if the product were defect free during or at the time of completion.

Quality costs are interrelated, and the objective in allocating resources for quality activities is that dollars spent on prevention and appraisal should prevent or reduce failure costs. In general, the optimal quality cost mix is the point at which an additional dollar allocated to prevention and/or appraisal will not reduce failure costs by more than one dollar. Beyond this point, although additional dollars spent on prevention and/or appraisal will continue to reduce failure and defects, overall quality costs will increase. However, particular industries which require stricter controls to achieve internal or external requirements may need to allocate more dollars to prevention and appraisal than initially indicated by the optimal mix point.

Therefore, each area or unit must determine its actual quality cost mix and its optimal cost mix, and consciously adjust quality activities (and their related quality costs) to reduce and erase any difference that may exist.

Quality Cost Process

The purpose of developing quality costs is threefold. The first objective of quality costs is to make management aware of the magnitude of quality costs and encourage proper managerial attention to quality and quality functions. The second objective is to help an area determine its current quality cost mix and whether or not adjustments should be made. The third objective of the quality cost study is to highlight any activities that have high failure costs associated with them. Activities with high failure costs can be identified as potential candidates for quality improvement projects. The successful completion of quality improvement projects can lead to substantial reduction in failure and rework costs, which ultimately results in higher productivity. Quality costs development can be broken into three parts: 1) data collection; 2) report preparation; and 3) analysis and feedback. To develop quality costs in a particular area, each activity done within the area is listed. Once the activities are listed, the supervisor enumerates how many people perform the activity, the salary level of each individual, and how much time is spent performing each activity. If time spent is not readily available, samples can be taken or work measurement can be used. Time can be specified as a percentage of a working day/week/month or in absolute hours.

The labor time shown on the report should balance to the total amount of labor time available in the area, including the supervisor's time. All the activities should be categorized into quality or non-quality activities. Non-quality activities are those activities which are considered strictly production or administration oriented. The identified quality activities are then categorized by specific quality cost category. It should be stressed that an activity can be part production/administration and part quality. An activity can be comprised of two or more quality components.

In preparing the report, each quality activity is first sorted according to type (prevention, appraisal, internal, and external failure), and is then costed out to determine each activity's cost, as well as each quality cost category's total. To determine the quality cost of a particular quality activity, the number of hours per month spent performing the activity is calculated. The total labor time is multiplied by the number of employees performing the activity to derive the total labor hours spent. The appropriate labor cost per hour is determined by multiplying the mid-point of the salary level by a factor of 1.xx which allows for fringe benefits associated with a basic salary. At Continental Bank, fringe benefits are 30% of basic salary, and therefore we use a 1.3 factor. A particular activity's cost is the product of multiplying the adjusted salary rate by the total labor hours spent. If more than one employee salary level performs an activity, the weighted average of the adjusted salary levels of all individuals performing that function are calculated when determining the cost of the activity.

An area's non-labor costs should also be reviewed and analyzed for their quality cost contribution. Examples of quality-related, non-labor costs are: having extra equipment on hand for possible machine downtime (an example of prevention), purchasing/ maintaining MICR testing equipment (an example of appraisal), destroying forms which were prepared incorrectly (an example of internal failure), and paying interest penalties (an example of external failure).

The calculations are completed when the quality costs of all activities are totaled by category to determine the percentage each quality cost category contributes to overall quality costs. Each category's total is also contrasted to the area's overall expense figure to provide further comparison of quality costs to overall area expense.

After the calculations are complete, the findings are analyzed and a summary with recommendations is prepared. The analysis highlights the magnitude of the quality costs, the area's quality cost mix and specific high failure costs. The analysis will suggest a future course of action. One potential course of action is a quality improvement project.

Generally, this is indicated when the optimum quality cost mix has not been attained. Specific quality improvement projects are indicated when a particular activity has high failure costs associated with it. The objectives in undertaking a quality improvement project are to reduce quality and total costs (or at least change the quality cost mix) while improving actual quality levels.

Quality improvement projects are not necessarily complex or difficult, and can yield a high payback without expending large amounts of money and/or time.

Case Study

A quality cost study was done in a loan processing section after developing quality costs for the area. The study showed that the activity "processing holdouts," which is defined as computer tickets that are rejected during daily processing, represented failure costs of more than $2,000 per month. Holdouts accounted for 30% of all tickets processed. In order to reprocess the holdouts, the reason for the reject had to be determined, and the tickets were held for next day processing. The delay prevented updating commercial lending information accurately and timely. This information is needed to determine a customer's credit availability, the bank's credit exposure and the bank's financial statements.

Prior to developing quality costs, the loan processing area had not been aware of the magnitude of quality costs, and in particular, the holdout problem as evidenced by its high monthly failure cost. The area was enthusiastic about initiating a quality improvement project in an effort to reduce holdouts, lower costs and increase quality. Since the cause of the excessive number of holdouts was unknown, the quality improvement project began with data collection.

In an effort to identify the problem, the area supervisor kept a holdout log. Each holdout was listed by the type of error that caused the reject, as well as the clerk who submitted it. Pareto analysis was done to determine which errors occurred most frequently and which clerks were responsible for the greatest number of errors. A matrix was constructed to see if there was a correlation between the two.

After completing the Pareto analysis, it was found that three types of errors were primarily responsible for the rejected tickets and that three clerks were responsible for the majority of the errors. The actual production operation was carefully monitored to determine the cause of the high error rates by certain clerks and the types of errors. Particular attention was paid to the quality of the incoming information and to how the clerk transferred it onto the computer input tickets. Further, the rejected tickets that were resubmitted were monitored to determine if they were again rejected. Since it was the responsibility of the clerk to determine the cause of the initial rejection and correct it, a twice rejected ticket might indicate that the clerk was unsure as to how to correct the error. It was found that few tickets were rejected twice which indicated that the clerks knew how to correct a rejected ticket.

Observations identified possible causes of the high error rates which were responsible for the rejected tickets. Frequently, incomplete or unclear information was received and some of the clerks were unsure as to how to proceed with the processing. Knowing the need for immediacy in processing, the clerks attempted to process the less than perfect information. It appeared that there was not a consistent method of completing input tickets since each clerk has his/her own method of completion. Further, since the supervisor had not been closely monitoring the holdouts, she was not giving the clerks necessary feedback or suggesting corrective action for the most frequently occurring errors.

These observations suggested multiple courses of action which were acted upon. Each clerk was required to participate in a training program in order to insure a uniform understanding of section and ticket processing procedures. A comprehensive procedures manual was written to accompany the training program and to serve as ready reference for the clerks. Clerks were encouraged to ask the senior clerk or supervisor questions concerning processing of non-routine items. By asking for clarification up front, prevention and appraisal activities were being performed in order to reduce failure costs later. Additional appraisal activities were now being performed by the clerks. They were instructed to reject input information that was incomplete or unclear and return it to the initiator, instead of attempting to process it and hoping it "would pass." By returning bad input to the user, the loan area was giving the user important feedback regarding their quality. This improved the quality of the input to the loan area.

The analysis had shown that the errors which occurred most frequently were caused by not matching two critical fields on the input ticket. As a result, the training course stressed that the clerks pay particular attention to correctly completing the matching fields.

The supervisor was encouraged to take greater initiative in measuring (sampling) the frequency and types of errors that were occurring as well as giving clerks frequent feedback. By monitoring the types of errors, the frequency and the person responsible, the supervisor could resolve or correct a potential problem before quality would be severely affected. The frequent feedback made the clerks more aware and more responsible for the quality of their work. Together, the supervisor and clerks worked to attack poor quality symptoms before they became quality problems.

Six months after initially developing quality costs, the costs were reviewed. The suggestions resulted in a shift of the area's quality cost mix, and a reduction in overall quality costs and total costs, as well as improved quality. The training and feedack increased the prevention and appraisal activities of both the supervisor and the clerks. Failure costs decreased significantly while quality improved, since rejects were virtually eliminated. The time the supervisor spent writing the procedures manual and developing the training were considered one-time costs; while the ongoing training and review of the clerks was considered part of the supervisor's basic responsibilities.

Impact on Productivity

The overall effects of quality costs and their improvement also can be observed by productivity measurement.

Productivity is measured by an output/input equation which yields an index. Tracking this index over time indicates the productivity trend. The numerator specifies the volume of the item processed during the month. The denominator includes the volume of resources expended to process the numerator volume.

Therefore, productivity can be positively affected by a change in the quality cost mix if failure costs are reduced.

A productivity measure was developed for the area. The numerator specifies the number of tickets processed during the month. The denominator lists the resources incurred to process the tickets, which included: labor hours, computer run time and ticket forms. Thus the measure appears as follows:

$$\frac{\text{Tickets Processed}}{\text{Labor + Systems + Forms}}$$

Since labor hours, systems and forms cannot be added together, a base period cost was assigned to each resource. The base costs do not change over time, only the amount of resources used. The base period cost of labor was $11.13 per hour, systems was a fixed $500 rate for the month, and forms were $.05 apiece.

At the time that quality costs were initially developed, approximately 2,080 tickets were processed per month at a cost of $7,753. Labor represented $7,123 ($11.13 × 640 hours), systems $500 and forms $130 ($.05 × 2,600 forms). The productivity index was calculated to be 26.83.

$$\frac{2{,}080}{(\$11.13 \times 640) + \$500 + (\$.05 \times 2{,}600)} = \frac{2{,}080}{\$7{,}123 + \$500 + \$130} = \underline{\underline{26.83}}$$

After the quality improvement project was completed, the productivity index was recalculated. Productivity had improved as evidenced by the increased index. Slightly more tickets were processed with less resources expended.

The new productivity measure appears as follows:

$$\frac{2{,}100}{(\$11.13 \times 546) + \$500 + (\$.05 \times 2{,}100)} = \frac{2{,}100}{\$6{,}077 + \$500 + \$105} = \underline{\underline{31.43}}$$

The new index represented an increase from the earlier productivity index, and a 17% increase in productivity.

This index confirmed that the process productivity had improved. Approximately the same number of tickets were processed, but the amount of labor and forms used decreased, since holdouts were eliminated.

Other divisions in the bank noticed the improvement of quality within the loan processing section because of the timeliness and accuracy with which the area posted critical information.

Conclusion

The importance of a comprehensive quality program cannot be overstated. When an operation's quality performance is poor, a quality standards program can alert an area to a potential quality problem. Some potential quality problems can be verified by developing quality costs. Quality costs can also alert an area to quality problems that may not be evident by traditional quality standards, as was seen in the case study. The inordinate amount of rejects was not apparent from measuring performance to standard or productivity. Only after developing quality costs was the problem identified and resolved. Once activities with high failure costs are identified, action can be taken to correct the specific problem. This will reduce failure activities and their failure costs, improve quality, and therefore, productivity. Hence, quality plus or minus quality costs equals productivity.

REDUCING FAILURE COST AND MEASURING IMPROVEMENT

William O. Winchell
Administrator, Service Section
General Motors Corporation
Detroit, Michigan

Abstract

This paper discusses two topics — reducing failure cost and measuring improvement — from the ASQC publication "Guide for Reducing Quality Costs." Four key steps are recommended for reducing the cost of failures. Some common pitfalls in using quality costs to measure improvement are also discussed. In addition, measurements that may be better than quality costs for tracking field failures are suggested.

Text

Reducing failure cost is a tough thing to do and once it is reduced it may be even tougher to maintain the new level of quality. Eliminating failure cost is something like a leaky tube of toothpaste. You fix the leak and force the toothpaste in the right direction — through the nozzle — but not for long. Soon it leaks in another spot. Frustrating as this may be, there is a very logical reason for this happening. The real cause of the problem has not been identified. Without knowing what is the real cause of a problem, how can we expect to fix it?

Too often, the quality manager sees his job as only informing management of symptoms of problems such as high rejects and returns from the field. Within his narrow scope of perceived activity, the quality manager adds more inspection to keep the bad parts from getting shipped until the problem is fixed. But without effective corrective action, the problem is never fixed. Management looks to the quality organization for more than this. The quality organization must not only report problem symptoms, it must also involve itself in the mainstream of defining the real problems and proposing the best possible solutions. The quality organization must develop an effective corrective action system that gets to the root cause of each problem so that failure cost can be permanently reduced.

The ASQC publication "Guide for Reducing Quality Costs" recommends following four key steps to reduce the cost of failures.

1. Communicate the problem to all affected persons.
2. Create a desire in others to mutually solve the problem.

3. Provide direction in the planning and carrying out a logical investigation by those in the company best able to resolve the problem.
4. Follow up to insure that the problem was permanently fixed.

Quality performance reports are a good tool for communicating problem areas to those that can solve the problem and to make top management aware of the progress. However the reports must be understandable, clear and to the point. They should summarize information and not overwhelm the user with needless detail. There is no one best format. The reports must be individually designed to meet the needs of those in your organization. If they are not, they will be useless documents on the desks of those people that the quality manager needs most — those that can find and fix the problems.

Creating a desire in others to mutually solve a problem is critical for most quality managers. It is directly reflected in the ability of the quality manager to eliminate the chronic quality problems existing in the company. This is because the chronic quality problems are not normally solvable by the quality manager alone. To get someone else to do something about the chronic problems is for the most part a selling job. This involves the quality manager selling his approach to corrective action and, most importantly, the fact that the problems are serious enough to justify spending the time and money necessary to fix them.

In the early stages of the investigation the cost of the problem, the amount of effort needed for the solution and the benefits that will be realized should be estimated and presented to management in your corrective action reports. Adequate justification will gain long lasting support when the problem is stated in terms others understand. For example, the product engineering department will be motivated to eliminate a reliability weakness if the amount of warranty can be determined. The production supervisor will join in solving a rework problem if it can be shown that the efficiency of the operation will be improved.

Key to your failure reduction efforts is a systematic corrective action system through which logical investigation and resolution of problems can be processed. The ASQC publication "Guide for Reducing Quality Costs" suggests several forms that can be used as a framework around which your system can be designed. Certain aspects need to be determined and documented early to insure the effective operation of your system. They are:

— A step-by-step plan for investigating and solving the problem. This should include the responsible individual and target date for completion of each step.
— Both tangible and intangible benefits in terms your management understands.
— The projected cost.

By having this plan, those involved will be confident that progress will be made and that the benefits will outweigh the costs. Also, priorities can be assigned to the various problems so that the important ones are worked on first while the others remain on the back burner.

Following up to insure corrective action really consists of a periodic measurement of the results. The initial measurement will determine if the plan achieved the desired results. If it did not, additional work is needed. Subsequent measurements will help insure that the gains are permanent. It is common to see improvement trends reverse themselves and head towards former levels. Measurements properly timed will help catch this early so that needed corrections can be made.

Being a quality cost paper, you might think that it would be stressed that quality cost is the best way to measure improvements. In contrast, only sometimes can quality cost provide the best evaluation and when it is used it must be done with a great deal of discretion. The ASQC publication "Quality Costs — What and How" is a good reference to help you in applying this technique.

For example, don't be misled in comparing percentages. One may assume that a failure cost decrease from 70 percent to 60 percent of total quality cost is an improvement. But if total quality cost increased from $2.00 to $2.50 a unit because of added inspection, the assumption is false. Quality cost is usually good for determining the magnitude of in-plant improvements such as inspection, scrap or rework costs. But, you must be careful because the changes could be really due to a major shift in the product mix to something easier to produce. Nothing substitutes for a walk to the production line for a first hand look at the situation rather than taking the cost data for granted.

Perhaps the greatest pitfall in using quality cost is measuring the effect of product improvements relative to field failures. There is often a long span of time between fixing a problem and an actual improvement in cost relative to field failures. This inherent lag makes other measurements better than quality cost for obtaining timely answers. The best measurement may be life cycle tests or outgoing quality audits that incorporate durability provisions. Such techniques can confirm that problems are fixed much more quickly than waiting for the product to fail in the field and the cost to be reported.

Other sources of data that are good indicators of field performance depending upon each individual situation are:

— Field trouble reports
— Market research surveys
— Spare parts sales
— Customer complaints
— Installation phase reports

There are many ways besides quality costs to track improvements in your company. In summary, be sure to evaluate all the possibilities in the context of your situation before selecting the measurement that you will adopt.

USING QUALITY COSTS IN PRODUCTIVITY MEASUREMENT

D. Scott Sink, PhD, PE
Associate Professor
John B. Keats, PhD, PE
Assistant Professor
School of Industrial Engineering and Management
Oklahoma State University
Stillwater, Oklahoma

Abstract

The measurement and evaluation of productivity and quality cost in one integrated system has been discussed by these authors in recent papers. This paper extends the multi-factor productivity measurement/quality cost integration by examining a reasonably comprehensive set of quality costs and categorizing each cost in terms of where and how they can be dealt with in the productivity measurement model. This development represents significant progress in terms of being able to effectively and efficiently integrate a variety of quality costs and to evaluate the relationships between changes in these costs and productivity. The paper presents the model, discusses entry of quality costs into the model, and presents a case scenario for clarification.

Productivity Basics

"You know you're getting smarter when you call things by their right names"

Productivity is an extremely abused and misused term and concept today. This occurs because there has been no disciplined attempt to develop a sound conceptual framework for the term. The "half-truth" rhetoric being written about productivity is amazing and at times overwhelming to those trying to better understand how to improve it. It has become such a significant buzz-word that almost every discipline and profession imaginable has begun to use the term in an attempt to further market and promote their own disciplinary and often myopic respective "solutions." The need for synthesis, clarification, disciplined definitions, and a generic conceptual framework is quite evident.

Productivity is a relationship between quantities of outputs and quantities of inputs for a given organizational system (i.e., work group, department, division, function, plant, firm, etc.). In order to better understand this definition in the context of organizations, it is useful to present the following conceptual framework and basic model of what is called the Productivity Management Process.

Figure 1 depicts a general system model an *organizational system*. By *organizational system* we could mean, depending upon the definition of the boundaries indicated by the dashed lines, a nation, a region, an industry, a state, a firm (multi-plant), a plant, a division, a function, a work group, or even an individual. Identifying the boundaries of the system we are interested in, defines the scope or *unit of analysis*. This step allows us to effectively discuss and develop *productivity measurement systems* because it allows us to accurately define inputs, transformations, outputs, and outcomes from the system we are interested in studying. Once we define major outputs and inputs we can begin to develop measures, ratios, and indexes with which to monitor productivity. The productivity measurement process is essentially the development of a managerial control system. Most organizations have control systems for behaviors, for costs, for prices, for information, for decisions, for financial performance, for production, for quality, etc.

There are many ways to classify control systems. We could classify or categorize them with respect to the resource they are intended to manage. Financial control systems, production control systems, behavioral control systems would be examples of this type. We could also classify control systems with respect to the type of "system" performance. A reasonably comprehensive set of system performance measures are: (1) Effectiveness, (2) Efficiency, (3) Quality, (4) Profitability (Benefit/Burden), (5) Productivity, (6) Quality of Work Life, and (7) Innovation. Every organization in one way or another, has systems designed to monitor, evaluate, control, and manage one or more of these seven measures of system performance. Note that productivity is only one measure of performance for a system. It is *not* clear it is the most important measure of system performance. We might consider these measures of system performance as a multi-attribute or multi-criteria measurement system.

In a way, one important job of a manager is to determine: (1) what the appropriate priorities or relative weights are for each performance measure; (2) how to measure, operationally, each performance measure; and, (3) how to link the measurement system to improvement. (In other words, how to most effectively use the control system to cause appropriate changes or improvements.) It is clear that the priorities or weightings for each of these performance criteria will vary according to several factors (i.e., size of the system; function of the system — marketing, manufacturing, R&D, etc.; type of system — job shop, assembly line, service, process industry, etc.; maturity of the system in terms of employees, management, technology, organizational structure and processes, etc.).

Productivity management is a management process. Precisely defined it is a subset of the larger management process. It involves the management functions planning, organizing, leading, controlling, adapting (POLCA). It focuses on controlling productivity, the relationship between quantities of output from a system and quantities of input from a system. It is not necessarily a more or less important process than the other control functions and processes. However, there is a growing concern that it has been overlooked and needs to be better developed. In other words, it is becoming apparent to some that the weights managers have placed on various system performance measures may need readjusting. The productivity management process necessarily includes productivity measurement and productivity improvement. The productivity measurement process is evolving and is beginning to have specific, unique techniques

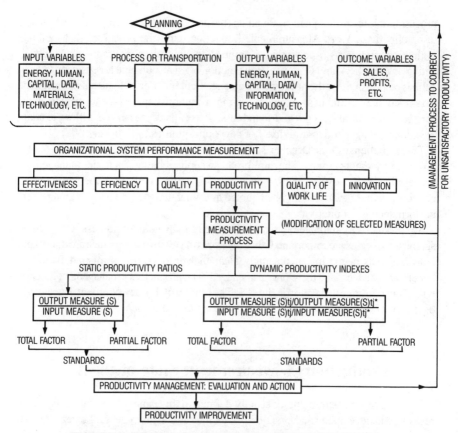

FIGURE 1. PRODUCTIVITY MANAGEMENT PROCESS

associated with it. The productivity improvement process is also evolving, however, the development is different from the measurement process. Peter Drucker's recent quote, "What we need to learn from the Japanese is not what to do, but to do it," is quite an appropriate description for the nature of the evolution of the productivity improvement process. Certainly we are learning more about how to use computer aided technology more effectively and efficiently, we are experimenting with participative problem solving (i.e., quality circles, productivity action teams, etc.), and we are reexamining the role that financial reinforcement (productivity gainsharing) can play in productivity improvement. But, there really are no significant new breakthroughs. Much of the Japanese success appears to be a result of more effective implementation. We are discovering the productivity improvement is 95% effective implementation and 5% technique.

Productivity management is the management of productivity. It is a direct focus on managing the relationship between quantities of outputs from a system and quantities of inputs into that same system. It is vitally interrelated with: Quality Management (the process that is to ensure quality), Planning (the process that determines what it will take to be effective), Work Measurement, Budgeting, etc. (the processes that facilitate evaluating efficiency), Accounting, Comptroller, etc. (functions responsible for evaluating profitability), and Personnel (the function often responsible for ensuring QWL). Who is responsible for the productivity management function? "In turbulent times, the first task of management is to make sure of the institutions capacity for survival, to make sure of its structural strength and soundness, of its capacity to survive a blow, to adapt to sudden change, and to avail itself of new opportunities." (Drucker, 1978)

These challenges that Drucker states for management are not just challenges for productivity management. They are broad sweeping challenges to the management process itself. However, Drucker goes on to say, "productivity is the source of all economic value." And, he has been quoted as saying that the first test of management performance is productivity.

So, productivity management is not the entire management process. Other management processes are important. Other measures of system performance are important. However, it is becoming increasingly clear that productivity is misunderstood by American managers and therefore mismanaged. The productivity management function will likely continue to develop in the U.S. The quality of this development process will be directly dependent upon our ability to understand the phenomena.

Productivity Measurement: State-of-the-Art

Productivity measurement, as strictly defined in this paper and as currently being researched, developed, and practiced, falls into at least two general categories: (1) economic, accounting based, generic and absolute measurement systems; and, (2) normative, participative, relative measurement systems. The latter focuses on obtaining "measurements" of productivity from participants in the organizational system through use of structured group processes such as the Nominal Group Technique and/or the Delphi Technique. The former focuses on absolute definitions and formulas for productivity ratios and indexes and utilizes system information (accounting, etc.) to drive a model that allows evaluation of changes in productivity, price recovery, and profitability from one period to the next. It is this model that will be briefly presented in this paper and that will be developed in terms of incorporation or integration with traditional quality cost measures.

The productivity measurement model to be developed in this paper is a multi-factor dynamic model. It is particularly applicable and appropriate at the firm, plant, company, or division level. It is, therefore, a reasonably macroscopic model. It is also primarily useful, at least easiest to apply, for systems that have measurable outputs and inputs. The model assumes ability to collect the data reflected in Figure 2. The model requires periodic data (weekly, or monthly, or quarterly, or annually, etc.) for output quantities and prices, by type of output and input quantities and prices, by class, type, and level. The model can be simplified by stopping at type for inputs if so desired. The designated type categories simply need to be collectively exhaustive and mutually exclusive.

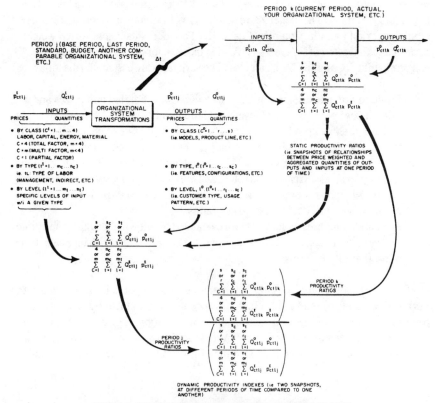

FIGURE 2. PRICE WEIGHTED AND AGGREGATED MULTI-FACTOR PRODUCTIVITY MEASUREMENT MODEL.

The model generates static productivity ratios as well as dynamic productivity indexes. That is for all comparisons of output to inputs (class, type, and level), the model makes within period calculations (ratios) as well as between period calculations (indexes). Figure 2 depicts this also.

The model utilizes relevant prices, p^I & p^O, to price weight inputs and outputs. For productivity calculations the prices are indexed to base period values so as to "partial out" the effect of changes in prices and just allow examination of changes in productivity. So, although use of prices and costs violates the basic definition of productivity, it does provide a relative weighting factor and a method for aggregation while base period pricing essentially indexes the effect of price recovery (inflation or deflation) out of the equation. The model also examines the independent effect of price recovery as well as the simultaneous effect of changes in productivity and price recovery on profitability.

The basic format for the model, which has been labelled "multifactor productivity measurement model — OSU version," is shown in the example presented in Table 1.

The basic productivity equations for ratios and indexes are depicted in Figure 2. As depicted in the table, the model works with these basic equations and calculates change ratios: for output quantities, input quantities, unit prices, unit costs revenues and costs (Columns 7, 8 and 9). The model then calculates cost/revenue ratios for each elemental input for period j and k (Columns 10 and 11). The model calculates price weighted performance indexes, see Figure 2 — Dynamic Indexes, for productivity, price recovery, and profitability (Columns 12, 13, and 14). The variations are essentially the following:

(a) Productivity indexes, column 12, are calculated by holding unit prices and unit costs at base period (i.e., K) levels and letting only quantity changes in outputs and inputs show up. This results in a between period rate of change calculation for price weighted and base price indexed ratios of output to input.
(b) Price recovery indexes, column 13, are calculated by fixing output and input quantity levels at current period values and thus looking only at changes in process compared to costs between periods. This results in a between period rate of change calculation of price recovery change ratios.
(c) Profitability indexes, column 14, are calculated by allowing quantities and prices to vary simultaneously. This results in a between period rate of change calculation of profitability margins.

With these productivity measurement fundamentals, the model to be utilized in the next section should be reasonably clear. It is of course important to realize that this has been a very brief presentation of a reasonably complex topic. The discussion will now focus on quality cost measurement and more specifically upon how some traditional quality cost measures can be integrated into this productivity measurement model.

Quality Costs in the Model

Table 1 presents, for selected levels (as obtained from the *Quality Control Handbook*) within each of the four quality cost categories the effects of these levels on the output and inputs of the multi-factor model. The effect is indicated as either being directly observable (D) or indirectly observable (I) in the model, D/I entries indicate that, depending upon the record-keeping practices of the company, the effect could either be directly or indirectly observable. Blank entries in the matrix do not necessarily indicate no effects in the model — only those levels that have a substantial effect have been marked.

With respect to output, internal failures will obviously have a strong effect. In a period with large amounts of scrap and rework, for example, output will be smaller than what it would have been had scrap and rework been under control. Returned items (external failure costs) can be subtracted from units produced either in the time period when these units were made, or in the time period when the returns occurred.

Space considerations for this paper will not allow comment on each of the input effects and thus only a few effects will be illustrated. Any activity which will reduce scrap and rework will also decrease labor, materials, and energy inputs. Repairs and adjustments under warranty will result in labor, material and energy changes which may or may not

	INTERNAL						EXTERNAL				APPRAISAL					PREVENTION						
	SCRAP	REWORK	RETEST	DOWNTIME	YIELD LOSSES	DISPOSITION	COMPLAINT ADJ.	RETURNED MAT'L.	WARRANTY CHGS.	ALLOWANCES	INCOM. INSP.	INSP. & TEST	ACCUR. OF EQUIP.	MAT'L. & SERVICES	STOCK EVALUATION	QUAL. PLANNING	NEW PRODUCTS	TRAINING	PROCESS CONTROL	DATA ACQ. & ANAL.	QUAL. REPORT	IMPROVEMENT
OUTPUT	I	I	I	I	I		D															
LABOR	I	D	D/I			I	I	D/I	D/I	D/I	D	D	D/I		D/I	I	D/I	D/I	D/I	D/I	D/I	D/I
MATERIAL	I	I	D/I	I			D/I	D/I			D		D		D							D
ENERGY	I	I	I	I			I	I			I		I					I				
CAPITAL							D	D					D							D	D	

TABLE 1. DIRECTLY (D) AND INDIRECTLY (I) OBSERVABLE EFFECTS OF THE MULTI-FACTOR MODEL

be directly measured. Capital expenditures are required for all quality equipment. These expenditures include depreciation, repairs, space charges, etc. Quality materials as appraisal costs would include x-ray film, materials used in printed reports and charts, as well as any disposables used in assessing the condition of the product. All labor used in prevention will be reflected in increased or decreased inputs although such labor may not be tracked according to levels such as planning, training and reporting.

To depict use of the multi-factor model with quality costs, an example is presented in Table 2. The example is extremely simplified for illustrative purposes. In practice, one would find hundreds of input variables and considerably more quality inputs than are presented here.

The quantities shown in columns 1 and 4 for output represent units sold in a period for Product X. Period 1 is the base period and Period 2 is any subsequent period chosen for comparison purposes. For input columns 1 and 4 are man-hours of labor expended, units of material consumed, and 1000 or 10000 BTU of electricity and gas, respectively. The prices of columns 2 and 5 are the selling price of Product X (output) and unit costs (all inputs). The values of columns 3 and 6 are the products of price and quantity. Column 7 represents the price weighted productivity change ratio, $P_1 Q_2/P_1 Q_1$. Column 8 is the quantity weighted price recovery ratio, $P_2 Q_2/P_1 Q_2$. Column 9 is the ratio $P_2 Q_2/P_1 Q_1$. The cost to revenue ratios of columns 10 and 11 represent input values (columns 3 and 6) divided by the output values of $1,691,200 and $2,166,000, respectively.

Column 12, change in productivity, is the result of dividing the total output quantity change ratio from column 7 (1.2983) by the corresponding input entities from column 7. Column 13 is the result of dividing the total output price change ratio in column 8 by the corresponding input price change ratios from column 8. Column 14 is the product of columns 12 and 13. It can also be thought of as between period rate of change profitability margins calculated by allowing quantities and prices to vary simultaneously.

Quality Costs: Ideas & Applications, Volume 2

	PERIOD 1			PERIOD 2			PRICED WEIGHTED CHANGE RATIOS			COST/REVENUE RATIOS			CHANGE IN		(14)	(15)	(16)	(17)
	(1) QUANTITY	(2) PRICE	(3) VALUE	(4) QUANTITY	(5) PRICE	(6) VALUE	(7) QUANTITY	(8) PRICE	(9) VALUE	(10) PERIOD 1	(11) PERIOD 2	(12) PRODUC-TIVITY	(13) PRICE RECVY.	PROFIT-ABILITY	CHANGE IN PRODUC-TIVITY	CHANGE IN PRICE RECVY.	CHANGE IN PROFIT-ABILITY
PRODUCT X	3020.	560.00	1691200.00	3800	570.00	2166000.00	1.2583	1.0179	1.2807								
TOTAL OUTPUTS			1691200.00			2166000.00	1.2583	1.0179	1.2807								
MGMT. NON-QUAL	5230.	20.00	104600.00	5360.	20.00	107200.00	1.0249	1.0000	1.0249	.0618	.0495	1.23	1.02	1.25	24415.91	2350.28	26766.19
MFTG. NON-QUAL	52600.	8.00	420800.00	53100.	8.80	467279.94	1.0095	1.1000	1.1105	.2488	.2157	1.25	.93	1.15	104683.50	-33024.94	71658.56
NON-QUAL LABOR			525400.00			574480.00	1.0126	1.0798	1.0934	.3107	.2652	1.24	.94	1.17	129099.38	-30674.63	98424.75
MGMT. QUALITY	1076.	18.00	19368.00	1240.	18.00	22320.00	1.1524	1.0000	1.1524	.0115	.0103	1.09	1.02	1.11	2050.33	435.18	2485.52
REWORK LABOR	2074.	9.00	18666.00	2120.	9.90	20988.00	1.0222	1.1000	1.1224	.0110	.0097	1.23	.93	1.14	440702	-1488.59	2918.43
INCOM INSPECTION	1080.	10.00	10800.00	1150.	11.00	12650.00	1.0648	1.1000	1.1713	.0064	.0058	1.18	.93	1.09	2089.40	-907.33	1182.07
INSPEC AND TEST	1522.	9.00	13698.00	1300.	9.90	12870.00	.8541	1.1000	.9396	.0081	.0059	1.47	.93	1.36	5535.89	-862.21	4673.68
QUALITY LABOR			62532.00			68828.00	1.0331	1.0654	1.1007	.0370	.0318	1.22	.96	1.16	14082.64	-2822.94	11259.70
TOTAL LABOR			587932.00			643308.00	1.0147	1.0783	1.0942	.3476	.2970	1.24	.94	1.17	143182.00	-3349.7.63	109684.38
RAW MTL.	10000.	25.00	250000.00	12200.	26.80	326960.00	1.2200	1.0720	1.3078	.1478	.1510	1.03	.95	.98	9569.56	-16342.69	-6773.13
MTLS. QUAL.	200.	2.00	400.00	230.	2.05	471.50	1.1500	1.0250	1.1788	.0002	.0002	1.09	.99	1.09	43.31	-2.51	40.80
TOT MATERIAL			250400.00			327431.50	1.2199	1.0719	1.3076	.1481	.1512	1.03	.95	.98	9612.88	-16345.25	-6772.38
ELEC MBTU	2700000.	.01	27000.00	3100000.	.01	34100.00	1.1481	1.1000	1.2630	.0160	.0157	1.10	.93	1.01	2973.51	-2493.33	480.18
GAS 10MBTU	980000.	.04	39200.00	1100000.	.04	46200.00	1.1224	1.0500	1.1786	.0232	.0213	1.12	.97	1.09	5324.50	-1319.20	4005.30
TOT ENERGY			66200.00			80300.00	1.1329	1.0707	1.2130	.0391	.0371	1.11	.95	1.06	8298.02	-3812.53	4485.48
CAP. NON-QUAL	1000000.	.20	200000.00	1050000.	.20	210000.00	1.0500	1.0000	1.0500	.1183	.0970	1.20	1.02	1.22	41655.63	4493.84	46149.47
INSPEC. EQUIP INC.	50000.	.20	10000.00	50300.	.20	10060.00	1.0060	1.0000	1.0060	.0059	.0046	1.25	1.02	1.27	2522.78	224.69	2747.47
INSPEC EQUIP	25000.	.20	5000.00	25300.	.20	5060.00	1.0120	1.0000	1.0120	.0030	.0023	1.24	1.02	1.27	1231.39	112.35	1343.74
TEST EQUIP	100000.	.20	20000.00	101000.	.20	20200.00	1.0100	1.0000	1.0100	.0118	.0093	1.25	1.02	1.27	4965.56	449.39	5414.95
COMPUTER QUAL	40000.	.20	8000.00	40500.	.20	8100.00	1.0125	1.0000	1.0125	.0047	.0037	1.24	1.02	1.26	1966.22	179.75	2145.98
SUBT QUAL EQUIP			43000.00			43420.00	1.0098	1.0000	1.0098	.0254	.0200	1.25	1.02	1.27	10685.96	966.18	11652.14
TOTAL CAP.			243000.00			253420.00	1.0429	1.0000	1.0429	.1437	.1170	1.21	1.02	1.23	52341.56	5460.06	57801.63
TOTAL INPUTS			1147532.00			1304465.50	1.0723	1.0601	1.1368	.6785	.6022	1.17	.96	1.13	213434.50	-48195.25	165239.25

TABLE 2. MULTI-FACTOR PRODUCTIVITY MODEL EXAMPLE

Comparing Period 1 and Period 2 with respect to quality labor, inspection and testing labor has been reduced while other labor hours have increased. The price of labor has increased except for management salaries. Quality materials usage has increased from Period 1 to Period 2 as has capital expenditures associated with quality equipment (inspection, testing, computer).

The price weighted change ratios of columns 7 through 9 are nearly all greater than 1, reflecting increased quantities, prices and values.

The cost/revenue ratios of columns 10 and 11 track the same type of information which is measured by quality cost studies. That is the model measures quality labor material and capital expenditures by each type as a fraction of sales (total revenue) for Period 1 and Period 2.

All of the productivity ratios of column 12 are greater than 1 indicating that the output change ratio of column 7 (1.2583) was higher than any of the input change ratios in the same column. Some of the price recovery changes of column 13 are below 1 and others are greater than 1. Those below 1 indicate that their costs in Period 2 relative to Period 1 were higher than the corresponding revenue ratio, i.e., costs increased more as a percent of their base period costs than did revenue as a percent of base period revenue. This can be thought of as a loss in price recovery. On the other hand, column 13 ratios greater than 1 are indicative of smaller cost increases relative to revenue increase (a gain in price recovery).

The total effects on profits are shown in columns 15, 16, and 17. These effects represent the dollar value impact on profits as a result of changes in productivity, price recovery, and profitability. All of the entries in column 15 are positive. They have been calculated by subtracting the appropriate input quantity price weighted change ratios of column 7 from the output quantity price weighted change ratio (1.2583) multiplying by the corresponding value of input from column 3. Each is a productivity gain since the output change ratio is larger than the input change ratio. Quality labor accounted for a $14,083 increase in profits. The price recovery figures in column 16 are obtained by subtracting column 15 from column 17. Column 17, dollar change in profitability is obtained by subtracting the appropriate input price weighted value change ratio in column 9 from the output price weighted value change ratio (1.2807) and multiplying by the corresponding value of input from column 3. Negative values in column 16 indicate that profits were adversely affected due to input costs increasing at a higher rate than the price of Product X. Column 17 reflects the simultaneous changes in productivity and price recovery.

The model may also be used for control purposes as well as for comparing outcomes with quality goals. If quality goals, standards, or desired outcomes are substituted for actual data in Period 1, then Period 2 could be used to record the actual quantities and costs incurred in a period of interest. The model will generate productivity, price recovery and profitability ratios and changes which may be used to verify the extent which quality goals have been achieved.

It is extremely difficult to explain the mechanics of the total factor model in a brief fashion. It is hoped that the reader will be stimulated to learn more about the model and will subsequently use quality cost data already being gathered as part of the total factor model.

References

Drucker, P. F., *Managing in Turbulent Times,* Harper & Row: New York, 1980.

Juran, J. M., Gryna, F. M., and Bingham, R. S., Editors, *Quality Control Handbook,* Third Edition, McGraw-Hill, 1974.

Keats, J. B. and Sink, D. S., "An Investigation of Quality Measurement Systems and a Methodology for Integrating those Systems into Productivity Measurement Models," IIE 1982 Annual Industrial Engineering conference *Proceedings,* (New Orleans), Institute of Industrial Engineers: Norcross, GA, May 1982.

Kogure, M., "Factors Required for Japanese Quality Cost System," *ASQC Quality Congress Transactions — San Francisco,* pp. 787-794, (1981).

Sink, D. S., "Productivity Management: Measurement and Improvement Strategies," A short course developed for the Oklahoma Productivity Center and the Institute of Industrial Engineers, 1980, 1981, 1982, Oklahoma State University: Stillwater, OK.

EUROPEAN ORGANIZATION FOR QUALITY CONTROL ANNUAL
CONFERENCE PROCEEDINGS

FOLLOW-UP AND ANALYSIS OF THE COSTS OF REJECTIONS

Miguel Arenas Lugea
Head of Testing Department
Industrial de Telecomunicación, S.A.
Spain

1. The Concept of Rejection

During the manufacturing of any kind of products, there are several factors having a negative influence upon the result of the process. Some of these factors are:
- tolerances in the characteristics of the raw materials.
- variations into the process itself.
- misadjustment of the machines/equipment.
- variations in the skill of the personnel.
- variations in the training.
- variations in the revision-state of the products.
- etc.

and because them the output of the process, includes a certain amount of pieces not fulfilling the established quality requirements.

The concept of *REJECTION* that we are dealing with means, parts, sub-assemblies, assemblies, or final products, that has been detected defective during the manufacturing process, being necessary to be repaired, reworked or scrapped.

The rejection has to be treated as an alarm signal, for the start of analysis of the factors causing the rejection, and for the determination of the actions to be taken for future prevention. The aim of these preventing actions is to improve process economy, but in this way it is of great necessity to know the economical costs caused by the rejection, as closely as possible.

2. The Cost of Rejection

The cost of rejection includes all costs resulting from products, processes and material not fulfilling the specified quality. These costs may be:
- Cost for scrapping.
- Cost for fault tracing, upon the rejection.
- Cost for repair and rework.
- Cost for articles, materials, etc., ordered to cover the shortage caused by scrapping.
- Cost for manufacturing the products to cover the rejected ones.
- Cost for re-inspection and/or retest of these products.
- Etc.

Those different costs can be grouped into three main groups, as follows:

a) *Material Cost*. It includes the cost of raw material, components, parts, etc., to be put out of the store, for the corresponding repair, rework or complementary manufacturing.
b) *Labour Cost*. Cost of the direct working hours, used to solve the scrapping.
c) *Overhead Cost*. Commonly associated to the labour Cost, includes for every working-group, the hour cost of the corresponding charges for indirect labour, staff, amortizations, supplies, etc.

Of course the first objective of the whole system is to know as clearly as possible for every rejection case, which is the cost that corresponds to each one of the three mentioned groups, and to go deeper in determining:

a) the defective article-code.
b) the type of fault.
c) the defective component, or detail (to the lowest possible level).
d) the operation-step where the fault has been detected.
e) the operation-step where the fault has been caused.
f) the cause of the defect.
g) the cost centre that should be debited.

in order to know the "high rejection-costing" machines, working groups/working-lines, parts, components, materials, and so on, to get the possibility of making a detailed study of every case and the corresponding proposals to improve the process quality and economy.

To meet this goal we group all rejection cost into one special cost-code, with some specific report documents we will deal with later on.

These documents give a double information flow:

a) *a fault-information* that is immediately handed over to those corresponding departments for discussion.
b) *a cost-information,* when the whole rejection case is ended, and that is computerized for its statistical analysis.

3. Sources for the Following of Rejection Costs

There are four documents from where all data for cost analysis, are obtained:
a) The Rejection Report.
b) The Requisition from store.
c) The Work-Cards.
d) The Manufacturing Order.

The Rejection Report, (see fig. 1), is the first document, and is issued at every cost-causing fault detection during the manufacturing process; (for practical reasons, this paper is produced, only when the defective quantity is higher than the 3% of total, or its value is higher than a certain amount of money). This Report is filled by the corresponding Quality Control, with all needed data, as we have spoken about before, so the Manufacture Engineering Dept. may calculate the costs and the Planning Dept. determines the need of completing or not the defective batch.

The Requisition from store (see fig. 2), is used to pick-up from store the material, parts, components, etc., to replace or repair the faulty units.

The Work-Card (see fig. 3) is assigned to every worker performing a rework operation with parts or material, found defective in store. It will collect at least, the following data:
- Code of the working place (to know overhead charges).
- Code of the worker itself (to know salary rates).
- Cost centre to be debited.
- Working hours consumed.

The Manufacturing Order (see fig. 4) is commonly used for updating of parts and products, for repair on products returned from customers, or when the work to repair or complete a defective batch in-process, is so big that it needs several operations and/or several workers to deal with; in this case, each worker needs a Work-Card for every operation, and all cards are related to the Mfg. Order.

4. Handling of Source Documents

In order to get a rejection cost information, as effective and deep as possible, all data from the mentioned documents are to be processed in a computer. For that, every fault is designed with a four digits alpha-numerical code. The two first digits correspond to the working group/working place origin of the fault, and the two last digits correspond to the fault cause. The figure 5 shows a cross-index with the codes for different fault-causes.

The alpha-numerical fault code must be written on each document issued at every rejection occasion, namely:
- Rejection Report
- Requisition from Store
- Work-Card
- Manufacturing Order

and the corresponding Quality Control has the responsibility of the right assignment of fault codes, for both, the causing working-group and the fault-cause.

The Manufacturing Engineering Dept. will then calculate the cost of the rejection, according to the number of rejected pieces, and the operation where the fault was originated. This cost will be debited the working-group origin of the fault, being preceptive the approval of this group responsible.

The computerized handling of these data is called SISTEMA VAEFA and has not a direct economical objective, but its goal is to provide information to the management for improving the intern quality and lower the processes economy. For these reasons, the results obtained from this system, have no accounting application, even though they may be compared to those of the General Cost Accounting.

The figure 6, shows the general diagram for the SISTEMA VAEFA and there may be seen its connection with other computerized economical systems of the company, as for example: MOCAM (store handling); INCOMAT (cost of inventories); COSTOS (general costs accounting); NOMINA (general pay-roll system). The system gives out monthly four data lists:
- Errors of input data (fig. 7)
- Summary of Rejection Reports (fig. 8)
- Fault-cause summary for each working-group (fig. 9)
- Working-group summary for each fault-cause (fig. 10)

Besides that, every cost-centre responsible may ask for a list of the rejection cost for each fault-cause in his cost-centre.

5. Handling of Results

Every month, when the lists are issued, the Quality Control Dept. performs the corresponding analysis, and distributes the information between all departments involved, not only in Production, but also in Design, Standardization, Economy, etc., in order to start the comments about the key points that for its weight are selected for the occasion.

From this discussion it is possible to make decisions for the improvement of the quality and economy (both are very related in between) of the manufacturing.

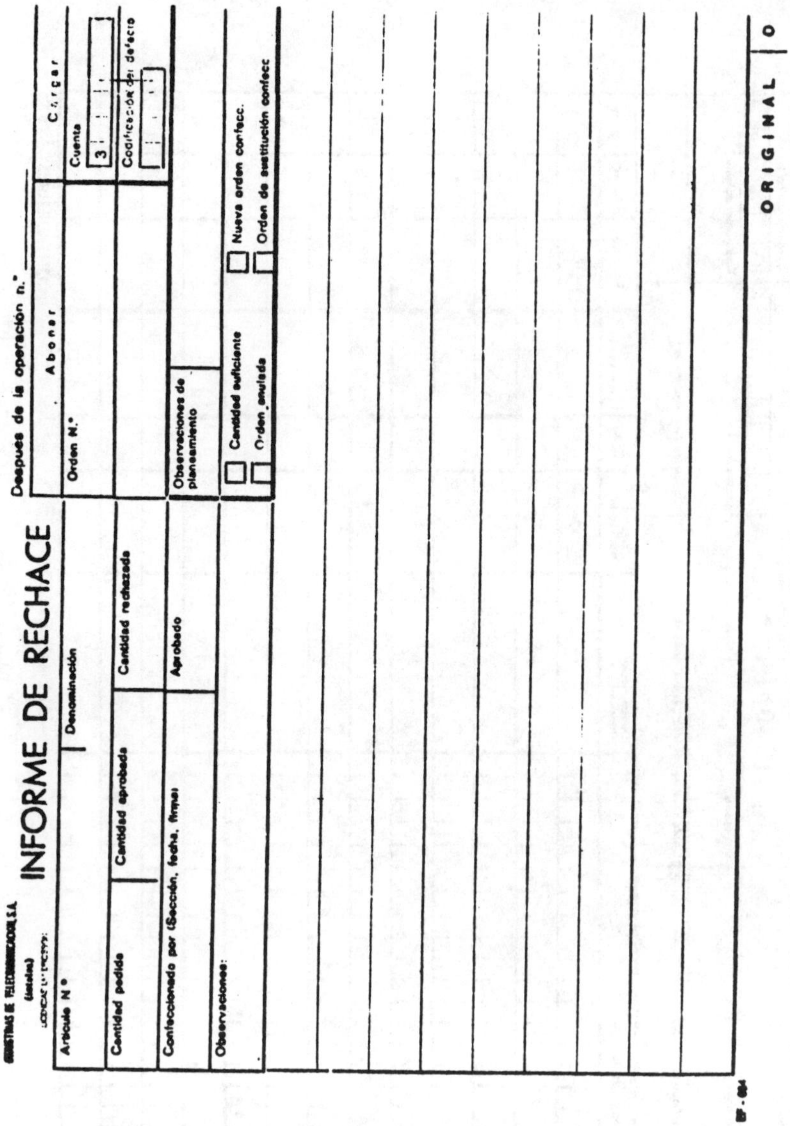

INFORME DE RECHACE

Despues de la operación n.º ______

Artículo N.º	Denominación		Orden N.º	Abonar	Cargar
Cantidad pedida	Cantidad aprobada	Cantidad rechazada			Cuenta 3 / Codificación del defecto
Confeccionado por (Sección, fecha, firma)		Aprobado	Observaciones de planeamiento		
Observaciones:			☐ Cantidad suficiente ☐ Orden anulada	☐ Nueva orden confecc. ☐ Orden de sustitución confecc.	

ORIGINAL

Figure 1

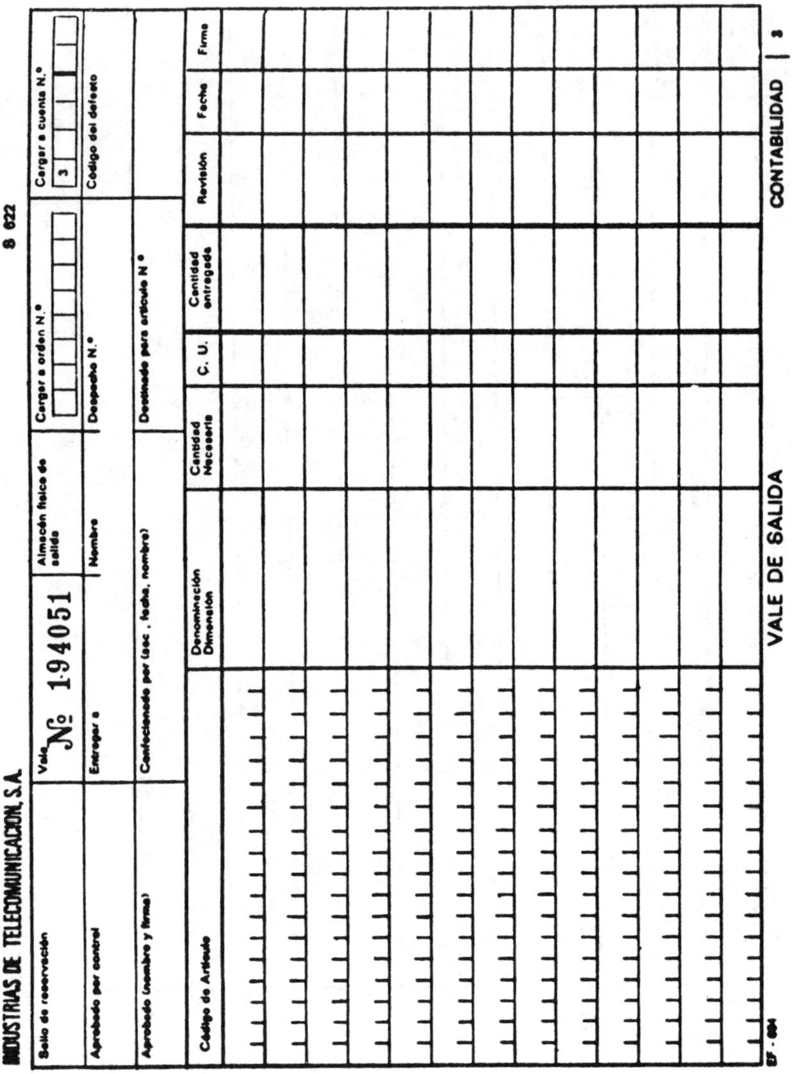
INDUSTRIAS DE TELECOMUNICACION, S.A.

S 622

Sello de reservación | Vale Nº 194051 | Almacén físico de salida | Cargar a orden N.º | Cargar a cuenta N.º 3

Aprobado por control | Entregar a | Nombre | Despacho N.º | Código del defecto

Aprobado (nombre y firma) | Confeccionado por (sec., fecha, nombre) | Destinado para artículo N.º

Código de Artículo	Denominación Dimensión	Cantidad Necesaria	C. U.	Cantidad entregada	Revisión	Fecha	Firma

VALE DE SALIDA

CONTABILIDAD | 3

Figure 2

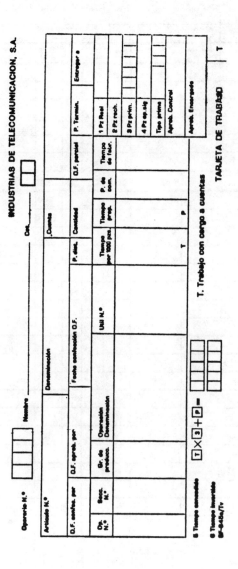

Figure 3

INDUSTRIAS DE TELECOMUNICACION, S.A.

Artículo N.º	Denominación	Cuenta	Orden N.º	O. F. Hoja N.º

O. F. confec. por	O. F. aprob. por	Fecha confección O. F	P. dist.	Cantidad	O. F. parcial	P. termin.	Entregar a	Gr. producto

Rechazo normal %	Util	Corr. del plano	Cantidad para entrega parcial Máx. .... Min.	F. de op. válida para cant. Máx. .... Min ....

A Op. N.º	A Secc. N.º	Artículo N.º N.º de material	Denominación Dimensión	P. de desp.	Cantidad total necesaria	Unidad	Cant. de mat. para prep	Almacén	Cantidad por 1000 pzas. Orden	Unidad cant. para almacén	P. de termin.
		2									
		3									
		4									
		5									
		6									
		7									

Op. N.º	Secc. N.º	Gr de preauss.	Operación Denominación	Util N.º	Tiempo por 1000 piezas	Tiempo prep.	P. de com.	Tiempo de fabr.	
			9						
			10						
			11						
			12						
			13						
			14						
			15						
			16						
			17						
			18						
			19						
			20						
			21						
			22						
			23						
			24						
			25						
			26						

Entrada	P. de termin.	P. de prep.	P. de ens.	P. de rutor.	Cantidad total		F. O. confec. por	F. O. Hoja N.º
Perfor.							F. O. aprob. por	F. O. Letra de corr.
Verif.								

[illegible]

B7–1 (10)

FICHA DE OPERACIONES | 0

Figure 4

1º \ 2º	0	1	2	3	4	5	6	7	8	9
0	STORE AND CARRYING								Loss		
1	NORMS, DRAWINGS AND DOCUMENTS	Non existent	Wrongs	Out of date	Storein floor 1 Complex	Storein floor 2					
2	QUALITY OF MATERIALS	Metalics	Plastics	Various	Details exterior	Details from store	Cables				
3	DIMENSIONS AND TOLERANCES	Metalics	Plastics	Various	Details exterior	Details from store	Cables				
4	MACHINERY AND TOOLS	Unsuitable	Incorrect Set-up	Incorrect function	Faulty design	Faulty construction	Breakage or failure				
5	INSTRUMENTATION	Unsuitable	Incorrect Set-up	Incorrect function							
6	STAFF	Bad training	Lack of								
7	SUPPLIES	Compressed air	Electricity	Gas							
8	RETURNS	Customs	Installations	Plant 1	Plant 2						
9	VARIOUS	Posterior endorsements	Production needs	Reparations						Omited or erroneous cause Don't use Generate by software	Unknown cause

Figure 5

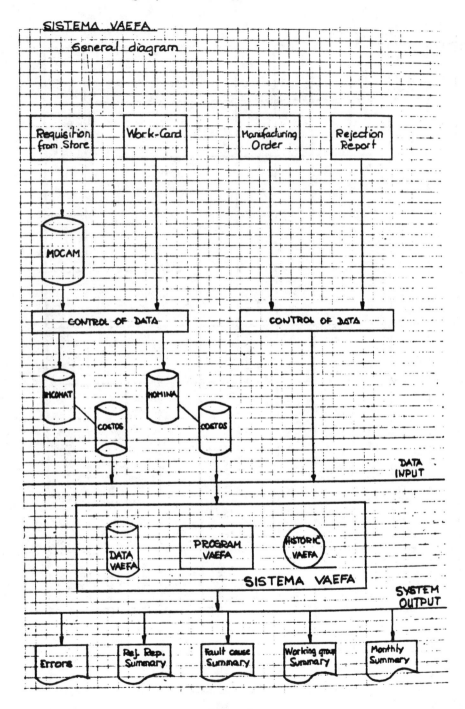

Figure 6

(VFC1020A1) * D E T A L L E D E L O S D A T O S T R A T A D O S P O R V A E F A E N E L M E S * 82/11

TIPO DOC.	SECC.	GRUPO CAUSA	NUM. DE ARTICULO	ORDEN	OPER.	CAN.RECHAZ.	C O S T E D E L R E C H A Z O MATERIALES	MANO DE OBRA	APLICACIONES	FECHA
TT	3540	JE-61		1184	00	0	0	670	0	82/11
TT	3540	CC-61		3913	00	0	0	616	0	82/11
TT	3550	CC-61		4207	00	0	0	41.676	0	82/11
TT	3550	CC-61		0153	00	0	0	91.718	0	82/11
TT	3550	CC-61		0350	00	0	0	49.775	0	82/11
MT	3551	HA-45		323785	00	0	15.030	0	0	82/11
MT	3511	AU-61		333659	00	0	1.056	0	0	82/11
MT	3572	JA-92		323103	00	0	1.904	0	0	82/11
MT	3572	JB-92		303795	00	0	2.469	0	0	82/11
MT	3572	HA-92		303829	00	0	3.325	0	0	82/11
MT	3572	JB-92		303795	00	0	3.420	0	0	82/11
MT	3572	HA-92		303830	00	0	3.385	0	0	82/11
MT	3572	HB-92		333794	00	0	69.800	0	0	82/11
MT	3572	HA-92		333348	00	0	3.000	0	0	82/11
MT	3572	JA-92		323102	00	0	3.180	0	0	82/11
MT	3572	JB-92		303821	00	0	1.840	0	0	82/11
MT	3572	HA-92		333348	00	0	2.370	0	0	82/11
MT	3572	JA-92		303812	00	0	1.218	0	0	82/11
MT	3572	JA-92		313792	00	0	19.035	0	0	82/11
MT	3572	HA-92		303830	00	0	3.240	0	0	82/11
MT	3572	HA-92		307150	00	0	198	0	0	82/11
MT	3572	HA-92		303830	00	0	24.000	0	0	82/11
MT	3572	JA-92		323102	00	0	1.142	0	0	82/11
MT	3572	HA-92		303785	00	0	3.905	0	0	82/11
MT	3572	HA-92		333348	00	0	14.000	0	0	82/11
MT	3572	JA-92		323437	00	0	12.600	0	0	82/11
MT	3572	HA-92		303829	00	0	8.325	0	0	82/11
MT	3572	JA-92		303812	00	0	10.200	0	0	82/11
MT	3572	JA-92		303812	00	0	1.098	0	0	82/11
MT	3572	HA-92		333348	00	0	2.500	0	0	82/11
MT	3572	JA-92		323102	00	0	812	0	0	82/11
MT	3572	JA-92		323103	00	0	1.494	0	0	82/11
MT	3572	JA-92		303812	00	0	1.388	0	0	82/11
MT	3572	JA-92		323103	00	0	3.260	0	0	82/11
MT	3572	HA-92		303830	00	0	15.200	0	0	82/11
MT	3572	JB-92		303823	00	0	3.240	0	0	82/11
MT	3572	JA-92		323102	00	0	906	0	0	82/11
MT	3572	JB-92		303795	00	0	2.976	0	0	82/11
MT	3572	HA-92		323441	00	0	33.000	0	0	82/11
MT	3572	JB-92		333917	00	0	2.400	0	0	82/11
MT	3572	JB-92		303795	00	0	2.841	0	0	82/11
MT	3572	JA-92		323102	00	0	1.156	0	0	82/11
MT	3572	JB-92		303795	00	0	1.800	0	0	82/11
MT	3572	JA-92		323102	00	0	934	0	0	82/11
MT	3572	JA-92		303812	00	0	1.313	0	0	82/11
MT	3572	HB-92		333924	00	0	18.148	0	0	82/11
MT	3572	HA-92		313622	00	0	1.050	0	0	82/11
MT	3572	HA-92		333348	00	0	2.250	0	0	82/11
MT	3572	HA-92		303830	00	0	8.600	0	0	82/11

Figure 7

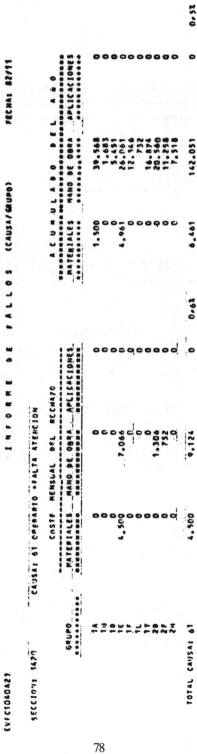

Figure 8

(VFC1142A1) INFORME DE FALLOS (GRUPO/CAUSA) FECHAS 82/11

SECCION: 3470   GRUPO: 20

CAUSA	COSTE MENSUAL DEL RECHAZO			ACUMULADO DEL AÑO		
	MATERIALES	MANO DE OBRA	APLICACIONES	MATERIALES	MANO DE OBRA	APLICACIONES
07 ALMACEN-TRANSP-EXTRAVIO	0	0	0	34.993	10.499	38.188
20 CALIDAD MATER.-METALICOS	0	0	0	0	1.573	0
22 CALIDAD MATER.-VARIOS	0	0	0	0	6.135	0
40 MAQ-UTILLAJE-INADECUADA	0	0	0	0	61.338	0
41 MAQ-UTILLAJE-MAL PREPARADA	0	0	0	0	5.056	0
42 MAQ-UTILLAJE-MAL FUNCIONAM.	0	0	0	0	5.048	0
44 MAQ-UTILLAJE-CONSTR.DEFICIE	0	0	0	0	3.543	0
45 MAQ-UTILLAJE-ROTURA/AVERIA	0	4.759	0	0	62.108	0
61 OPERARIO -FALTA ATENCION	0	0	0	0	3.743	0
72 SUMINISTROS-GAS	0	0	0	0	3.654	0
TOTAL GRUPO: 20	0	4.759	0	34.993	162.697	38.188
		0,2%				0,8%

Figure 9

CAUSA	COSTE MENSUAL DEL RECHAZO			ACUMULADO DEL AÑO			
	MATERIALES	MANO DE OBRA	APLICACIONES	MATERIALES	MANO DE OBRA	APLICACIONES	
03 ALMACEN.TRANSP.ALM. NAVE L4	0	0	0	45.150	6.282	40.111	0.2%
07 ALMACEN.TRANSP.EXP.PAVIO	1.009	0	0	979.507	48.680	0	3.6%
10 NOR.PLANOS-DOC.INEXISTENTES	0	0	0	3.260	0	0	0.0%
11 NOR.PLANOS-DOC.ERRONEAS	0	0	0	0	13.160	0	0.1%
13 NOR.PLANOS-DOC.COMPLEJAS	0	0	0	0	47.793	0	0.2%
20 CALIDAD MATER.-METALICOS	0	0	0	44.203	4.826	100.443	0.8%
21 CALIDAD MATER.-PLASTICOS	0	0	0	24.925	16.440	0	0.3%
23 CALIDAD MATER.-DET.EXTERIOR	0	0	0	26.312	44.322	0	0.3%
24 CALIDAD MATER.-DET.ALMACEN	0	0	0	129.104	187.826	0	1.6%
25 CALIDAD MATER.-CAILES	0	0	0	12.295	65.656	0	0.3%
30 DIMENS.TOLERAN-METALICOS	0	0	0	41.964	4.897	0	0.2%
31 DIMENS.TOLERAN-PLASTICOS	0	0	0	45.726	0	0	0.2%
33 DIMENS.TOLERAN-DET.EXTERIOR	5.148	0	0	770	1.457	0	0.0%
34 DIMENS.TOLERAN-DET.ALMACEN	0	0	0	32.531	144.967	0	0.7%
40 MAQ.UTILLAJE-INADECUADA	0	0	0	13.600	70.198	0	0.5%
41 MAQ.UTILLAJE-MAL PREPARADA	0	0	0	127.440	20.439	0	0.7%
42 MAQ.UTILLAJE-MAL FUNCIONAM.	0	0	0	3.228	5.048	0	0.0%
43 MAQ.UTILLAJE-DISEÑO DEFICIE	0	0	0	0	37.264	0	0.4%
44 MAQ.UTILLAJE-COVSTR.DEFICIE	0	0	0	78.254	3.343	0	0.4%
45 MAQ.UTILLAJE-ROTURA/AVERIA	4.759	0	0	0	103.857	0	0.4%
50 INSTRUMENTAC-INADECUADA	0	0	0	15.370	23.551	0	0.1%
51 INSTRUMENTAC-MAL PREPARADA	1.770	0	0	0	5.650	0	0.0%
60 OPERARIO .FALTA EXP/FORMAC	53.660	0	0	45.491	379.467	0	1.4%
61 OPERARIO .FALTA ATENCION	5.556	418.587	0	1.083.988	3.556.904	143.092	16.1%
62 DESCONOCIDA	0	0	0	136.484	899.389	17.672	3.5%
72 SUMINISTROS-GAS	0	0	0	0	3.454	0	0.0%
80 DEVOLUCIONES-CLIENTES	65.479	317.114	219.917	355.402	162.536	0	1.7%
81 DEVOLUCIONES-INSTALACIONES	0	0	0	1.980.747	3.096.877	3.015.093	33.9%
90 VARIOS-RODOS POSTERIORES	13.405	100.692	0	16.183	402.974	0	-0.5%
91 VARIOS-NECESIDADES PRODUCC.	435.102	71.923	0	1.576.285	4.795.159	0	21.4%
92 VARIOS-REPARACIONES	4.972	422.805	0	81.370	3.546.231	742	12.2%
99 VARIOS-CAUSA DESCONOCIDA	0	0	0				19.8%
TOTAL	547.557	1.389.540	219.917	6.737.202	19.709.883	3.317.153	100.0%

Figure 10

THE SIGNIFICANCE OF USER-CONSUMER QUALITY COSTS IN THE WORLD OF UNLIMITED RESOURCES

Professor Bronislaw Oyrzanowski, DSc/Econ
Tadeusz Wawak, DSc/Econ
Jagellonian University
Poland

1. Producer Quality Costs

The meaning of Quality Costs is fairly well-known but for our purposes it needs some further elucidation. Quality costs are the costs covered by the producer on account of the quality of conformance not achieving the quality of design. They would be equal to zero if the product corresponded to the quality designed. For this reason the French usually call them "Coûts de non-qualité." They do not concern the costs of the improvement of the quality of design. They are composed of the prevention costs, appraisal costs and internal and external failure costs. The latter are the costs covered by the producer and paid to the consumer under the warranty system because the product sold does not correspond to the quality of design.

Besides the producer's quality costs, there was introduced a concept of *quality costs in a branch* concerning the cases in which the costs were transmitted either from one producer to another or from the producer to a salesman; in this case the two firms had to be treated together in calculated quality costs. (See: "Quality costs in a branch," B. Oyrzanowski, J. Skrzypczak, Proceedings of XXI EOQC Conference, Varna 1977.)

2. User Quality Costs

The third type of quality costs was introduced by Mr. Frank M. Gryna (Quality Progress, November 1972). This type of quality costs called by him "user quality costs" are the costs which are not covered by the producer-vendor under warranty and which are born by the buyer — of a producer good or of a consumer durable good. It means that in this case for the bad quality of conformance pays the user and not the producer-vendor. Usually they are due to the low reliability and short durability after the guarantee period.

According to Frank M. Gryna user quality costs fall under five main categories:
(1) cost of repairs,
(2) cost of maintaining extra capacity because of expected failures,
(3) cost of effectiveness loss,
(4) cost of damages caused by failed item,
(5) lost income.

At first sight one could say that they are no concern of the producer because he does not cover them. However, the buyer, a producer, consumer who covers user quality costs compares the user quality costs of two appliances manufactured by two different producers and in the future quite obviously he would be buying the machine with lower user quality costs. Under these circumstances, although these user quality costs are not covered by the manufacturer of a given machine, he cannot stay quite indifferent to them because competitiveness of his products depend on them, and so do his profits.

This problem has acquired a new significance explained by Dr. J. M. Groocock in his paper "Quality to the Customer and Quality Costs the Two Measures of Quality Improvement" (Proceedings of XXVI EOQC Conference, Amsterdam 1982).

Dr. Groocock believes that the customer judges the quality of a product in comparing the same product of two different manufacturers under four different headings: 1) product quality, 2) support quality, 3) delivery performance, and 4) price/basic price, discounts, payment terms. The most important for him is product quality which can be subdivided into design quality, conformance to design at time of delivery and performance after delivery (reliability, maintainability, durability). The two last items are of special importance in case of comparing products of the similar quality of design.

At present, in many countries as a result of consumers associations and the publications of periodicals like "Test," "Which," "Que Choisir," etc., the consumers have at hand the exact comparisons of the quality characteristics of competing products. For this reason the manufacturer must seriously take into consideration the user quality costs which although not covered by him but by the buyer have indirect influence on the buyers preferences and so on the sales and profits of the manufacturers firm. They related mainly to Dr. Groocock's "Performance after Delivery (reliability, maintainability, durability, safety, etc.)."

This problem is of great importance not only in the free market economy where the producers must take it into consideration on account of the existing competition but also in the centrally planned economies where the lowering of user quality costs increased the social quality costs and decreases real income and satisfaction of consumers.

3. A Survey of User-Consumer Quality Costs

Taking all these problems under consideration, the Economic Institute of the Jagellonian University, backed by the Polish Committee for Standards, Measures and Quality undertook the research into user quality costs of consumer goods or shortly consumers quality costs of some goods. The novelty of this approach lies in the fact that considered commodities are not only limited to consumers durable goods but the consumers perishable goods are included.

The following articles were chosen:
(1) washing machines,
(2) washing powders
(3) bread,
(4) flats/houses.

A. Washing Machines

In case of durable consumer goods, the consumer usually takes all the opportunities of warranty system and tries to make the producer pay for the lack of conformance between the product and its design. In the majority of the cases they result from the lack of reliability. Thus, what are the bad quality costs which have to be covered by the consumer in case of a durable commodity:

(1) Costs connected with warranty:
   a. consumer's lost time taken for bringing the machine to be repaired or exchanged and additional costs born by doing it,
   b. costs of services to be paid for (washing) during the time the machine is being repaired.
(2) Losses covered by consumer due to the malfunction of the machine:
   a. costs of eventual damages to the washing,
   b. eventual costs of damages to the health of the operator,
   c. costs of depreciation of the machine during the period of the machine being repaired under warranty system.
(3) Costs of repairs after the warranty period.

The special survey prepared by the Economic Institute of the Jagellonian University and carried by Committee for House Economics gave the following results:

Consumers quality costs amounted to 8.3% of the price of the washing machine during average 4.5 years of use. The costs of repairs after the warranty period amount to 40% of the quality costs. Indirect costs of the failure of washing machine amounted to 24%, and the costs of lost time to 22%. It is quite obvious that the costs of repairs in the 4th, 5th, 6th were higher than in the 1st, 2nd, and 3rd year and lower than in the 7th, 8th and 9th year.

B. Washing Powders

Washing powder as a non-durable consumer good causes different kind consumers quality costs, because there are no costs of repair and no costs related to repair. Under warranty you can only exchange a packet of washing powder for a new one. Theoretically speaking, the consumers quality costs of washing powder fall under the following categories:

(1) eventual damages of the washing powder to the washing:
   a. washing powder damages the colour of washed materials,
   b. washing powder makes the washed items grayish,
   c. washing powder damages washed items in other ways,
   d. washing powder causes colouring of colour textiles by each other;
(2) eventual damages done to the washing person:
   a. washing powder dries excessively hand skin of the operator — the cost of the special greasy hand-creams to be used after washing by the operator,
   b. eventually causes allergy — both to the operator and person using washed linen,

the cure of the allergy (the doctor's fee, medicines, etc.) has to be paid,
c. the costs of protection, special gloves to be used;
(3) the costs of insufficient packing of the washing powder:
   a. damages caused by the package breaking, i.e., the damages to other products carried by the buyer in the same bag,
   b. underweight of the washing powder in the package;
(4) Lack of exact instruction how to use given powder printed on the packet may be the cause of some damages.

Some findings of this survey caused that the production of low quality washing powder was abandoned.

C. Bread

Bread is another non-durable product taken under consideration. Consumer quality costs may be divided into following items:

(1) value of bread bought and not consumed by the buyer on account of:
   a. taste and nutritive value being not sufficient,
   b. getting stale and dry in too short a time,
(2) time lost for complaints of the low quality of the product,
(3) eventual cost of damages to health caused by the low quality of bread.

The results of the survey carried by the Economic Institute and the Committee of the House Economics showed that more than 95% of the costs resulted in not consuming the bread previously bought. Only a little more than 4% were the cost of lost time for complaining. The main reason for these losses was the fact that the bread was getting stale in a relatively short time. This discovery induced a technological research to lengthen the period during which the bread is kept fresh.

D. Flats/Houses

In Poland the great majority of houses are generally constructed by the Building Cooperative. The greater part of the town population live in big buildings composed of a number of flats. After the construction of these buildings is finished the members of a cooperatives get the keys of their respective flats and they can occupy them.

Before the members move into the flats, in a number of cases they find it necessary to make certain corrections and eventual improvement. Thus, the user quality costs fall under two headings:
(1) the cost of additional work required to make the flat ready for occupation,
(2) the cost of improvement of the standard of the flat.

Some costs of the first category are covered by the construction firm under warranty system. The majority, however, befall on the inhabitant of the flat. All costs of the second category are born by the member:

(1) the cost of additional work required to make the flat ready for occupation may be divided into following items:
   a. correction of the badly laid electrical, heating, gas water and draining installations,
   b. corrections of the badly laid floor,
   c. corrections of the plastering/ceilings and walls,
   d. corrections of the badly fitted doors and windows,
(2) cost of improving the standard of the flat:
   a. additional built-in cupboards,
   b. additional tiling,
   c. laying of parket-floors,
   d. additional electrical fittings,
   e. eventual shifting of walls between rooms,
   f. eventual shifting of doors;
(3) the cost of rent during the period between the moment a tenant gets the keys of his flat and the moment he can move in after necessary corrections and improvements have been made.

The consumers quality costs of a flat during the last few years amounted to around 7.5% of the value of a cooperative flat (the costs of construction). Out of this 5% concerned the costs of improvement of the standard of the flat and only 2% concerned the quality costs caused by the differences between the quality of conformance and the quality of design. As it was mentioned before, part of these latter costs were covered by the building constructor. As the majority of corrections and improvements are in the kitchens and bathrooms the consumers quality costs per square meter are much higher in small flats rather than in big flats.

4. Some General Conclusions of the Undertaken Survey

The first conclusion of the survey is that consumers quality costs are quite significant because they reach 7-8% of the value of the commodity. Although they are much lower than the producers quality costs which quite often reach 12% in case of a number of goods in different countries, one could say that the consumers quality costs are considerable well above 50% of the producers quality costs.

The second conclusion is that being so high the consumers quality costs must be taken under consideration by the buyer who tries to find commodities with lower quality costs.

The third conclusion is that in free market economies a producer wishing to increase his sales must try to reduce consumers quality costs especially by increasing reliability in case of durable goods in order to increase the competitiveness of his goods as compared with these of other manufacturers.

The fourth conclusion is that in centrally planned economies there must arise a strong tendency to lower consumers quality costs in order to diminish the social costs of the state and raise the standard of living of the population.

The fifth conclusion, however far reached, may be stated as follows: the other side of lowering the producer and the consumers quality costs is the reduction of the use of materials, capital equipment and labour which in the world of scarce resources becomes of ever greater significance.

CONSUMER PRODUCT QUALITY CONTROL COST REVISITED

(April Issue)

Harold L. Gilmore
Head and Professor
Division of Business Administration
Pennsylvania State University
Middletown, Pennsylvania

Quality does not cost; it is free[1]: that idea has provoked a debate that goes on and on. Nevertheless, no one can deny that every organization, in one way or another, expends resources in the pursuit of product quality.

Some time ago a study of the quality control cost experience of a sample of U.S. manufacturing firms was reported in this journal.[2] To see what change, if any, could be detected in the expenditure experience of the same or similar corporations, the author decided to revisit the quality cost scene. Since the product cost experience of manufacturers is considered a measureable reflection of corporate quality control behavior, any change in managerial perspective and decision-making toward quality should be evident in expenditure data. Leonard and Sasser[3] recently reported finding a shift, at the general mangement level, from an inspection-oriented, manufacturing-focused approach toward quality, to a defect-prevention and company-focused strategy. They noted a second important change in the view held by management of quality personnel. They were seen as managerially focused, planning- and prevention-oriented, assertive, powerful, responsible for preventing failures, and well-respected. This was in sharp contrast with earlier views where quality personnel were viewed as technically- and problem-oriented, defensive, powerless, responsible for inspection and "fixing" failures, and not well-respected. These changes should be reflected in the allocation and expenditure of corporate resources devoted to the quality control task.

Research Method

As in the previous study, a structured cost data collection form was used to request quality cost data from manufacturers. In fact, the form used was the same one employed earlier. Furthermore, clarification was provided if requested as to the use of the form in the interest of obtaining compatible cost data.

From the outset there was no intention of conducting a statistically designed study. Earlier experience suggested there would be little likelihood of doing so even though such a study would be highly desirable from a research perspective. This study like most studies of this type, is dependent upon voluntary participation. While the investigator may carefully select the sample, respondents provide information "voluntarily." Quality cost information has the additional characteristic of being regarded as "confidential" or proprietary. It is not freely given, if it is given at all. The net result is that these study results are subject to the bias that such voluntary and reluctant participation is bound to produce, however carefully the research is designed.

What the survey attempted to do was to draw upon past experience and current knowledge with regard to corporate effort in the collection and analysis of quality cost data. Thirty-five consumer product manufacturing corporations were selected to receive the data collection instruments. All of those corporations were known to be collecting quality costs in a form both consistent with generally accepted practice and amenable to analysis and publication.

Seventeen survey instruments were returned. Not all contained complete or completely useful data. This report makes maximum use of the information provided consistent with study integrity and in the interest of achieving as much confidence in the statistics presented as possible. Obviously, with only 17 respondents distributed over 10 industrial classifications, there will be a wide confidence range in the statistics presented.

Table 1 is a brief description of the participating organizations that responded to the survey. Because of the confidential nature of the data, companies are not identified by name. Instead Standard Industrial Classification (SIC) code numbers are used to identify respondents.

The respondents can also be identified by size in terms of employment level. (See Table 2.) The number of employees reported by the respondents reflected to a certain extent the structure of the industry within which the firms were classified. For example, the SIC 20 participants (food and kindred products, a highly concentrated industry) reported employing over 2,000 people.

Even though participants were not selected according to size, 42% of the respondents were major corporations employing more than 2,000 people. This happened largely because quality cost systems are more frequently utilized in large firms or in major business units of complex organizations.

Most respondents reported sales, profits, and output to be satisfactory to good. Few rated these characteristics as unsatisfactory. All reported the product price trend was up over recent years and all attached major importance to quality in the sale of their products.

The survey results clearly indicated that quality control management established the quality budget which was, in turn, approved by top mangement. The basis for the budget, however, was not clear from the data. Several companies used multiple bases, but most based their budgets on factors other than sales, manufacturing costs, or direct labor hours. The expense budget and process capability were mentioned by two respondents as alternatives to the above.

TABLE 1
Survey Respondents (N = 17)

SIC		Number of Business Units
20	Food and Kindred Products	2
22	Textile Mill Products	1
25	Furniture and Fixtures	1
26	Paper and Allied Products	3
28	Chemicals and Allied Products	3
34	Fabricated Metal Products, Except Ordnance Machinery and Transportation Equipment	2
35	Machinery, Except Electrical	1
36	Electrical Machinery Equipment and Supplies	2
38	Measuring, Analyzing and Controlling Instruments, Photographic, Medical and Optimal Goods, Watches and Clocks	1
39	Miscellaneous Manufacturing Industries	1

TABLE 2
Company Size (N = 12)

Number of Employes	%
< 200	17
201-500	25
501-2000	17
> 2000	42

Production System

The production system employed by the respondents resulted in either discrete products or mixed, i.e. continuous production or discrete products. Most firms established production levels based on a sales forecast; some used a sales forecast in conjunction with market research and incoming orders. The resulting volume of production for 66% of the respondents involved either large batches or mass production of well-established, stable, slowly changing products.

Two other related characteristics bear on quality control and its cost: labor cost and the number of production workers. The respondents reported that their labor costs as a percent of manufacturing costs ranged from 13-25%. Table 3 shows the ratio of production to quality control personnel employed by the respondents.

TABLE 3
Production to Quality Control Personnel Ratio (N =10)

Ratio	%
<10:1	30
10-100:1	60
>100:1	10

Most of the respondents identified "appearance" type characteristics as the most important product characteristic affecting the quality control activity. Relatively few cited "functional" characteristics of the products as most important. Concern for appearance has an effect on both manufacturing and quality control: the processes having the greatest impact on appearance become the most important from a quality control perspective. It should be noted that the respondents in this survey manufacture products for the consumer market, which may account for the significance attached to appearance. Informed buyers of commercial products would most likely have cited functional specifications as most important.

Industrial orientation, operational climate, budgetary responsibility, and quality-sensitive product characteristics individually and collectively influence the allocation and expenditure of effort on product conformance quality control. The nature of that influence is presented below as it is reflected in selected expenditure data.

Findings

These survey data depict manufacturer quality control behavior and priority as reflected in expenditure practice. The quality cost data in Table 4 are organized into the categories of prevention, appraisal, and failure. Their total makes up what is commonly identified as conformance quality control costs.

Theoretically, one would expect to see a distribution of expenditures on prevention and appraisal activities skewed to over 50%; however, only 25% reported spending over 50% of their total quality costs on prevention, and only 38% reported spending over 50% on appraisal. Failure costs exceeded 50% of total quality costs in nearly one-fourth of the companies that responded. Based on these data — if they could be generalized — there is clearly much to be done to shift the focus of attention towards defect prevention activity. It should be noted that the previous study reported that over one-half of the manufacturers devoted in excess of 50% of their quality costs to the area of failure and failure-related activities.[4]

The expenditure practice presented by industry classification gives additional insight into the behavior or organizations with regard to product conformance quality. The reported data are depicted in Table 5, which illustrates a substantial variation in practice across the spectrum of industries represented.

Again a note of caution is appropriate. Even though the cost data collection instrument was designed and tested to obtain compatible information from all participants, semantics could have influenced the responses. Definitions of prevention, appraisal, and failure may differ within a single company and among people even when they are presumably fully informed. Such differences, should they exist, could account for some of the variation reported here.

It is widely accepted in theory that quality activity should be oriented toward defect prevention. Table 5 shows two noteworthy examples of this theory in practice, namely paper (SIC 26) and fabricated metals (SIC 34). However, it should be noted that the two SIC 34 participants reported cost data at wide variance with each other. As noted above, there may be differences in what each classifies as prevention, appraisal, and failure costs. In this instance, the reported average data are not likely to be typical for the industry in general.

TABLE 4
Expenditure on Prevention, Appraisal, and Failure Activity
(Percent of Total Quality Cost and Percent of Respondents)
N = 8

Prevention		Appraisal		Failure	
Cost (%)	Respondents (%)	Cost (%)	Respondents (%)	Cost (%)	Respondents (%)
<5	38	<20	12	<5	38
5-9.9	25	21-50	50	5-50	38
10-50	12	>50	<u>38</u>	>50	<u>24</u>
>50	<u>25</u>		100		100
	100				

Chemical products (SIC 28) and electrical machinery (SIC 36) report the next highest percentage of prevention costs, although both are some distance behind the leading categories. The chemical-industry respondents reported the highest expenditures for appraisal activities; when those costs are added to the ones associated with prevention, a strong emphasis can be seen on reducing failures. Non-electrical machinery (SIC 35) and instrumentation (SIC 38) appear to be a completely opposite picture. The largest portion of their costs are incurred in the failure area. Such a dichotomy might be attributed to the fact that each industry is represented by only one company, but the author believes that each properly represents its industry class in general.

TABLE 5
Total Quality Cost Allocation by Industry
(SIC Code)

SIC	26	28	34	35	36	38
Quality Activity (Percent of Total Quality Cost)	N=2 %	N=2 %	N=2 %	N=1 %	N=1 %	N=1 %
Prevention	40	22	48	8	16	3
Appraisal	48	60	28	13	45	34
Failure	7	20	16	79	40	62

Note: SIC Nos. 20, 22, 25, 39 did not provide useable cost allocation data.

Indexing

A final statistic of interest is the amount expended on quality in relation to some widely employed index of corporate activity. Three such indices are manufacturing costs, gross sales, and value added.

Table 6 clearly shows that gross sales is the most popular base against which quality costs are measured. Manufacturing cost is the next most popular. Unadjusted value added is not as widely used as the former two. Total quality costs as a percentage of gross sales ranged from less than 1% to over 8%. Over one-half of the companies providing the data reported expending less than 5% of gross sales on conformance quality control. The previous study disclosed the same expenditure experience: 58% expended less than 5% on quality. The manufacturing cost and value added bases show quality costs representing a much larger percentage of each. The significance of these various bases lies in the use that is made of the information. Obviously sales and marketing personnel

TABLE 6
Total Quality Cost as a Percentage of Selected Bases

Industry (SIC Code)	% of Manufacturing Cost TQC	% of Gross Sales TQC	% of Unadjusted Value Added TQC
20	*	.45	*
26	1	1	1
28	7	2	8
34	11	3.3	.5
35	7.4	5.3	18.6
36	*	6.5	*
38	12	8.2	25.6

*Data not available.
Note: May not add to 100% due to averaging or rounding.

find sales relationships most meaningful. Manufacturing management would find either manufacturing cost or value added measures more useful.

Table 7 depicts the components of total quality costs — prevention, appraisal, and failure — expressed as a percentage of the same bases used in Table 6. It follows that prevention and appraisal account for a substantially smaller portion of the measurement bases that failure costs.

Company conformance quality control expenditure practice for some respondents conforms to the theoretical framework underlying quality control management allocation decisions. The paper-industry respondents offer an example of the theory in practice. Two responses from the fabricated metal industry — which when averaged together showed high prevention costs — showed a wide variation in reported prevention costs; it is unclear what the practice of similar business units of the same industry might be. Although the chemical-industry respondents reported a lower percentage of prevention costs, the total for prevention and appraisal was much higher than for failures.

Total quality costs as a percentage of gross sales is still the most popular measure. Companies probably should not rely on only one indicator. Multiple measures are highly recommended to meet the decision-making needs of the various members of management as they relate to product quality. Discussions with several respondents revealed other measures including quality cost per unit of output. The specific measures employed reflected what the users felt would result in the most effective decisions.

One final observation: the level of detail and precision with which quality costs are collected suggests the degree of significance attached to this information within the organization. A question that must be continuously asked and answered is whether the cost of quality cost data collection and analysis is justified by the benefits derived

TABLE 7[5]
Prevention, Appraisal and Failure Cost as a Percentage of Selected Bases

Industry (SIC Code)	% of Manufacturing Cost			% of Gross Sales			% of Unadjusted Value Added		
	P	A	F	P	A	F	P	A	F
20	*	*	*	*	*	*	*	*	*
26	.35	.25	.10	.25	.18	.07	1.35	.95	.40
28	1.7	4.2	1.2	.5	1.3	.4	*	*	*
34	.9	3.7	6.2	.3	1.1	1.9	.04	.17	.28
35	.6	1.0	5.9	.5	.7	4.2	1.6	2.4	14.6
36	*	*	*	1	3	2.5	*	*	*
38	.4	4.1	7.5	.3	2.8	5.1	.8	8.7	16.1

*Data not available.
Note: May not add to 100% due to averaging or rounding.

therefrom. To many it is of little importance. On the other hand, a sizable body of advocates is making effective use of this data in the management of the organization as a whole, as well as of the quality function in particular.

Footnotes

1. Philip Crosby, *Quality Is Free,* (New York: McGraw-Hill, 1979), p. 1.

2. Harold L. Gilmore, "Product Conformance Cost," *Quality Progress,* VII, 6 (June 1974), pp. 16-19.

3. Frank S. Leonard and W. Earl Sasser, "The Incline of Quality," *Harvard Business Review* (Sept.-Oct., 1982), pp. 163-171.

4. Harold L. Gilmore, "Product Conformance Cost."

5. Tables 5 and 7 are not provided for comparative purposes. Cost data involved in each are from different respondents where there is more than one study participant.

DON'T BE DEFENSIVE ABOUT THE COST OF QUALITY*

(April Issue)

Philip B. Crosby
Chairman and Chief Executive Officer
Philip Crosby Associates, Inc.
Winter Park, Florida

In the 30 years I have been a working quality professional nothing has disappointed me as much as the way the "cost of quality" is not used. I am reminded of the chemistry sets we used to receive for Christmas when I was a youngster. The set came equipped with all the materials necessary to perform certain experiments. The bigger the set, the more experiments.

During the dreary winter days following Christmas I would sit entranced in the basement working my way through the experiments. I performed each one with the intensity of a Curie — without the slightest idea of what I was doing or why I was doing it.

It was a lot of fun, and it kept me out of trouble. I probably even learned some self-discipline and new words. But it wasn't a great deal of help in the chemistry I took in school since they didn't use the same book or the pictures of a boy pouring things into things.

My big problem was I never really understood WHY I was taking all those steps; I never knew the purposes of the effort.

The same is true with the cost of quality. No subject has received more attention from the quality profession over the past several years than COQ. At each convention, the best attended sessions are on cost of quality. Every ASQC Section worth its salt conducts at least one class a year on the subject.

The headquarters staff will confirm, I am sure, that the most requested literature refers to the cost of quality. "Do-it-yourself" books and exhaustive reference material are available. Certainly the "cost of quality" is an item well-known. This would lead you to believe that everyone and their brother is doing cost of quality. You would then think that the corporations of America were well-versed in the subject. You might imagine Wall Street brokers discussing the relative COQ of the Fortune 500 companies. You might feel that the reputations of Chief Executives could rise or fall based on their personal capability for getting the most out of COQ.

You might think that, but you would be wrong.

I have never seen a company that had its COQ figured out right or used it properly. Now before you mail me a copy of your report, hear what that means to me.

"Figured out right." The COQ has two aspects: the price of conformance (POC), and the price of non-conformance (PONC). (I like to call them the "price" rather than the cost since they are not inevitable.)

---

*Copyright ©1983 by Philip B. Crosby

Every cost incurred because things were not done right the first time goes into PONC. Every cost incurred to make certain things are or were done right the first time goes into POC. These prices include all the benefits, overheads, and whatever that are associated with the real costs of the company.

It should take no more than a couple of days to figure this out if the company has any kind of an accounting system at all.

"Used it properly." The purpose of calculating the cost of quality is to give quality professionals something to talk to management about in order to get corrective action.

Corrective action is what it is all about.

All the time I was a line quality manager I was never invited to an important meeting. Today as a Consultant I am invited to go to corporations and meet with top management. They very rarely invite their quality person to that meeting. Why not?

The reason is simple: they don't think we have anything to talk about. And for the most part we don't.

When the COQ takes its place on the corporate agenda along with inventory, compensation, accounts receivable, general and administrative, sales, profits, taxes, and other normal measurements, then the life of the quality executive becomes useful. There is something to talk about. They will get invited everywhere.

Taken as a part of the management system of the company, the COQ suddenly becomes a star. We normally see figures like 25% of sales for the COQ. We see companies reducing that number by half in a year or two. Once they know it exists they can identify the causes and go after them. It becomes routine.

Why isn't all this happening if it is so easy?

I never said it was easy; I said it was practical and only took a couple of days. But it isn't happening for two not-so-very-good reasons:

First, quality professionals are still gathering all the data so they can make absolutely sure they get every single little nickel of the cost documented, and at the same time not hurt anyone's feelings.

Second, quality professionals resolutely refuse to include the study of management in their work on how to succeed. The principles of management, and of running a company, are not mysterious. They just have to be learned. They need to be included in our arsenal if we are to communicate with other members of management. Everything you ever needed to know about what to collect has been documented by author after author and committee after committee. It is not all that important what you choose to put in as long as you are ready to do it the same every time.

All of the waste, all of the appraisal, all of the field service, all of the computer verifying of software, all of the purchasing and engineering changes — you know what categories count. If you are in doubt, put it in.

What you need is a cost accumulation that will permit you to develop a trend and then show the areas that need corrective action. For the most part this will only provide verification of what everyone already knows. After all you don't really need the sales figures to know when sales are down — you can see it.

But remember that COQ is a communications tool, not an accounting process. It is meant to help you in your job, not to point out that you are inadequate.

The quality profession has the opportunity to bring our industrial empire back to profitability by causing the elimination of waste in our service, administration, and production operations. No other function has that possibility. I suggest that you go to work on your COQ in practical terms. You owe it to your company and to yourself.

Buy the Controller a lunch, and get two of you with two of them to take a serious look at the COQ. (If you "figure it out" and you come up with less than 20% of sales in a manufacturing company, or 25% of the cost of operations in a service company, then you may be looking at your company in too limited a fashion.) But don't worry about what it is in percentage or actual numbers. The purpose is to spend your time reducing the COQ through honest corrective action oriented around prevention.

Don't let anyone set a specific cost reduction as a goal; that's too easy: for starters, just eliminate the inspection department — that would save money, at least for a while. To eliminate costs, you have to remove the causes of problems. Many causes cannot be eliminated unless management wants them eliminated.

Your tool for getting management to want to eliminate the causes of the problems is the COQ status. This brings us back to you. Go to it.

MANAGEMENT TEAM SEEKS QUALITY IMPROVEMENT FROM QUALITY COSTS

(April Issue)

Joseph J. Tsiakals
Quality Engineering Manager
Travenol Laboratories, Inc.
Cleveland, Mississippi

The quality cost concept has been developing for several decades. There are many books, articles, talks, and seminars on this subject. Many companies have attempted to start quality cost programs; there have been some fine success stories but there have been many more failures.

It is clear that a formal quality cost program is not necessary to achieve high product quality at competitive prices. To attain these goals, a company must have effective management that continually strives for improvement. Quality costs can be viewed as a method that provides new, exciting insights to the management team. A quality costs program is more likely to succeed when there is an effective management team in place working in an atmosphere of "managerial breakthrough"[1] and when quality cost information is made easily available in a form that pinpoints areas of opportunity.

A program that has proven particularly successful began in 1981 at one of the Travenol Laboratories, Inc. manufacturing facilities in northern Illinois. The plant was selected because, among other reasons, it had a strong management team in place and a history of high product quality and good cost performance. The program was designed to incorporate three ideas: the quality cost concept; Juran's "managerial breakthrough" techniques; and the decision support system concept.

Steering Committee

The key to the program's success is a plant quality cost steering committee. The steering committee consists of the plant manager, the production manager, the quality control manager, the engineering manager, and the controller. The committee has a continuous existence, meeting once a month. Its primary function is to use quality cost analysis to identify areas of opportunity for improving quality and reducing costs. Once the committee has found an area of opportunity, it assigns a diagnostic team to investigate. These teams have a limited life: the duration of the investigation. The teams consist of those individuals on the technical staff and in operations who are most knowledgeable about the area being investigated.

The structure of the team depends upon the nature of the problem. The diagnostic team seeks to discover causes and remedies. It evaluates the quality and economic impact of various alternatives and recommends a course of action to the steering committee. It is not uncommon to have eight or more such teams in existence at any time.

To identify areas for diagnostic teams to tackle, the steering committee requires quality cost data presented in a variety of ways which are difficult to define in advance. Data may be expressed in dollars, may be adjusted for production volume, may be adjusted for the value of production, or may be expressed in some other terms. To isolate a particular area for investigation, the committee must first take a broad view, then consider alternative views, then examine selectively narrower perspectives. It may look at quality cost categories, then cost elements, and then cost sub-elements. The committee may look plant-wide, or at a department or a product line. Additionally, it is desirable to observe performance over time in order to identify trends and track improvement. It was felt to be important that the managers themselves conduct these analyses; that from their positions in the organization, the managers have the perspective to identify significant areas of opportunity for diagnostic teams to investigate.

To allow the steering committee to perform those kinds of analysis, a decision support system was designed for the committee. The system is hierarchical, interactive, and menu-driven with color graphic screen and copy output. Since the users were to be managers, the system was designed to require no computer language training. Graphs are defined from selections offered on the screen. It is relatively easy to go quickly from a plant summary graph through layers of intermediate information to very specific detail. Managers can work their way through the hierarchy of data, easily selecting and displaying the information important to them. Typically, in an hour session, 20 or so graphs will be generated from the thousands of combinations available. Hierarchical data provides the ability to go quickly from summaries to specifics. Interactive menus allow quick designation of the output. Graphical displays allow managers to spot trends and disparities immediately. The color feature — at first thought to be somewhat unnecessary — turned out to be an important attribute of the system. Much more data can be analyzed and displayed in terms of trends when each line is a different color than when all lines are the same color. This is additional to the advantages obtained from using different patterns of lines.

All data is presented in trend chart fashion. A legend on the chart defines each line. The initial system consisted of six menus. After entering an identification number and password, the user sees a menu on which the performance measure must be specified. Users may select month or year-to-date data; dollars, dollars divided by the value of production, or dollars divided by machine hour. Machine hour is a good measure of the level of production when operations are automated and machine dominant. In more manual production environments, Travenol prefers to measure production volume in earned labor hours. Units produced is not used as an indicator of production volume because of the typically large variety of product types and sizes.

Finally, if they so desire, users can define on the first menu their own performance measure by selecting "ratio analysis." The users would then be required to enter on a subsequent screen the desired ratio (i.e., prevention/internal failure, product A/product B, prevention + appraisal/internal failure + external failure, etc.). Ratio is a unitless measure since terms in both the numerator and denominator cancel.

Breakdown of Information

Following screens permit the user to select plant totals, departments, and products. Additional screens contain the quality cost hierarchy: cost category (prevention, appraisal, internal failure, external failure); cost element (a further breakdown of a specific element).

A typical analysis might proceed like this:

A first graph shows the dollar amounts in each of the four quality cost categories. The user notes that internal failure costs have increased and calls up quality cost totals in terms of dollars per machine hour. This graph reveals that production has increased and that internal failure costs have fallen, rather than risen, when corrected for the increased production; the graph also shows an upward trend for appraisal costs — a trend that was not apparent on the first graph. By calling up a third graph showing dollars per machine hour for each element of the appraisal cost category, the user discovers that in-process inspection costs are rising. The user therefore selects a fourth graph — dollars per machine hour for in-process inspection, broken down by department — and finds that in-process inspection costs are going up in Department B. The user then prints that fourth graph.

At this point, the manager might point out that a major new product has been introduced into Department B and that higher-than-normal shakedown inspections are planned. He might look into this further with additional graphs or assign a diagnostic team to more thoroughly evaluate it. The diagnosis might include a comparison of specific cost categories and elements for the new product, use of technical data, action plans, and recommendation of new performance goals.

The analysis system uses TSO on an IBM 370 mainframe. We use an IBM 3279 color graphics CRT and associated printer. Initially, the main program was FORTRAN with SAS and GDDM subroutines. As data became more extensive and as enhancements were added, we found it advantageous to incorporate RAMIS databases into the structure. Enhancements completed or in progress include adding a master menu, providing Pareto diagrams, and providing additional levels in the internal failure category — linking the support system to the scrap reporting system and capturing reason codes.

The plant controller is responsible for maintaining data. No additional staffing has been required to maintain this system. The great majority of the quality cost data was available from existing accounting systems. The small fraction that was new was easily incorporated into the current reporting mechanisms.

This program has expanded from its initial location. The continual reductions in total quality costs have provided strong incentives for adopting the program. Even so, claims are avoided as to what a quality cost program can achieve. Furthermore, it must be emphasized that certain conditions need to be present before the program is adopted. Quality costs reporting alone does not improve quality and reduce costs. Neither does quality cost analysis. A company will get results only from the hard work and dedication of a strong management team that takes the fullest advantage of available resources.

Used in a way that helps make quality improvement routine, quality costs can serve as a valuable means of achieving cost-effectiveness and quality.

Reference

1. Juran, J. M., "Managerial Breakthrough," McGraw-Hill Book Company, New York, 1964.

PRINCIPLES OF QUALITY COSTS

(April Issue)

Jack Campanella
Director of Quality Assurance
Saab-Fairchild 340 Program
Fairchild Republic Company
Farmingdale, New York
Frank J. Corcoran
Quality Assurance Specialist
Singer, Kearfott Division
Little Falls, New Jersey

The concept of the economics of quality can be traced back to the early 1950s. Chapter I of J. M. Juran's first "Quality Control Handbook,"[1] published in 1951, was titled "The Economics of Quality" and contained discussions of the "Cost of Quality" and his now famous analogy with "gold in the mine." However, most other papers and articles of the time were based on specific applications, such as "The Economic Choice of Sampling Systems in Acceptance Sampling" by J. Sittig in 1951.[2]

Among the earliest articles on quality cost systems as we know them today were Harold Freeman's 1960 paper, "How to Put Quality Costs to Use,"[3] and Chapter 5 — "Quality Costs" — of A. V. Feigenbaum's famous text "Total Quality Control"[4] in 1961. They are among the earliest writings categorizing quality costs into the costs of prevention, appraisal, internal and external failure.

In December 1963 the Department of Defense issued MIL-Q-9858A, "Quality Program Requirements,"[5] making "Costs Related to Quality" a requirement on many Government contracts and subcontracts. From that point on, the subject boomed. In 1964 the Industrial Engineering Department of Stanford University did a research study for the Air Force Systems Command and published the QUICO (Quality Improvement Through Cost Optimization) System[6] ultimately leading to the May 1967 release of Quality and Reliability Assurance Technical Report TR8, "Guide to Quality Cost Analysis,"[7] by the Office of the Assistant Secretary of Defense for Installations and Logistics.

The ASQC Quality Cost Technical Committee was formed in 1961, and in 1967 published "Quality Costs — What and How,"[8] the most popular document on the subject and one of the largest sellers of any ASQC publication. Other popular committee publications are the 1977 "Guide to Reducing Quality Costs"[9] and more recently, the "Guide for Managing Vendor Quality Costs"[10] published in 1980.

The more than 300 excellent articles and papers listed in the current "Bibliography of Articles Relating to Quality Cost — Concepts and Improvement,"[11] maintained by the Quality Cost Technical Committee, further attest to the popularity of the subject and to the recognition given to it throughout industry.

Applicable Specifications

As mentioned above, military specification MIL-Q-9858A makes quality costs a requirement on many government contracts. Paragraph 3.6 of that document requires the contractor to "Maintain and use quality cost data as a management element of the quality program." Except for requiring the identification of the costs of "prevention and correction of nonconforming supplies," this paragraph is not very definitive or restrictive. It permits a very wide interpretation and allows "the specific quality cost data to be maintained and used" to be "determined by the contractor."

A more recent military standard, MIL-STD-1520B[12] (original issue MIL-STD-1520, May 1, 1974), requires Air Force contractors who have MIL-Q-9858A on contract to collect "the costs associated with nonconforming material" with an objective of providing "current and trend data to be used for contractor and government management review and appropriate action." This document is much more specific than MIL-Q-9858A and lists the actual costs that must be collected and summarized.

In addition, more and more contracts, both government and commercial, are spelling out quality cost requirements... from the basic collection of scrap and rework costs, to full-scale quality cost programs.

The real value of a quality program is ultimately determined by its ability to contribute to customer satisfaction and to profits. That is the environment in which quality management exists, and is the principal reason why quality costs should be an integral part of an effective quality management system.

To understand the concept of quality costs, it is necessary to establish a clear picture of the difference between quality costs and the cost of the Quality Department. It is important that we don't view quality costs as the expenses of the Quality function. Fundamentally, each time work must be redone, we are adding to the cost of quality. The most obvious examples are the rework of a manufactured item, the retest of an assembly, or the rebuilding of a tool because it was originally unacceptable. Other examples may be less obvious: the repurchasing of parts, for instance, or the response to a customer complaint.

In short, any cost that would not have been expended if quality was perfect contributed to the cost of quality. Unfortunately, many such costs are overlooked or unrecognized simply because most accounting systems are not designed to identify them. It is for this reason that the system of quality costs was created. It was designed to demonstrate that the cost of "doing things over" is a significant addition to the cost of quality management, and to show that these costs collectively offer an otherwise hidden opportunity for profit improvement.

The most costly condition exists when a customer finds defects. Had the manufacturer, through much inspection and testing, found the defects himself, it would have saved money. And if the manufacturer's quality program had been geared towards preventing defects, defects and their resulting costs would have been minimized; obviously the most desirable condition.

The basic problem is to strike the correct balance between the cost of quality and the value of quality. Figure 1 shows that total cost is at a minimum when the cost of quality is at an optimum point. It is at this optimum point that the correct balance between the cost and value of quality has been reached.

Quality costs must be categorized if they are to be managed. The three major categories commonly used are prevention costs, appraisal costs and failure costs, defined as follows:

• *Prevention Costs* are those costs expended in an effort to prevent discrepancies, such as the costs of quality planning, supplier quality surveys, and training programs;

• *Appraisal Costs* are those costs expended in the evaluation of product quality and in the detection of discrepancies, such as the costs of inspection, test, and calibration control;

• *Failure Costs* are those costs expended as a result of discrepancies, and are usually divided into two types:

Internal Failure Costs are costs resulting from discrepancies found prior to delivery of the product to the customer, such as the costs of rework, scrap, and material review;

External Failure Costs are costs resulting from discrepancies found after delivery of the product to the customer, such as the costs associated with the processing of customer complaints, customer returns, field services, and warranties.

Total Quality Cost is the sum of these costs, i.e., prevention, appraisal, and failure. It represents the difference between the actual cost of a product, and what the reduced cost would be if there were no possibility of failure of the product nor defects in its manufacture. It is, as decribed by Juran,[13] "gold in the mine" waiting to be extracted. Quality costs can run as high as 15 to 20 percent of sales in some companies and as low as 2.5 percent of sales in others.

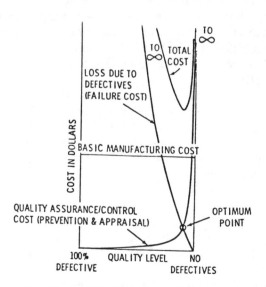

Figure 1. Cost of Quality Versus Value of Quality

The objective is to bring the total quality cost to a minimum while maintaining required quality levels. The basic concept is that an increase in the cost of prevention should result in a larger decrease in the cost of failure, thereby reducing the total quality cost. (An ounce of prevention is worth a pound of failure.) When this no longer happens, prevention costs have been "saturated," and no further dollars should be invested in prevention until conditions change or a breakthrough[14] is achieved. It is at this point, the *saturation point,* that further costs of prevention become larger than the savings afforded.[15]

It may be determined, for example, that the cost of a quality training program (a prevention cost) was responsible for a large decrease in failure costs. Future programs of this type would therefore be indicated since the value of the program (the decrease in failure costs) exceeded the cost of the program. When the cost of training exceeds the savings it produces, the saturation point has been reached. Properly analyzed, quality costs will let you know when you have reached the saturation point.

An increase in the cost of prevention should reduce the cost of failures; it may also cause appraisal costs to decrease somewhat. For example, the reduction in failure costs may justify an increase in sampling inspection due as you become confident that you have improved quality, thereby reducing the amount of inspection performed.

An increase in appraisal costs may also cut failure costs, because a higher proportion of discrepancies will be found in-house. External failures — those found by the customer — inevitably cost more than internal failure. Returned material costs, retrofit costs, and possible loss of future business due to unhappy customers: these are examples of expensive external failure costs.

Appraisal costs and failure costs should change in opposite directions, appraisal costs going up as failure costs go down, *until an optimum point is reached.* An investment in appraisal costs beyond the optimum point is uneconomical. The determination of the optimum point should be part of the quality cost analysis.

Total quality costs, compared to an applicable base, may be plotted and periodically analyzed in relation to past indices. The base should be representative of and sensitive to fluctuations in business activity. Some bases commonly used are manufacturing direct labor, net sales billed, and cost input.

It should be obvious that increases in expenditures for prevention and appraisal will not show immediate reductions in failure costs because of the time lag between the cause and effect.[16] This lag can be observed on a quality cost index trend chart. It may therefore be desirable to indicate on the chart when major changes to the quality program were made. Figure 2 is a quality cost index trend chart marked to indicate a change to the quality program. We can see the reason for the steady improvement over the last five months: back in April, the company initiated operator training programs.

The first effect was an increase in the quality cost index to reflect the cost of the programs (prevention costs increased but failure costs remained the same). After a "cause-and-effect lag" of about two months, the value of the training began to become evident: there is a steady reduction in the quality cost index (failure costs decreased while prevention costs remained the same). By November, a 45% reduction was achieved.

Obviously, the training programs were a worthwhile investment. But if the index had not shown an improvement within a reasonable period of time, the company would have

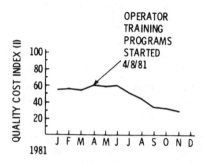

Figure 2. Quality Cost Trend

had to take some action. The programs would have been re-evaluated and either revised or dropped in favor of some other course of action.

How to Start

The first step in starting a quality cost program is to determine the need for such a program. Then make a presentation to management in a way that will justify the effort and will interest them in participating in the program.

One way to do this is by establishing a trial program. The trial program can be simple: only the major costs need be gathered, and only data readily available need be included. You may find that much of the data required for the program is presently available. You may even estimate some of these costs, if necessary. Select a program, division, facility, or area of particular interest to management. The results should be sufficient to sell them on the need for the program.

Most trial runs will show eye-opening results — spectacular enough to make management sit up and take notice. They'll see that quality costs can run as high as 20% or more of net sales dollars; they'll see opportunities for significant savings (Juran's "gold in the mine"). Once top management in convinced, getting the all-important cooperation of the Accounting Department should be easy.

Defining Quality Cost Elements

With management sold, and accounting ready to go, you must then determine the specific quality costs to be collected. To determine prevention costs to be collected, list the tasks your company performs in order to prevent discrepancies, together with the departments responsible for those tasks. Remember that quality costs are not only incurred by the Quality Department. In a like manner, appraisal cost elements are determined by listing those tasks associated with the inspection and test of product for the detection of discrepancies.

For failure cost elements, you need to determine those costs which would not have been expended if you had no discrepancies. If you had no discrepancies, you would not have rework, nor would you have to respond to customer complaints, nor take corrective action. Remember to divide failure costs into "internal" and "external" categories.

Quality costs elements may differ from company to company, and especially from industry to industry; however, the overall categories of prevention, appraisal, and failure costs are always the same.

Cost Collection

Now that you have determined the specific costs to be collected, you must develop a method for collecting them. Collection of quality costs should be the responsibility of the Controller. If top management was properly sold on the program, the Controller will have been charged with the task of heading this effort.

With the help of the Quality Manager, the Controller should review the list of costs collected, determine which of these are available under the present accounting system, and decide where additions to the existing system need to be made. Occasionally, the mere addition of new cost element codes to the present charging system is sufficient. However, if necessary, the present system may be supplemented by separate forms designed especially for the quality costs program.

Ideally, a complete system of cost element codes could be generated and coded in such a way that the costs of prevention, appraisal, and internal and external failures could be easily distinguished and sorted. These codes would be entered on a labor distribution, charge, or time card together with the hours expended against the cost element represented by each code. The labor hours would be converted to dollars by Data Processing. (An exception might be scrap. In many companies, the existing scrap reporting documents are forwarded to the Estimating Department, which estimates the labor and material costs expended to the stage of completion of the scrapped items.) Accounting then provides all collected quality costs to Quality in a format suitable for analysis and reporting.

Of course, training programs will be necessary to inform all personnel about the method to be used to report their quality cost expenditures. The training should be repeated periodically and the collection system should be audited on a regular basis.

Summary and Analysis

Quality costs may be summarized by company, by division, by facility, by department, or by shop. They may be summarized by program, by type of program, or by the total of all programs. The decision must be predicated on the individual needs of *your* company.

In your analysis, you can compare total quality cost to an appropriate measurement base such as net sales, cost input, or direct labor. This comparison will relate the cost of quality to the amount of work performed.

Quality costs normally increase in proportion to increases in the base. It is the nonproportionate change that should be of interest. The index [Total Quality Cost/Measurement Base] is the factor to analyze. Your goal is to bring this index to the point of most economical operation — the optimum point. The index may be plotted so that you can analyze how well you're doing in comparison to past performance and future goals may be analyzed.

Another method of analysis is to study the effect that changes in one category (prevention, for example) have on the other categories, and on the total quality cost. This technique can help you see where the quality dollar can most wisely be spent. You must investigate increases in failure costs so that you can decide which preventive actions — and prevention costs — can reverse these trends and reduce the total quality cost. You must define losses, identify their causes, and take preventive action to make sure it doesn't happen again.

You can use other existing quality systems, such as that for defect reporting and analysis, to identify significant problems. While the losses are distributed among numerous causes, they are not uniformly distributed. A small percentage of the causes will account for a high percentage of the losses. This is an adaptation of "Pareto's Principle."[17] These causes are the "vital few," as opposed to the "trivial many." By concentrating on prevention of the "vital few" causes, you will achieve maximum improvement at minimum cost. Your goal is to determine and attain an optimum level, where return for the effort expended is greatest. By its very nature, this strategy will have the effect of improving quality while reducing costs.

There are almost as many ways to report quality costs as there are companies reporting them: how they are reported depends to a large extent on who they are being reported to and what the report is trying to say. The best way for you is the way that's best for your company and purpose. The examples provided as Figures 3 through 6 are samples of reporting techniques that have worked in companies with established quality cost programs.[18, 19]

The amount of detail included in the quality cost report generally depends upon the level of management the report is geared to. To top management, the report may be a scoreboard, depicting, in a few carefully selected trend charts, Quality program's status — where it's been and the direction it's heading. The report might also identify savings afforded over the report period and point out opportunities for future savings. To middle management, the report might provide quality cost trends by department or shop to enable these managers to identify areas in need of improvement. Reports to line management might provide detailed cost information, perhaps the results of a Pareto

QUALITY COST - By Program
FARMINGDALE

Month AUGUST 1978

CODE	ELEMENT DESCRIPTION	PROGRAM (In Thousands)					TOTAL ALL PROGRAMS
		A-10	SSVT	F-4	747	Misc	
K	MATERIAL REVIEW ACTIVITY	71.8	—	.7	7.7	—	80.2
L	CORRECTIVE ACTION	249.8	—	.9	2.7	—	253.4
X	TROUBLESHOOTING / FAILURE ANALYSIS	47.2	—	—	.8	—	48.0
R	REWORK / REPAIR	128.6	.4	6.3	26.1	—	161.4
P	SCRAP	19.2	—	—	0.5	—	19.7
V	RWK / RPR / SCRAP - VENDOR RESP.	27.1	—	.1	2.6	—	29.8
U	PROCESSING OF CUSTOMER COMPLAINTS	5.4	—	—	1.4	—	6.8
I	PROCESSING OF CUSTOMER RET'N'D MAT'L	23.3	—	—	—	—	23.3
J	FIELD SERVICES	—	—	—	—	—	—
Y	WARRANTY COSTS	.1	—	—	—	—	.1
	TOTAL "UNQUALITY" COSTS	572.5	.4	8.0	41.8	—	622.7
	QUALITY PREVENTION AND APPRAISAL COSTS	718.2	.9	8.3	69.8	38.6	835.8
	TOTAL QUALITY COSTS	1290.7	1.3	16.3	111.6	38.6	1458.5
	MFG DIRECT LABOR COSTS	6973.8	7.5	116.5	882.8	396.5	8377.1
	SCRAP / REWORK / REPAIR AS % OF MFG D/L	2.1	5.3	5.4	3.0	—	2.2
	COST INPUT	20943.4	66.1	978.4	1946.6	104.1	24038.6
	TOTAL QUALITY COSTS AS % OF COST INPUT	6.2	2.0	1.7	5.7	37.1	6.1

Figure 3. Quality Costs by Program

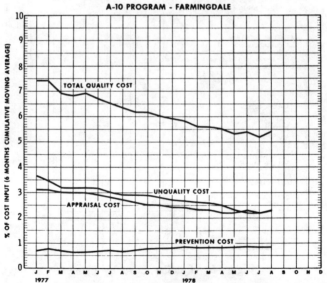

Figure 4. Quality Costs by Category

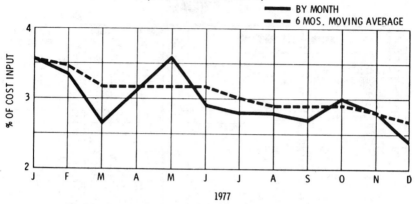

Figure 5: Failure Costs by Month With a 6 Months Moving Average for Trending

QUALITY COST REPORT
FOR THE MONTH ENDING _____
(In Thousands of U.S. Dollars)

DESCRIPTION	Current Month			Year to Date		
	Quality Costs	As a Percent of Sales	Other	Quality Costs	As a Percent of Sales	Other
1 PREVENTION COSTS						
1.1 Product Design						
1.2 Purchasing						
1.3 Quality Planning						
1.4 Quality Administration						
1.5 Quality Training						
1.6 Quality Audits						
TOTAL PREVENTION COSTS						
PREVENTION TARGETS						
2 APPRAISAL COSTS						
2.1 Product Qualification Tests						
2.2 Supplier Production Inspection and Test						
2.3 In Process and Final Inspection and Test						
2.4 Maintenance and Calibration						
TOTAL APPRAISAL COSTS						
APPRAISAL TARGETS						
3 FAILURE COSTS						
3.1 Design Failure Costs						
3.2 Supplier Product Rejects						
3.3 Material Review and Corrective Action						
3.4 Rework						
3.5 Scrap						
3.6 External Failure Costs						
TOTAL FAILURE COSTS						
FAILURE TARGETS						
TOTAL QUALITY COSTS						
TOTAL QUALITY TARGETS						

MEMO DATA	Current Month		Year to Date		Full Year	
	Budget	Actual	Budget	Actual	Budget	Actual
Net Sales						
Other Base (Specify)						

Figure 6. Quality Costs by Element

analysis, identifying specific areas where corrective action would afford the greatest improvement. Scrap and rework costs by shop are also effective charts, when included in reports to line management.

Charts are used to present data and trends pictorially. Properly done, they can bring home a point or tell a story more simply, better, and more interestingly than the raw data they represent. However, they should rarely be designed to portray more than one idea. Charts with several ideas usually do not present any of them effectively.

Charts can be designed in various ways, depending on their purpose. Line graphs are best for depicting trends, while bar charts are useful for showing proportions and comparisons. Circle or "pie" charts are also used for showing proportions, and are particularly effective in illustrating the slice of each silver dollar expended for prevention, appraisal, and internal and external failures. However, circle charts are limited in that they cannot be plotted over time as can bar charts.

Use by Management

Once the quality cost program is implemented, it should be used by management to justify and support improvement in each major area of product activity. Quality costs should be reviewed for each major product line, manufacturing area, or cost center. Management can then look at the improvement potential that exists in each individual area and can establish meaningful goals. The quality cost system then becomes an integral part of quality measurement. It can be used to establish improvement efforts at whatever level is necessary to reduce total quality costs, and then, as progress is achieved, can be adjusted to reduce quality costs to the lowest level possible. This prevents unheeded growth in quality costs and creates improved overall quality performance, reputation, and profits.

A quality cost program, based on the concept and methods of implementation endorsed by the ASQC Quality Cost Technical Committee[20] and broadly presented here, can be used by management as an aid to achieving its goal of an optimum quality program at a minimum quality cost. The program will measure the value of the quality effort, identify the strong and weak points of the quality program, indicate how the quality dollar can be spent most effectively, and provide quality improvement while reducing costs.

An additional benefit of a quality cost program is its ability to be used as a budgeting tool. Once quality cost elements have been established and costs are being collected against them, you can generate a history with which to determine the average cost per element. These averages can serve as the basis for future quotes and "estimates to complete." Budgets can be established for each element. Then, going full circle, the actuals collected against these elements can be used to determine budget variances and, as with any good system of budget control, action can be taken to bring variances into line. Figure 7 illustrates a report providing actual-versus-budget figures for each quality cost element.

```
                                          DATE
                                          MONTH
                    QUALITY COST REPORT
                                      MONTH                YEAR TO DATE
PREVENTION                        ACTUAL    BUDGET      ACTUAL    BUDGET

P1.1 ENGINEERING EFFORT            20.0      21.0        200.0     200.0
P1.2 QUALITY ADM., PLNG. & SERV.    5.0       5.0         50.0      50.0
     P1.  TOTAL                    25.0      26.0        250.0     250.0
APPRAISAL

A1.1 INCOMING INSPECTION           20.0      25.0        200.0     250.0
A1.2 IN-PROCESS & FINAL INSPEC.    50.0      55.0        600.0     700.0
A1.3 TESTING                       50.0      50.0        600.0     600.0
A1.4 PRODUCT AUDIT                 10.0      12.0        100.0     150.0
A1.5 OTHER TESTS-RELIAB. ETC.      10.0      11.0        100.0     100.0
A1.6 EQUIP. CALIBRATION            10.0      12.0         50.0      50.0
     A1.  TOTAL                   150.0     165.0       1650.0    1850.0

INTERNAL FAILURES

N1.1 DEFECTIVE WORK/JUNK           30.0      35.0        350.0     400.0
N1.2 REMAKES, REWORK & RETEST      30.0      20.0        400.0     300.0
N1.3 INVESTIGATION                 10.0       5.0        100.0     150.0
     N1.  TOTAL                    70.0      60.0        850.0     850.0

EXTERNAL FAILURES

E1.1 WARRANTY - COMPLAINTS         10.0      20.0        100.0     200.0
E1.2 FIELD REPAIR - (IN
             WARRANTY)              2.0       5.0         50.0      50.0
E1.3 FIELD REP. - (OUT OF WARRANTY) 1.0       5.0         50.0      75.0
E1.4 OTHER                          2.0       5.0         50.0      25.0
     E1.  TOTAL                    15.0      35.0        250.0     350.0

Q1.  TOTAL                        260.0     286.0       3000.0    3300.0

- RATIO TO TOTAL SALES              3.2%      5.0%        3.0%      3.2%
- RATIO TO TOTAL STANDARD
      COST OF OUTPUT                5.1%      4.6%        4.6%      4.7%
```

Figure 7.   Quality Costs Actuals Versus Budget

Reducing Quality Costs

"An ounce of prevention is worth a pound of cure." That old maxim is as true as ever when it comes to reducing quality costs. An increase in the cost of prevention should result in a larger decrease in the cost of failures, thereby reducing total quality costs.

Quality improvement means cost improvement. Designing and building a product right the first time always costs less. You will see measurable savings if you find the causes of problems with existing products and then eliminate these problems. To cash in on these savings requires that the quality performance on the past be improved. The ASQC publication, "Guide for Reducing Quality Costs,"[21] describes ways to do this. The publication is intended for general management and professionals engaged in quality program management; it will enable them to structure and manage programs for quality cost reduction. It describes techniques for using quality cost data in programs to reduce costs and improve profits.

Corrective action is a key factor in the reduction of quality costs. Quality costs do not reduce themselves; they are merely the scorecard. They can tell you where you are, and where your corrective action dollar will afford the greatest return.

Once you have identified a target for corrective action — through Pareto or other methods of quality cost analysis — you must carefully determine the action to be taken. It must be individually justified on the basis of an equitable cost trade-off (e.g., a $500 rework problem versus a $5000 solution). At this point, experience in measuring quality costs will be invaluable in estimating the payback for individual corrective action investments. Cost-benefit justification of corrective action is a continuing part of the program.

Some problems have fairly obvious solutions. They can usually be fixed immediately (e.g., replacement of a worn bearing or a worn tool). Others are not so obvious (such as a marginal condition in design or processing) and are almost never discovered and corrected without the benefit of a well-organized and formal approach. Marginal conditions usually result in problems that can easily become lost in the accepted cost of doing business. Having an organized corrective action system justified by quality costs will bring such problems to the surface, where management can see and take action on them. The true value of corrective action is that you have to pay for it once; whereas failure to take corrective action may be paid for over and over again.[22]

Auditing Quality Costs

A quality cost program will be effective only as long as it continues to accurately measure true quality costs within an organization.[23] The financial establishment has long recognized that setting up sound and reasonable procedures is, in itself, insufficient to maintain an accurate reporting system. Periodic audits are required to determine if the system is functioning as it was designed to and if it is still conceptually adequate. Some companies use their financial auditors to review the cost collection system, and their Quality Auditors to review the balance of the program; however, there is no method applicable to all companies. Each company knows best how to audit its own cost systems.

Often, however, the major emphasis is on auditing to determine whether the system is functionally adequate and there is too little, if any, emphasis on seeking to determine whether the system is still conceptually adequate. An annual "conceptual" review of the total quality cost system with more frequent (monthly or quarterly) "functional audits" of major portions of the system is usually sufficient.

Conclusion

The concept of quality cost has become a principal management tool. Definitions and standards have been developed and refined along with techniques and methods for implementation. The quality cost program is the bridge between line and executive management. It provides a common language, measurement, and evaluation system which proves that quality pays in increased profits, productivity, and customer acceptance.

References

1. Juran, J. M., *Quality Control Handbook*, First Edition, McGraw-Hill, Inc., 1951.
2. Sittig, J., *The Economic Choice of Sampling Systems in Acceptance Sampling*, Bulletin of International Statistics Institute, 1951.
3. Freeman, H. L., *How to Put Quality Costs to Use*, Transactions of the Metropolitan Conference, Metropolitan Section, ASQC, 1960.
4. Feigenbaum, A. V., *Total Quality Control*, McGraw-Hill, Inc., 1961.
5. Department of Defense, *MIL-Q-9858A, Quality Program Requirements*, December 16, 1963.
6. Morgan, E. D. and Ireson, W. G., *Quality Cost Analysis Implementation Handbook*, Stanford University, Department of Industrial Engineering, March 27, 1964.
7. Office of the Assistant Secretary of Defense (Installation and Logistics). Quality and Reliability Assurance Technical Report TR8, *A Guide to Quality Cost Analysis*, May 31, 1967.
8. Quality Cost Effectiveness Technical Committee, ASQC, *Quality Costs — What and How*, American Society for Quality Control, 1967.
9. Quality Costs Technical Committee, ASQC, *Guide for Reducing Quality Costs*, American Society for Quality Control, 1977.
10. Quality Costs Technical Committee, ASQC, *Guide for Managing Vendor Quality Costs*, American Society for Quality Control, 1980.
11. Quality Costs Technical Committee, ASQC, *Bibliography of Articles Relating to Quality Cost Concepts and Improvement*, January 1, 1981.
12. Department of Defense, *MIL-STD-1520B, Corrective Action and Disposition System for Nonconforming Material*, July 3, 1980.
13. Juran, J. M. *Quality Control Handbook*, Third Edition, McGraw-Hill, Inc., 1974.
14. Juran, J. M., *Managerial Breakthrough*, McGraw-Hill, Inc., 1964.
15. Campanella, J., *A Simplified Approach to the Use of Costs Related to Quality*, Transactions of the 13th Annual All Day Conference on Quality Control, Long Island Section, ASQC, 1975.
16. Office of the Assistant Secretary of Defense (Installations and Logistics), op. cit.
17. Juran, J. M., op. cit., *Handbook*, Third Edition.
18. Campanella, J., *The Fairchild Republic Company Quality Cost Program*, Transactions of the 33rd Annual Technical Conference, ASQC, 1979.
19. Hagan, J. T., *Quality Costs*, ITT, N.Y., 1981.

20. Quality Cost Effectiveness Technical Committee, ASQC, *Quality Costs — What and How,* Second Edition, American Society for Quality Control, 1971.

21. Quality Costs Technical Committee, ASQC, *Guide for Reducing Quality Costs,* American Society for Quality Control, 1977.

22. Hagan, J. T., op. cit.

23. Quality Costs Effectiveness Technical Committee, ASQC, op. cit., Second Edition.

Acknowledgement

The material used in this article was obtained from many sources. References were provided whenever possible. The authors wish to give special thanks to the many members of the Quality Costs Technical Committee who submitted material.

QUALITY COSTS: CURRENT APPLICATIONS

(April Issue)

Edward Sullivan
Editor, Quality Progress
American Society for Quality Control
Milwaukee, Wisconsin

There is no such thing as *the* quality costs system. Even after a firm has put a quality cost system in place, it may find that it wants to modify that system, adjusting it to fit the company's needs a little better. As a result, there is currently a wide range of approaches to the basic task of gathering quality cost information in order to help improve quality. Some of those approaches are well-established; others are experimental. Some work with the basic categories of prevention, appraisal, and failure; others go beyond those categories; still others disregard the categories. Some may be unique to individual firms; others are used more widely.

To find out how quality costs are currently being measured and reported, *Quality Progress* spoke to five top quality professionals who have worked with quality costs systems. This article is based on those discussions

Keeping Management in Mind

A finely tuned quality cost system won't get your company anywhere if top management doesn't use it to improve the company's performance. But it may not be management's fault if the high-powered numbers in a quality cost system are overlooked.

"I have a very strong prejudice that people in our field don't communicate very effectively on this subject," observes Dana Cound, Vice President of Quality Assurance at American Can Company.

American Can's quality costs system is set up to provide an accurate picture of the cost of quality in terms that top management can easily relate to. Both the accuracy of the figures and the way they are presented are important. In fact, the closer you look at the two, the harder it is to separate them.

To develop the most useful quality cost figures possible, Cound has set up a system that is "a little bit different than what you see in the literature" on quality cost, he explains. His company divides all quality costs into two basic categories: discretionary and consequential. In the discretionary slot, Cound puts costs that can be directly controlled: prevention, appraisal, and administration. Consequential costs, quite simply, are internal and external failure costs.

"The purpose of doing this is simply to try and keep in the front of everybody's mind that you can make unilateral and arbitrary reductions in your prevention, appraisal, and administrative costs. You can't really do that with your failure costs. Yet we often say that we are going to budget a 20% reduction in failure costs next year. That's just out of your power to unilaterally edict. You must put in place something to make it happen."

In the discretionary category, Cound is careful to separate administrative costs from prevention and appraisal. Too often, he fears, administrative costs get thrown into the prevention account. "The prevention category is typically the smallest category in quality costs and when we start dumping things into it that are only remotely related to prevention, we swamp that category," Cound observed. "I think we paint pictures for ourselves that frequently aren't true." The American Can system makes clear what the prevention costs really are.

To make reports as useful as possible, Cound is "pretty conservative" in deciding what expenses should go into the discretionary slot. His rule is this: if the activity is devoted primarily to maintaining or improving quality, the cost goes into the discretionary cost. "If quality is an incidental aspect," the expense isn't counted as a quality cost.

"Sometimes I feel that the reason for including so many splinter activities is to make the number look bigger. It is a great attention-getter," Cound admits. "But it can also have an adverse effect on the credibility of the information."

Cound sees other problems with reaching out too far for discretionary costs. For one thing, he observes, "you find yourself spending dollars to account for dimes." For another, you may make large errors if you try to account for small fragments of many people's time. Finally, Cound says, "a particular action may be clearly preventive in nature but also very modest in impact."

Companies can run into a more subtle problem if they reach out too far for quality costs. If a quality program is successful, failure costs fall. It is then time to review discretionary costs. If you have reached out too far, you may find that you have included activities that are highly internalized, marginal in impact — and nonreducible. "You cannot tell a manufacturing foreman or designer to think 10% less about quality next year," Cound observes. "As a consequence, the burden of the reduction falls disproportionately on the primary, heavily weighted quality activities."

Once a company knows which expenses it wants to keep track of with its quality cost system, it has to decide how to report to management. Cound tracks quality costs in terms of standard manufacturing costs, but he believes that he gets more attention by emphasizing quality costs in terms of cost avoidance. "I think that quality costs systems ought to reflect the return on management's investment in the quality program." Cound declares. "ROI ought to be as important to a quality manager as any other manager. And the ROI is cost avoidance.

"By putting it in cost-avoidance terms, I can talk about millions of dollars that we are not spending this year or this month that we would have spent last year. This has an impact."

But monthly cost-avoidance reports aren't all that Cound does with his quality cost data. For example, in his quarterly and annual reports, Cound figures out quality cost per average share of common stock outstanding. An executive "knows very well what his earnings per share of common stock were last year," Cound explains. "You lay that beside quality costs per share of common stock and he gets a very vivid picture of the impact on the bottom line performance of his company."

Cound also reports cost avoidance per share so that everyone can see how much lower earnings per share would have been without an improvement in quality costs. He calls that "a very good way to get your message across."

Cound uses reserves as another example of the need to keep management in mind when putting together a quality cost system. Suppose a company receives notice of a claim and puts aside a reserve to cover that claim. Should the quality cost report reflect an expense when the reserve is established or when the claim is paid? For awhile, Cound says, he did it the second way — taking what he calls "the conservative approach." But that approach creates a management problem. The claim "hits the P and L statement the month they reserve for it." Cound says. If it doesn't show up in the quality costs reports until it is paid, "you are kind of out of synch with life going on around you." Your report will be "ancient history."

At American Can, therefore, "external failure costs monthly input becomes the algebraic sum of changes to claim account and net change in the reserve," explains Cound. "That way, we don't count anything twice.

"It is part of playing management's music."

What Costs Count?

Francis X. Brown believes that the only important quality costs are failure costs."We don't even consider prevention and appraisal," explains Brown, Quality Manager of the Westinghouse Low Voltage Breaker Division.

The idea of ignoring prevention and appraisal cost is heresy to most people who work with quality costs. And in the system at the Low Voltage Breaker Division, "failure costs" is a very broad category that includes appraisal costs.

Even when the Division defined failure costs in a very traditional way — as losses attributable to scrap, rework, and warranty work — it found that, in terms of profitability, "failure costs are the only thing that counts," according to Brown. In a paper presented at the 32nd Annual Technical Conference in 1978,[1] Brown and Roger W. Kane described the Westinghouse experience with quality costs. They found that "true failure costs" were generally much higher than costs usually reported: "our experience indicates that a multiplier effect (on profitability) of at least three or four is directly related to the hidden effect of quality failure." As a result they concluded, "*minimum* TQC (total quality cost) is hardly even optimum... Reported TQC is at a minimum when prevention plus appraisal cost (P and A) constitute about half of the total quality cost, but *profits* continue to improve until (P and A) contributes 70% - 80% of the reported TQC."

"The basic literature is erroneous," contends Brown today. "Prevention and appraisal really had nothing to do with anything." Brown believes that those costs are "irrelevant to the problem" of increasing profitability.

Since the 1978 paper some Westinghouse units — including the Low Voltage Breaker Division — have adopted an even broader definition of failure costs. "Whatever it is that we haven't done well by the standard of the time is a failure cost," explains Brown. He divides failure into two categories: failure in execution and failure in planning. Into

---

[1]"Quality Cost and Profit Performance," F. X. Brown and R. W. Kane, 37th ATC Transactions, pp. 505-511.

one of those two categories, Brown puts "any deduction from financial performance that results from some kind of failure or error."

The most costly failures usually fall into the second category. "The problem is most often that we decide to do the wrong thing," explains Brown. For example, if the Division plans for 20% return on investment on a new piece of equipment but achieves only an 18% ROI, the difference — 2% — is treated as a failure cost. Similarly, if the Division has to shut down a line because a part is out of stock, Brown includes those expenses in failure costs. "Any aspect of the financial statement is fair game for failure costs." The goal of the system, Brown says, is "to reduce failure to a random event."

Appraisal costs also wind up as failure costs. Brown sees appraisal as "an insurance policy against failure. That's a failure cost."

But Brown sees no reason to consider prevention a failure cost. "That is not a quality cost. That is just doing the job." For example, Brown believes that "an ongoing training program is part of the cost of doing the business." But it is a different story if Design and Quality Engineers have to overhaul a process or product that has been having problems. In cases like that, "we didn't do the job right the first time around the track."

"It is how you look at life. With the benefit of 20/20 hindsight, could we have done better?"

Westinghouse's definition of failure takes the quality cost system beyond the usual boundaries of the Quality Department's responsibilities. Every department has to participate to make the system work. What the Quality Department brings is "the ability to collect and analyze the information in some statistical fashion" — what Brown calls "the ability to diagnose the problem."

Brown is excited about the quality costs system his Division uses. "It expands your horizons," he says. "It gives you more things to work on.

"If you expand the problem statement, the opportunity for improvement is a whole lot bigger."

Staff Time Is Company Money

If a quality manager tried to tell the top management of a $60,000,000 operation that they ought to be concerned about $40,000 in quality costs, "you would probably get them to fall asleep on you," jokes Frank Scanlon, Director of Quality/Education, The Hartford Insurance Group.

The Hartford has 52 profit centers. A single office can generate $60 million in gross income and have a $3.5 million expense budget — much of it for salaries. "What we monitor very closely in our company is staff," says Scanlon. With that in mind, the company decided to measure quality costs in terms of staff instead of dollars. For example, rather than reporting that quality costs come to $40,000 in a year, the quality program identifies that quality costs amounted to two or three staff people. And in an office of 75 people, the idea of two more staff members commands a lot of attention.

Measuring quality costs in staff equivalency solves another problem for a service firm like The Hartford. A manufacturer can compare quality costs to the cost of manufacturing or to value added. That won't work in an insurance firm, and The Hartford had to find a substitute. The company settled upon staff equivalency because that measurement attracts the most attention.

One total that The Hartford does measure in dollars is "escape costs" — money lost because defects were never caught. Scanlon uses billing mistakes as an example. If a company makes a mistake in its own favor on a bill, the customer is likely to complain. But if the mistake is in the customer's favor, the company may never hear about it. Those losses can be especially significant in a service industry because those industries have so many direct customers — often many more than a manufacturer would have.

The Hartford does not measure escape costs on an ongoing basis, explained Michele Redman, Manager of Quality Improvement and Systems. Instead, the firm measures escape costs on a project-by-project basis. The company then considers that information when making financial commitments to project development.

Scanlon reports that some improvements have saved The Hartford significant — and sometimes enormous — amounts of money. For example, to avoid the problem of "underpricing," the company developed an automated policy issuance system. The system paid for itself — and much more: it has saved the company over $7 million.

Measuring Opportunities for Quality

Celanese Fibers Operations is trying an experiment: it wants to find out how well it can measure opportunities for quality.

The company still keeps track of the classical quality costs. "You don't want to de-emphasize the standard cost of quality categories," cautions Ed Ewald, Director of Quality Management for Celanese Fibers Operations. "That basic cost of quality is our bread and butter. There is no question about that."

Into the basic cost of quality category, Ewald puts expenses that are "in our ability to control if we follow effective operating practices." But the operating procedures themselves do not provide the whole picture: a certain amount of other loss may have been designed into the processes. The opportunity for quality category is intended to show how large those other losses are.

For example, most chemical processes require a certain minimum amount of raw material. Even if the process is run exactly as it was designed to, "in essence, losses are designed in," Ewald notes. Those design losses go into the opportunity for quality slot. Any other raw material loss in excess of design goes into the standard quality cost categories.

The opportunity for quality totals might allow you "to move your process technology closer to perfection," Ewald points out. If a plant has outdated equipment, the opportunity for quality category can give a company an idea of how much it could save by investing in new equipment. "It can be eye-opening," Ewald says. "I'm not sure that is true in all cases, but it *can* be eye-opening."

The opportunity for quality category may also give a corporation some indication of the effect improved quality could have on the company's profit margin. Suppose a company makes 100 units, scraps one, and sells the other 99. If it had done everything right the first time, it could have had an extra unit to sell. If it could then have sold that extra unit, it would have improved its profit margin. The opportunity for quality figures give the company an idea of the profit potential if it could sell the extra units.

The opportunity for quality category adds "an interesting dimension" to a quality cost system, says Ewald. The category reaches out into areas like Design and Marketing and "gives them something they can identify with."

Further Quality Cost Refinements

After a firm has established its operating quality costs system covering prevention, appraisal, and failure costs, it may decide to track certain detailed quality cost areas "to further help in its quality improvement program," observed A. V. Feigenbaum. Those measurements can provide valuable information, notes Feigenbaum, though he points out that they are "later refinements."

Feigenbaum described four types of "refinements":

• Indirect quality costs are "those quality costs which are hidden in other manufacturing costs," explained Feigenbaum. An important part of this category is the cost of "operations made standard because of uncertain quality." For example, if a company puts a reaming operation after a drilling operation simply because it isn't sure of the quality of the drilling, the cost of the reaming operation would go into the indirect quality costs category.

• Equipment quality costs cover the capital investment in quality information equipment like automated testing equipment. These costs become more significant as the Quality Department obtains increasingly sophisticated equipment.

• The loss of a customer's goodwill as a result of a quality problem is an intangible quality cost, but is "nevertheless real," according to Feigenbaum. This category is broader than it might seem at first: it includes customers who refuse to buy a product because of problems they themselves have had in the past; but it also includes consumers who turn away from a product because of publicity about a lawsuit or recall. "Modification of these costs will never be possible comparable to operating quality costs," Feigenbaum notes, "but general awareness of them can be of critical importance."

• Life cycle and use oriented costs represent "the cost of maintaining quality over the reasonable life of a product," Feigenbaum explains. The category includes the cost of service, repair, and replacement. A company cannot ignore these costs, even if it does not specifically track them, Feigenbaum adds.

All of these categories represent real quality costs. Feigenbaum worries, however, that quality professionals get "sidetracked from the basic job of installing and measuring operating quality costs by being diverted into too many refinements."

What the Quality Department must do is decide how it's going to use quality costs, set up a system to measure those costs, and then "get them done," Feigenbaum emphasizes. "The big job is getting the operating quality costs and putting them to work as the basis for making quality improvement a habit throughout the organization."

QUALITY COSTS: CURRENT IDEAS

(April Issue)

Edward Sullivan
Editor, Quality Progress
American Society for Quality Control
Milwaukee, Wisconsin

More companies are keeping track of them. A wider range of expenses are being considered. Reports about them are now starting to reach upper management.

Despite all the recent excitement about quality costs, however, one bit of old advice is still news: the goal of a quality costs system is to help improve quality. Simply measuring quality costs won't do that.

• "Quite a few companies have gone through an extensive broadening of their accounting systems in order to capture these costs," comments J. M. Juran, one of the first to see that a quality program could reduce companies' losses — "gold in the mine," he called it. "All of that work hasn't gotten them very far.

"To get things improved, you need the organizational machinery for improvement."

• "It is always a mistake to try to do quality costs in a vacuum," contends A. V. Feigenbaum who originally divided quality costs into the categories of prevention, appraisal, and failure. Top management *has* to understand the cost of failures, but it *wants* to know what Quality is doing about them, Feigenbaum argues. "The direct management of quality costs represents a central program for obtaining quality improvement throughout the company."

• "Whatever it is that we haven't done well by the standard of the time is a failure cost," explains Francis X. Brown, Quality Manager of the Westinghouse Low Voltage Breaker Division. "The idea is not to account for it so much, but to recognize conceptually what it is" — a missed opportunity for better performance, "and opportunity costs do not even appear in the financial statements."

• "In the U.S. particularly, top management is primarily concerned about costs" and reducing losses, observes August B. Mundel, a consultant and member of the Quality Costs Technical Committee. "There is too great an emphasis on trying to fix the problem rather than trying to identify the actual cause and correct the system." By concentrating on losses and quick fixes, companies can miss an important part of the value of a quality costs program — quality improvement. "A small percentage increase in yield can have a tremendous effect on profit."

Quality costs play an important role in successful quality programs. There is sometimes a difference of opinion about which costs should be included, but there is no dispute about the importance of measuring quality costs. "One of the things not to do in quality costs is not to do quality costs," insists Feigenbaum.

Feigenbaum believes that quality costs systems are leading more and more corporations to view quality as "fundamentally a way of managing a business, just as marketing is a way of managing a business, just as production is a way of managing a business."

As a result, more and more companies have installed systems for "the direct management of quality costs." They manage, measure, and control quality costs, and use them for strategic planning and budgeting, just as they would production or marketing costs. "In some companies, that comes as something of a revolution."

Feigenbaum calls the "direct management of quality costs as much of a corporate breakthrough, if you will, as direct management of production costs."

The direct management of quality costs systems represents a "step forward" for quality management, according to Feigenbaum. "There is an old myth that better quality costs more," he says. The fact is, however, that "good quality means good costs." Some companies do spend too much on quality, but that's only because they spend money in the wrong areas. "A fortune may be going down the drain because of failure costs," says Feigenbaum, President of General Systems Co. "Another large sum may be spent for a sort-the-bad-from-the-good appraisal stream to keep too many bad products from going to customers. Very little may be spent for the true defect prevention technology that could do something about reversing the vicious upward cycle of higher quality costs and less reliable quality."

Despite the success of many quality costs systems, however, Feigenbaum points out that there is not yet "general corporate use" of quality costs.

For one thing, he observes, "the messenger syndrome still exists." Some managers may be reluctant to measure quality costs for fear of being the bearers of bad news. "But properly handled," Feigenbaum argues, "the reverse is true because the proper identification of quality costs provides the groundwork for a strong quality improvement program."

The success of many quality costs systems has brought these systems — and the quality profession — into new areas. For example, there is "a big move toward administrative, clerical, and software costs," reports Mundel. These costs occur in manufacturing as well as service industries. There are important differences between service and manufacturing companies, and those differences affect the ways some quality costs in each area are measured. But almost every company devotes time to administrative matters, and many companies use computers. In those areas, the difference between a big service company and a big manufacturer is one of scale: "we just do more of it," says Frank Scanlon, Director of Quality/Education at The Hartford, a major insurance firm.

Accounting

Quality costs systems are also putting Quality in touch with Accounting. If a company wants to use quality costs figures for a single quality improvement program, it can usually estimate its scrap and rework costs. Feigenbaum calls that "an entry-level use of quality cost." By contrast, formal quality costs systems must be able to stand up to scrutiny by management and accounting. Once a company sets up a system of cost controls, "you are in conventional accounting," says Juran. Management will need "complete documentation."

Quality costs systems are therefore putting Quality in touch with Accounting. "The emphasis in quality costs is to get the accounting people involved," Mundel reports. To that end, the Quality Costs Technical Committee has been in contact with representatives of professional associations of accountants to develop better ways of capturing quality costs.

If the system is properly explained, finance and accounting departments are generally glad to help set up a quality costs system, according to Feigenbaum. "This gives them the opportunity to measure major costs that just haven't been measured."

Quality costs systems are also trying to capture what Mundel calls "the most pernicious, least obvious" costs, like time spent on answering complaints or fighting fires. He agrees that it is "very difficult to get a handle on" those costs, but adds that improvements in those areas can have a dramatic effect on morale and productivity, as well as losses.

But Juran cautions against trying to include too many areas in the quality costs system. He worries that companies may find themselves "all wrapped up in purifying the accounting system and meanwhile you're not getting results."

Quality costs systems have given management a better idea of the value of quality systems; but they have also given quality professionals training in the language of top management: the language of money. Dana Cound, Vice President of Quality Assurance for American Can Company, analyzes quality costs in terms of earnings per share. "You have to be more than a little bit of an accountant to play this game," he says, and adds that "it is a very good way to get your message across."

Computers have also contributed — and will continue to contribute — to the growth of quality costs systems, according to Mundel. Computers originally made it easier for Accounting to keep track of an extra set of costs — those associated with quality. Now, computers may allow companies to track costs that are harder to quantify.

Feigenbaum sees another role for computers. He believes that Quality Departments will turn to small computers dedicated to quality costs. As it stands now, quality may have to wait in line for access to the corporate mainframe. On the other hand, with a small dedicated computer, Quality "can pinpoint just where costs are going up because quality is going down." Instead of waiting in line for quality costs data, the Quality Department will be able to identify "corrective action areas immediately."

If the Quality Department is to do its job, it needs support from top management. A successful quality costs system is one way to get that support. Feigenbaum sees a unique opportunity for the quality profession. Management today is interested in better quality, he argues; the challenge now is to "convert management commitment to quality into management participation in quality. Quality costs is an ideal vehicle."

COST AND MANAGEMENT

MEASURING QUALITY COSTS

(July-August Issue)

Wayne J. Morse
Professor of Accounting
The University of Tennessee

As manufactured goods become more complex and customers' expectations grow, increased attention is being paid to maintaining and improving product quality. The short-run costs of a lack of quality include the costs of warranty repairs, legal liability, and product recalls. Long-run costs may include lost sales or even bankruptcy. To avoid these costs and improve competitive position, many businesses are investing large sums of money in quality assurance.

Because of the increasing significance of quality and the growing magnitude of quality costs, a number of business firms have established or are in the process of establishing special accounting systems to measure quality costs. It is important that management accountants actively participate in the development and implementation of quality cost systems. Such participation is necessary to help ensure the success of both the quality cost system and the firm's quality assurance program.

The purposes of this article are to:
1. Identify the important types of quality costs and the relationships existing among them.
2. Indicate the potential uses and limitations of quality cost data.
3. Outline some important steps involved in designing and implementing a quality cost system.

Quality of Design and Conformance

Quality is often defined as "fitness for use," "grade of excellence," or "an essential character of something." None of these definitions is precise enough to guide the measurement of quality costs.

Juran and Gryna make a useful distinction between quality of design and quality of conformance:[1]

Quality of design refers to the planned quality of a product. The designed quality of a product may include specifications for product life, reliability, operating costs, and so forth.

Quality of conformance refers to the degree of correspondence between the customer's actual experience with a product and the product's designed quality. A manufacturer often issues a warranty guaranteeing that a product will conform to certain design specifications. If the product does not conform, the manufacturer must take action to remedy the situation. The cost of this corrective action is one type of quality cost.

Types of Quality Costs

Quality costs are related to quality of conformance rather than quality of design. Quality costs are costs incurred because actual quality may not, or does not, conform to designed quality.

Some persons inappropriately talk about the "cost of quality" rather than "quality costs." The problem with the former phrase is that it implies a trade-off between cost and quality. When accountants refer to the "cost of quality" they evoke a negative reaction in quality assurance personnel who then think of cost versus quality. . ."you get what you pay for." Once quality assurance people have adopted this attitude, they are unlikely to cooperate in quality cost measurement — to the long-run detriment of their purpose and the organization's profits.

While there may be a trade-off between cost and quality of design, there is no trade-off between cost and quality of conformance. An automobile or a television set that is designed to be energy-efficient will likely cost more to produce than a less efficient competing product. Once the design is agreed upon, however, quality costs are associated with both the ensurance of conformity and the lack of conformity.

By restricting quality costs to costs associated with quality of conformance, we avoid the trade-off between cost and quality. Here the manufacturer incurs costs if the product fails to conform. These costs include such things as warranty costs, product liability, and lost sales from a poor quality reputation. To ensure quality, the manufacturer incurs costs for inspection, process controls, training, and so forth. Hence, there is a trade-off between two types of quality costs: those incurred because poor quality of conformance *can* exist, and those incurred because poor quality of conformance *does* exist. Each of these quality costs is further classified into two sub-categories:

1. Costs incurred because poor quality of conformance can exist;
   Prevention costs
   Appraisal costs
2. Costs incurred because poor quality of conformance does exist;
   Internal failure costs
   External failure costs.

Prevention and appraisal costs are incurred because poor quality of conformance can exist. *Prevention costs* include the costs of planning and designing the production process to ensure conformance. Also included in this category are the costs of developing quality standards, the costs of quality training, the cost of forming and operating quality circles, and the cost of evaluating the quality control system. *Appraisal costs* include the cost of testing and inspecting both purchased materials and the firm's own product. They also include the cost of routine field tests and the cost of outside product endorsements.

Internal and external failure costs are incurred because poor quality of conformance does exist. *Internal failure costs* include the cost of rework on defective items, the net cost of scrapped items, the cost of downtime due to failed products and materials, and other costs for adjusting machine settings, retesting failed products, and so forth. *External failure costs* include the cost of warranty service and replacement, the cost of product liability, and the opportunity cost of lost sales.

Economics of Quality Costs

The relationship among the types of quality costs is illustrated in Exhibit 1. Expenditures for prevention and appraisal can reduce the costs of internal and external failure. While the graph indicates an optimal level of quality of conformance, it is unlikely that an equation could ever be developed to determine this optimal level. There are two reasons for this. First, quality relationships are dynamic rather than static. Exhibit 1 indicates quality relationships at a particular point in time. By the time sufficient information has been accumulated to determine these relationships, they likely will have changed. Second, it is difficult precisely to measure quality costs. This latter point is discussed subsequently in this paper.

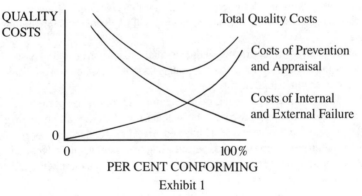

Exhibit 1

Despite these limitations, Exhibit 1 is useful as a general guide to action. Actual quality cost information might be accumulated and presented to management along with this graph to convince them of a maldistribution of quality costs. It the firm has large failure costs and small prevention and appraisal costs, the graph suggests that investments in prevention and appraisal may reduce total quality costs.

A business that spends little or nothing on prevention or appraisal is likely to have low internal failure costs and very high external failure costs. This is true for the simple reason that all the defects are going out the door. To improve the quality of goods leaving the factory, management may institute an inspection program. The effect of the program is to increase appraisal costs and reduce external failure costs. However, internal failure costs may also rise. The defects, which are now identified before they leave the plant, must be reworked or scrapped.

The ultimate solution to the firm's quality problem is to invest in prevention. Only prevention will reduce all failure costs. Prevention will also reduce the need for high levels of inspection. Many people think of increased inspection (appraisal) as the cure for product defects. This is not the case. The ultimate objective of a quality assurance program is to prevent rather than to find defects.

Purpose of a Quality Cost System

Because of the growing significance of product quality and the increasing size of quality costs, it is important that management have some systematic means of planning and controlling quality costs. This is the ultimate purpose of a quality cost system.

Quality costs reports constitute the backbone of a quality cost system. Quality cost reports are periodic (weekly or monthly) summaries of quality costs accumulated by plant or product or other appropriate unit. A quality cost report is illustrated in Exhibit 2.

The potential uses of the information contained in such a report are limited only by the imagination of management. When a quality cost system is first established, it will have its greatest usefulness in revealing the magnitude of quality costs. Because most accounting systems are not designed to measure quality costs, most managers have no idea of the size of these costs. The initial reports will likely be both enlightening and alarming.

The initial quality cost reports may indicate a maldistribution of quality costs — high failure costs and low prevention costs. This suggests that investments in prevention and appraisal might reduce total quality costs.

After quality cost information has been accumulated for several periods, it may be possible to budget certain quality costs. Management may, for example, be able to predict internal failure costs as a percentage of budgeted direct labor dollars and external failure costs as a percentage of some previous period's sales revenue. Appraisal may be a fixed cost and prevention may be subject to negotiation between the plant superintendent and the head of quality assurance.

The next stage is the establishment of goals for the reduction of quality costs. Management might target a 50 per cent reduction in external failure costs to be achieved by a 20 per cent increase in appraisal and internal failure costs. Quality assurance personnel might set a goal of a 30 per cent reduction in total quality costs to be achieved by several new defect prevention programs.

Finally, quality cost reports provide the information needed to evaluate the success of investments in prevention or appraisal or both. This can be done by comparing actual quality costs with budgeted quality costs or with the initial level of quality costs (see Exhibit 3).

SAMPLE QUALITY COST REPORT

XYZ Corporation: Product 1
Quality Cost Report
For the month of September, 19x3

	Current Month	Per Cent Direct Labor	Year to Date	Per Cent Direct Labor
Prevention:				
Quality engineering	$ 2,500	1.25	$ 20,000	1.67
Quality circles	3,000	1.50	15,000	1.25
Quality training	5,000	2.50	5,000	0.42
Supervision	1,200	0.60	11,000	0.92
Total	$ 11,700	5.85	$ 51,000	4.26
Appraisal:				
Inspection	$ 12,000	6.00	$ 96,000	8.00
Testing	3,000	1.50	21,000	1.75
Supervision	2,000	1.00	16,000	1.33
Total	17,000	8.50	133,000	11.08
Internal failure:				
Net cost of scrap	$ 20,000	10.00	$180,000	15.00
Rework labor and overhead	50,000	25.00	420,000	35.00
Downtime	5,000	2.50	50,000	4.17
Other	4,300	2.15	36,000	3.00
Total	79,300	39.65	686,000	57.17
External failure:				
Warranty	$ 8,000	4.00	$ 50,000	4.17
Out of warranty	5,000	2.50	38,000	3.17
Total	13,000	6.50	88,000	7.34
Total Quality Costs	$121,000	60.50	$958,000	79.85

Exhibit 2

Limitations of Quality Cost Information

It should be emphasized that quality costs are only a general guide to the overall planning and control of a quality assurance program. Quality cost reports seldom provide the detailed information needed to suggest specific actions which will reduce quality costs. Industrial engineers, quality assurance personnel, and product personnel must use the less aggregated data of things (as opposed to aggregated dollars) to guide them in making specific decisions.

Current quality cost information is more like a meat axe than a scalpel. It is not precise. But, given the state of the art in quality cost measurement and the magnitude of quality costs, a meat axe is useful. As Rowland Caplen has observed:

QUALITY COST TREND ANALYSIS

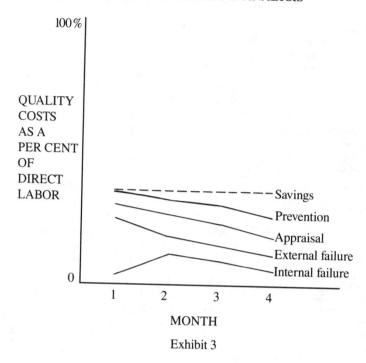

Exhibit 3

The fact that accurate reliability costs are so difficult to estimate is of course no excuse for neglecting them. As technology advances, reliability costs continually increase so that in many cases even the roughest of calculations will leave no doubt about the need for intensive control.[2]

Five specific problems with quality cost reports should be noted:

1. Much of the information is subjective. The cost of the time a production supervisor spends on appraisal and prevention activities, for example, might be based on a rough estimate of the time devoted to each activity.

2. Important costs are omitted from the report. One of the most significant costs of external failure is the opportunity cost of lost sales. Unfortunately, it is almost impossible to develop a reasonable estimate of these costs. Consequently, they are almost always omitted from quality cost reports.

3. Overhead cost assignments to scrap and rework may be imprecise. Scrap and rework costs are often the largest internal failure costs. The direct materials and direct labor costs associated with these items are fairly easy to measure accurately. Such is not the case with overhead. Many persons working with quality costs feel that overhead allocations made for external reporting purposes should not be included in the costs of scrap and rework. Sample reports developed by a committee of the American Society for Quality Control include only material and labor costs.[3]

The often-cited reason for the omission of overhead costs is that they are not needed to support the argument for increased investments in prevention. Therefore, why muddy the reports with questionable items? Others argue that, only the imprecise (meat axe?) nature of quality cost information is acknowledged, all relevant costs should be included no matter how imprecise their measurement.

4. Variations in activity may reduce the comparability of quality costs from different periods. A reduction or an increase in total quality costs may merely reflect a reduction or an increase in production or sales volume. To overcome this problem, quality costs from different periods should be compared on the basis of some activity measure such as their percentage of direct labor dollars of direct materials costs.

5. Effort and accomplishment are probably not matched in a single reporting period. It may take months, or even years, for investments in prevention to pay off. Consequently, too much emphasis on the significance of individual reports may be counter-productive. Total quality costs may, in fact, go up before they start moving toward their optimum level.

Quality assurance personnel are particularly sensitive to the matching problem. Because of concerns about matching and pressures for instant results, many quality assurance personnel are actually opposed to the establishment of a quality cost reporting system. They are interested in selling improved quality and are concerned that accounting reports will lessen top management's willingness to make the necessary long-term commitments. The counter-argument is that an improved awareness of the magnitude of quality costs will lead management to support long-run programs to reduce these costs.

Implementing a Quality Cost System

At least ten distinct steps are involved in establishing and implementing a quality cost system:
1. Obtain management commitment and support.
2. Establish a quality cost team.
3. Obtain the cooperation and support of users and information sources.
4. Operationally define quality costs (to limit the scope of the system).
5. Identify specific quality costs.
6. Determine sources of quality cost information.
7. Set up a code system and forms to accumulate information.
8. Design quality cost reports.
9. Accumulate information.
10. Distribute reports.

Obtaining management commitment and support is a prerequisite to the successful implementation of any quality assurance program. Management must take this commitment seriously and understand that it will probably not produce immediate results. Management is, however, more likely to support a program that can be justified in terms of dollars, and has established procedures for planning and measuring program costs.

Once managers have committed themselves and the firm's resources to a quality assurance program, the next step is to establish an organization to design and implement the program. This organization may, for example, include quality circles or an inspection department. Statisticians may be hired to evaluate the current variability in product quality. A team of engineers might be instructed to start considering ways to prevent product defects. One group should be assigned the task of designing and implementing a quality cost system. This group, often called the quality cost team,[4] should include at least one person familiar with the organization's cost accounting records and, in situations where records are stored in electronic media, at least one person familiar with the organization's electronic data processing system. The quality cost team should also include representatives of the major suppliers and users of quality cost information — perhaps a production supervisor, an engineer, and a product manager.

Successful implementation of a quality cost system requires the active cooperation and support of the users of quality cost information and the suppliers of this information. Obtaining this support should be one of the quality cost team's first assignments. Extreme care should be taken to indicate both the users and limitations of quality cost information. The quality cost team needs to ensure that expectations are reasonable. Nothing is more detrimental to the system than unrealistic expectations that cannot be achieved. Quality cost information is simply another tool to help people do their job. Another benefit of talking to information users and suppliers at any early stage is that they can offer suggestions that are of immense value as the quality cost team gets into the details of system design and implementation.

To ensure that everyone understands exactly what is and what is not being measured, the quality cost team should develop, or adopt, a working definition of quality cost. This definition might refer to the major categories of quality costs; it might indicate if overhead costs are to be included or excluded; and it might include a statement about whether or not opportunity costs are included. Since the definition will be designed to meet the information needs of a specific organization, it may vary significantly from the definition given in this paper. There may, for example, be three, four, and even five categories of quality costs. As the operational definition of quality cost is developed, the team should look ahead to subsequent implementation steps.

Going from the general to the specific, the next step is to identify and classify specific quality costs that should be included in quality cost reports. The quality cost team should identify specific costs under each of the previously established categories — preventional, appraisal, and so forth.

It is also necessary to determine the specific source of information for each quality cost. Is the information currently available in the accounting records? Can accounting data be used without modifications? Will it be necessary to accumulate additional raw data? If so, who will provide them? Will subjective data be needed? Again, who will provide them? In this step the team should constantly think of the costs and the benefits of accumulating additional or more refined data. If report users understand the state of the art and the uses and limitations of quality cost information, the data do not need to be precise; precision probably cannot be achieved in any case.

After identifying the sources of quality cost information, the next step is to develop a systematic means of accumulating and storing this information. This stage may involve computer programming or the design of forms to be completed by production workers, supervisors, bookkeepers, quality inspectors, and others. It is recommended that an expandable coding system be developed to facilitate the expansion of the quality cost system and the accessing of quality cost information stored on electronic media.

Quality cost reports must also be designed. These reports will be used to report quality cost information to responsible managers. It is likely that several different reports will be developed for reporting to different levels of management. Reports to top management should be more aggregated than reports to first-line supervisors. As a minimum, the reports should indicate the total quality costs for the period in each major category. They should also relate quality costs to some standard measure, such as direct labor, that changes with volume.

The last two steps — accumulating the data and distributing the reports — are self-explanatory. However, the quality cost system should not be initially implemented on a firm-wide basis. Implementation should start with a single plant, product, or even operation. Only after the "bugs" have been worked out should the system be implemented on a broader basis. Think small; develop a pilot system and then expand.

Conclusions

Quality costs are costs incurred because actual quality may not or does not conform to designed quality. Increases in product complexity and customer expectations have dramatically increased the magnitude of quality costs.

To improve product quality and reduce quality costs, many businesses have established quality assurance programs. A quality cost measurement system is a vital element of such a program. A quality cost system can provide information on the magnitude of quality costs and a systematic basis for planning and controlling such costs. While quality cost information is imprecise, this should not be an excuse for neglecting it. Given the magnitude of quality costs, even the roughest of calculations will prove useful.

References

[1] J. M. Juran and F. M. Gryna, *Quality Planning and Analysis* (New York: McGraw-Hill, 1971), pp. 38-47.

[2] Rowland Caplen, *A Practical Approach to Reliability* (London: Business Books Limited, 1972), p. 59.

[3] Quality Cost — Cost Effectiveness Technical Committee, *Quality Costs — What and How,* American Society for Quality Control, 1971, p. 41.

[4] The phrase "quality cost team" appears to have been coined by the Quality Cost — Cost Effectiveness Technical Committee of the American Society for Quality Control.

LET'S HELP MEASURE AND REPORT QUALITY COSTS

(August Issue)

Harold P. Roth
Assistant Professor of Accounting
University of Tennessee
Knoxville, Tennessee
Wayne J. Morse
Professor of Accounting
Chairman of the Faculty of Accounting and Law
Clarkson College

In recent years, U.S. industry has become more concerned with product quality and its relationship to productivity. Although their involvement in this area so far has been minimal, it's time for accountants to use their expertise in measuring and controlling costs to help management in the quality area.

The quality costs discussed here deal with costs associated with quality of conformance as opposed to costs associated with quality of design. Quality of design refers to variations in products which have the same functional use.[1] For example, a Cadillac and a Chevrolet automobile may be used for the same function, transportation, but their design qualities are different. For this type of quality, higher quality generally means higher costs. The same relationship, however, does not exist when cost associated with quality of conformance is considered.

Quality of conformance refers to the degree with which the final product meets its specifications. In other words, quality of conformance refers to the product's fitness for use. If products are sold and they do not meet the consumers' expectations, the company will incur costs because the consumer is unhappy with the product's performance. These costs are one kind of quality costs that will be reduced if higher quality products are produced. Thus, higher quality may mean lower total costs when quality of conformance is considered. Accountants, therefore, should be aware of the costs associated with quality of conformance and how they can be of assistance to management.

The costs associated with quality of conformance generally can be classified into four types: prevention costs, appraisal costs, internal failure costs, and external failure costs.[2] The prevention and appraisal costs occur because a lack of quality of conformance can exist. The internal and external failure costs occur because a lack of quality of conformance does exist.

Prevention costs are the costs associated with designing, implementing, and maintaining the quality system. These costs include engineering quality control systems, quality planning by various departments, and quality training programs.

Appraisal costs are the costs incurred to ensure that materials and products meet quality standards. These costs include inspection of raw materials, laboratory tests, quality audits, and field testing.

Internal failure costs are the costs associated with materials and products that fail to meet quality standards and result in manufacturing losses. They include the cost of scrap, repair, and rework of defective products identified before they are shipped to consumers.

External failure costs are the costs incurred because inferior quality products are shipped to consumers. They include the costs of handling complaints, warranty replacement, repairs of returned products, and so forth.

These quality costs are incurred in many different departments within an organization, and most cost accounting systems are not designed to measure and report the costs in these four categories. This lack of a quality cost accounting system means that most companies do not know the size of their quality costs although they can be great. As Philip Crosby notes, the costs of quality ". . .are all a result of not doing things right the first time. You can spend 15 to 20 percent of your sales dollar on such expenses without even trying hard."[3] Other writers also have noted the high quality costs American businesses are incurring.

The Magnitude of Quality Costs

The costs of quality (COQ) can be a major cost item for many companies and represent an area for substantial savings. Schmidt and Jackson note, "Recent studies indicate the typical COQ for U.S. companies is from 10 to 20 percent of total sales. It has been suggested by Philip Crosby and others that COQ should be about 2.5 percent of total sales."[4] They suggest that the Japanese automobile industry lends creditability to this estimate by showing cost of quality in the 2.5% to 4% range.

The quality costs for the U.S. automobile industry have been reported to be high. One source, for example, notes, "Auto industry sources estimate that as much as 25% of the price of a car is attributable to poor quality: scrappage, reject parts, inspection and repair, and warranty costs."[5] Another example of quality costs incurred in the automobile industry is given by Robert M. Reece regarding the external failure costs associated with recalls of defective products. He notes:

"To fix the pollution control systems on 270,000 of its 1976 cars, American Motors Corp. figures it spent around $3 million — including close to $40,000 just for first-class postage to notify the car owners. For Firestone, the cost of replacing 7.5 million steel-belted radial tires in its notorious recall case, in which 41 deaths and 675 injuries were allegedly connected with the tires, ran upwards of $135 million after taxes — more than the company's net income in fiscal 1977."[6]

Quality costs are not limited to manufacturing industries. They also can be a major cost item for service industries. The banking system provides an example:

> "All the checks issued in this country today include the magnetic-ink character recognition (MICR) code, known as E-13B. Yet not all these checks can be used. When the attempt is made, computers reject them as unreadable. The problem is serious enough so that the Bank Administration Institute estimates that [in] 1980, faulty checks...cost the country's banks a whopping $435 million a year in rework. This figure...represent[ed] approximately one-half of all check-processing costs."[7]

These examples indicate that the cost of quality in American industry is high. Several companies also have reported information supporting this observation. For example, a recent *Business Week* article on TRW notes that "a new accounting system underpins the effort to reduce quality costs. Everything from scrap to warranty costs must be examined with an eye to achieving an average savings of 10% per year. For TRW as a whole, that would amount to hundreds of millions of dollars." TRW "...is nearly on schedule in meeting its goal of cutting total quality costs by $24 million — even though many units are not yet fully participating."[8]

The size of another company's quality costs recently was volunteered by F. Timothy Plummer, director of manufacturing services at Borg-Warner Corp. He notes "...the company has been spending an average of 20% of sales to correct poor quality, and trimming that 20% represents the most significant opportunity we have to increase profitability."[9]

To effect savings in quality costs, it is helpful to identify poor quality as soon as possible in the production process. Richard W. Anderson, general manager of the Computer Systems Division of Hewlett-Packard, for example, describes the damage a faulty 2¢ resistor can do. "If you catch the resistor before it is used and throw it away, you lose 2¢. If you don't find it until it has been soldered into a computer component, it may cost $10 to repair the part. If you don't catch the component until it is in a computer user's hands, the repair will cost hundreds of dollars. Indeed, if a $5,000 computer has to be repaired in the field, the expenses may exceed the manufacturing cost."[10]

These examples show that quality costs represent an area where substantial savings can be achieved. Before the savings can be measured, however, the quality costs must first be determined. Accountants would seem to be the logical choice for measuring these costs although the concept is rarely mentioned. Despite this omission, many quality control personnel believe accountants should be involved in this area.

Why Accountants Need to Be Involved

Management accountants need to be involved in the quality cost area because they generally design the accounting systems and provide the reports for measuring and controlling costs. While quality control generally focuses on item data rather than costs, most managers are more concerned with cost data than with statistical data on number of defects. Former ITT Chief Executive Officer Harold S. Geneen summed up this attitude: "Operating managers know they cannot get into deep trouble for creating nonconformance products or services. They will be frowned at for these difficulties, but they can be really put down only for profit loss. Therefore, they concentrate on financial and schedule matters. Quality is third."[11]

Because managers relate to financial data, it seems appropriate for the quality cost data to originate in the accounting department. Such a procedure would produce reports that are perceived by managers to be more objective than reports originating in other departments such as quality control. Crosby notes that "having the comptroller establish the cost of quality removes any suspected bias from the calculation. Most important, a measure of quality management performance has been established in the company's system."[12]

As Juran and Gryna write: "The accounting department becomes unhappy whenever anyone undertakes to prepare cost reports. The accountants argue that two sources for such reports inevitably come up with inconsistent figures, create duplication of effort, etc. (The accountants also strongly prefer to have a monopoly on plying their trade within a company.)"[13]

Although these arguments support a role for accountants in the quality cost area, the contributions made to date by management accountants in the quality area have been limited. The first instance of using a company-wide quality measurement, actually calculated and reported by the comptroller, may have taken place within the ITT program instituted in the mid-60s. Other companies also may have involved their accountants in this area since then, but their involvement has apparently gone unreported.

To fill this gap, quality control personnel often prepare quality cost reports from other accounting reports, and, occasionally, subjective estimates by managers. Unfortunately, these reports are lacking in four areas: consistency, confidence, accuracy, and cost efficiency.

The quality cost reporting systems installed by quality control personnel vary in format and content between plants. Because these personnel are not used to company-wide reporting systems, these systems lack consistency, and it is difficult for top management to obtain accurate estimates of quality costs and the trade-offs that exist among the various types of quality costs.

Moreover, because quality cost information is often used to justify additional quality control expenditures, its accumulation by the same persons responsible for such expenditures is suspect. Management would have more confidence in such data if it were accumulated by accountants.

Quality cost reports prepared by quality control personnel vary in accuracy. The reports are derived from information prepared for other purposes, and the cost data are often subjectively adjusted. Of more concern is the fact that they tend to *understate* quality costs. In an effort to improve the reports' credibility, and increase management's confidence in them, quality control personnel tend not to include questionable overhead costs. After all, they argue, even their conservative reports indicate an obvious need for increased expenditures on prevention and appraisal.

Even if all of these problems can be overcome, quality control personnel are not accountants. Their time could be better spent doing the things they are trained for rather than learning how to operate and then operating a specialized accounting system. If management accountants measured quality costs, report preparation costs would likely fall while consistency, confidence, and accuracy would increase.

Measuring and Reporting Quality Costs

If accountants are going to make contributions in this quality cost area, they need to develop accounting systems for measuring quality costs data and reports for summarizing the data for management. Geneen suggests that management:

> "Ask the comptroller to calculate the cost of rework, scrap, warranty, unplanned service, qualification test, inspection test, and calibration. Make sure he includes the overhead or other burden change. Relate this as a percentage of sales, shop cost, or other measurements that are meaningful to you. Establish a ratio that you want to achieve and publish it. Set an achievement standard of 2.5% of sales."[14]

Some quality costs are available from standard accounting data, including:

1. Quality direct labor and associated fringes, overtime, cost of living, and shift premiums,
2. Quality overhead (including indirect labor),
3. Total scrap (vendor and production),
4. Policy and warranty expense,
5. Product liability, and
6. Maintenance, repair and calibration effort on all gauges and test equipment.

Other quality costs, however, will probably have to be estimated, including:

1. Service effort,
2. Remedial engineering effort,
3. Production rework effort,
4. Production in-process inspection effort, and
5. Remedial engineering change losses.

Because these latter quality costs will probably require estimates, the data might be viewed with less creditability than data derived from the accounting system. The use of estimated data, however, will provide information that accountants can use in modifying the accounting system to produce quality cost information and, in addition, it will help management control its quality costs. "Those who do the calculating (of quality costs) will still miss a third of the costs, but they will develop a reasonable enough estimate to provide direction for cost elimination. Nothing is quite as effective as having cost data to show competing areas that one department has more effective methods of reducing costs than others."[15]

Once the quality costs have been assembled from the accounting system or estimated from available data, they need to be classified into prevention costs, appraisal costs, internal failure costs, and external failure costs. This breakdown into categories will help show where quality costs are being incurred and indicate areas where increases in one type of quality costs may reduce the total costs. Table 1 illustrates one method that can be used for presenting these data.

To illustrate how total quality costs might be reduced by redistributing costs from one area to another, let's take a look at the computer resistor example presented earlier. If the company has no prevention, appraisal, or internal failure costs, the defect would not be detected until there was an external failure, and the entire cost would be classified as an external failure cost. As noted in the example, this repair may cost hundreds of dollars or more. Thus, spending money on internal detection will reduce the total cost as long as the amount incurred is less than the difference in the cost to repair externally and internally. If external repair costs $200 and internal repair costs only $10, up to $190 could be spent on internal failure detection before it becomes more costly. Likewise, if the defective resistor is caught in the appraisal stage, only a 2¢ loss occurs. Thus, as long as the incremental cost of detecting the bad resistor is less than the savings, the total cost can be reduced. If we assume the faulty resistor was acquired externally, prevention costs would have to be incurred by the producer. The user may want to relate data on the defect to the producer, however, so appropriate correction can be made.

Once quality costs have been collected for several periods, graphs also can be used to illustrate the trends in these quality costs over time and indicate if total quality costs are within the ranges set by management.

Table 1 is only one report accountants could prepare to report on quality costs. They doubtless will be able to develop other kinds of reports once they become more involved in measuring and reporting quality cost data.

Quality costs are becoming an area of increasing concern as managers try to reduce costs and increase productivity. Management accountants can help by designing accounting systems to measure and report quality costs. This involvement will help ensure that quality cost data are consistent, accurate, and cost effective.

Quality Cost Report

	Current month's cost	% of total
Prevention costs:		
Quality training	$ 2,000	1.3%
Reliability engineering	10,000	6.5
Pilot studies	5,000	3.3
Systems development	8,000	5.2
Total prevention	$ 25,000	16.3
Appraisal costs:		
Material inspection	$ 6,000	3.9
Supplies inspection	3,000	2.0
Reliability testing	5,000	3.3
Laboratory	25,000	16.3
Total appraisal	39,000	25.5
Internal failure costs:		
Scrap	$ 15,000	9.8
Repair	18,000	11.8
Rework	12,000	7.8
Downtime	6,000	3.9
Total internal failure	51,000	33.3
External failure costs:		
Warranty costs	$ 14,000	9.2
Out of warranty repairs & replacement	6,000	3.9
Customer complaints	3,000	2.0
Product liability	10,000	6.5
Transportation losses	5,000	3.3
Total external failure	38,000	24.9
Total quality costs	$153,000	100.0

Table 1

References

[1] J. M. Juran and F. M. Gryna, *Quality Planning and Analysis*, McGraw-Hill, New York, 1971.

[2] *Quality Costs — What and How*, Quality Cost-Cost Effectiveness Committee: American Society for Quality Control, Milwaukee, 1971.

[3] Philip B. Crosby, *Quality Is Free*, New American Library, Inc., New York, 1980.

[4] Jack W. Schmidt and Jerry F. Jackson, "Measuring the Cost of Product Quality," *Proceedings of the February 1982 Meeting of the Society of Automotive Engineers*, February 1982.

[5] "Quality: The U.S. Drives to Catch Up," *Business Week*, November 1, 1982.

[6] Robert M. Reece, "QC as an Inflation Fighter," *Quality Progress*, August 1980.

[7] William J. Latzko, "Quality Control for Banks," *The Bankers Magazine*, Autumn 1977.

[8] "TRW Leads a Revolution in Managing Technology," *Business Week*, November 15, 1982.

[9] Quality: The U.S. Drives to Catch Up."

[10] Jeremy Main, "The Battle for Quality Begins," *Fortune*, December 29, 1980.

[11] H. S. Geneen, "Fourteen Steps to Quality," *Quality Progress*, March/April 1972.

[12] Crosby.

[13] Juran and Gryna.

[14] Geneen.

[15] Crosby.

1984

ASQC ANNUAL QUALITY CONGRESS TRANSACTIONS

APPLICATION OF ECONOMIC PRINCIPLES TO QUALITY

Charles W. Bradshaw, Jr.
Western Electric Company
Richmond, Virginia

Introduction

The present paper departs from the traditional approaches to controlling quality of manufactured product. In the past, the attention of the quality control (QC) professional has been focused on the rather myoptic goal of eliminating defective product. Although this is an important objective of QC, it should not be the primary one. The primary goal of QC or any other manufacturing function, should be to provide a contribution to the overall profit of the firm. Therefore, the focus of the QC professional should be on profit maximization rather than defect minimization. Although defect minimization will certainly reduce certain types of quality costs, it will increase others. Some QC professionals have attempted to account for this phenomenon by adopting a cost minimization approach to controlling quality. The present author takes this trend a step further by advocating a profit maximization approach. It is a well known economic fact that cost minimization doesn't necessarily lead to profit maximization. Therefore, the profit maximizing firm cannot be satisfied with the blind application of cost minimizing QC techniques.

Since the bulk of the procedures used in the QC field during the past half century have concentrated on reducing defect rates, the main line of attack has been from a statistical perspective. Process checking intervals and product sample sizes have been chosen to maintain certain quality levels with specified probabilities of exceeding prescribed limits of process performance or product specifications. Since these approaches entail the use of statistical and probability theories, they have come to be known as statistical quality control (SQC) techniques. As a result, the discipline which involves the application of these techniques to controlling quality is commonly known as statistical quality control.

The present paper advocates a different approach to attacking quality control. Instead of choosing sample sizes and intervals to maintain processes within certain stochastic limits to minimize defects, the present paper recommends the use of a true economic approach to the task of controlling quality of manufactured product. Such an approach requires an entirely different perspective than the SQC approach. For this reason, I will refer to the body of techniques touched upon in this paper as *economic quality control (EQC)*.

The Controlled-Process Production Function

The author has developed a specific production function to describe the situation where the production process is controlled by checking product. This is a common quality control procedure. The process is assumed to generate a continuous flow of product. The individual pieces of product are all of good quality until such time as the process ceases producing good quality product and commences producing bad quality product. From that point on, it is assumed that the process continues to produce bad product until a correction to the process is made. The process is said to be in control when it is producing good product and out of control when it is producing bad quality product. This is the simple, but not uncommon, case of dichotomous output quality. A bad quality condition, or defect, which continues to occur on every piece of product once the process goes out of control will be called a *repeating defect*. Repeating defects are often associated with problems with dies, tools, or molds. Although repeating defects may not represent all of the defects produced by a process, they are generally the most prolific because of their tendency to persist until corrective action is taken to rectify the process.

Based upon the type of process control system just described, the author has arrived at the following controlled-process production function

$$Q = f(X,W) = X^a(1-(((X^a/W)-1)/2M_0)).$$

The output, Q, is an explicit function of the process inputs, X and W, while M_0 is a parameter. A parameter is an exogenous variable which is determined externally from the model. The variables X and W are considered to be endogenous to the model.

The proposed production function focuses on three variables which affect output. The endogenous variables can be treated as economic factors of production in an analogous manner to the way such classical factors as capital and labor are handled in microeconomics. The factor X is a measure of the amount of raw material used by the process during a given period of time. In many cases, this raw material is simply the unfinished product entering the process. (Labor is assumed to be held constant. Increased raw material entering the process is accommodated by raising the production rate rather than increasing man-hours). When a product requires several processes to complete production, its degree of completion will increase as it passes from process to process. The factor X will often be nothing more than a count of the number of pieces of partially completed product entering the process being studied during a given period of time. Clearly, the production output of the process during that time period will be influenced by this factor.

The other factor of production included in the model, represented by W, is a measure of the process quality control efforts which involve periodically checking a piece of product from the output stream to determine whether the process is in control. If the process is found to be out of control, it is assumed that an immediate correction will be made to bring it back into control. Specifically, W is intended to be a count of the number of pieces of product receiving a quality check during the given time period.

It is assumed that checks are made at a constant interval where there is always the same number of unchecked pieces between every two consecutively checked pieces.

The parameter M_0 is a measure of the inherent ability of the process to produce good quality product. It can be called the *process capability*. More specifically, it is the average number of pieces produced between the time a process correction is made to bring it into control and the time that it goes out of control again. This can be thought of as the expected value of the in-control intervals, where the interval length is measured in terms of consecutive good pieces produced.

The only remaining unexplained quantity in the production function is the exponent a of the variable X. The purpose of this exponent is to account for the inability of the process to maintain a throughput which keeps constant pace with increases in raw material input. Here, throughput is taken to be the number of pieces of product, both good and bad, emerging from the process during the given time period under study. Clearly, a continuous increase in the number of pieces loaded on the process cannot continually generate corresponding increases in the process throughput. The process has physical limitations. This characteristic has to do with what economists call decreasing marginal productivity. This is a manifestation of the well known law of diminishing returns. This law often comes into play when production is increased by increasing a factor or factors when certain other potential factors are held constant. In our case, capital and labor are assumed to be constant. We are not considering adding more processes or man-hours in order to increase production output. Our model is only designed to account for production changes resulting from changes in product input and process quality checking for given levels of process capability. To ensure decreasing marginal productivity of raw material (quantity of input pieces) the exponent a is taken to be a quantity less than one (i.e., $a < 1$). It is assumed that constant returns ($a = 1$) and increasing returns ($a > 1$) are not possible.

Optimum QC Sampling

The author has derived formulas for the optimum QC sample sizes for both the constrained and unconstrained profit maximization cases based on the controlled-process production function described above. In the constrained case, the profit maximizing QC sample size depends on the constrained level of product input. In the unconstrained case, product input to the process is allowed to vary along with the QC sampling. This yields unconstrained profit maximization formulas for both the optimum product input level and QC sampling level. Rather than belabor the reader with the mathematical derivations and resulting models and formulas, they will not be presented here but will be used in the following hypothetical case example to illustrate the application of the technique to a practical situation. The interested reader should refer to the appendix to this paper for an elaboration of the technical aspects behind the formulas used below.

A Case Example

The ACME Widget Company advertises that it sells a line of high quality widgets. However, in order to ensure their customers receive only quality widgets, the firm has become involved in an extensive inspection and repair operation at the end of the production line. The plant manager has become concerned that too much money is being spent on culling out the bad product and not enough attention is being given to preventing the defects from occurring. He feels that some sort of process control using QC techniques might be the answer. However, he is afraid that mandating a wholesale introduction of process sampling will not justify the cost involved. He remembered from his experience as a QC supervisor that traditional statistical quality control techniques tend to focus on defect rates and give little attention to the cost/benefit relationships associated with the process quality control efforts. He took his problem to ACME's quality engineering department and asked them to come up with a plan that would take account of the economic aspects of the situation. "Remember," he said, "our primary objective is to maximize profits. Quality control is fine, but somebody has to pay for it. I'm not as much interested in percent defective as I am in the dollar figures on our balance sheets and income statements."

Due to management's concerns, the quality engineering department elected to adopt an *economic quality control* approach to the problem. A single critical process was chosen as a pilot process to test the planned scheme which involved checking the process at various intervals to determine whether it was in control and making adjustments to the process to bring it back into control when necessary. The process chosen for initial implementation of the new EQC plan was the crinkling process.

Widgets are made in a series of operations involving several separate machines. The raw material is purchased from a supplier in pieces. Each piece is referred to as a blank trogeal before any assembly or other processing is performed. Nadles are mounted onto the trogeal at the nadling machine. After that, varvits are attached to the nadled trogeal. The varvitting process is followed by the crinkling operation. It is very important that a good crinkle be made in each trogeal. The final process is the mepler which protects the crinkle and adds to the widget's distinctive cosmetic appeal.

We will focus our attention on the critical crinkling process. Product specifications only allow a very small tolerance window for the depth of the crinkle. ACME owns the most modern, sophisticated, and highly regarded crinkling machine available on the market today. Yet, the state of the art in crinkling is such that the crinkle depth characteristic tends to drift out of tolerance. No design has been perfected to automatically correct for this drift. The only method for determining when the process drifts out-of-tolerance has been to require constant operator attention. Whenever the final inspection department detects a large increase in crinkle depth defects, they write a memorandum to the process supervisor recommending that the operators pay closer attention to the crinkle depth tolerance. The operator increases checks, but the problem may have already been cleared up by routine maintenance or other factors.

The quality engineering department decided to provide the operator a constant interval for sampling product using EQC techniques. This is equivalent to stating the required number of pieces that must be checked for a given amount of product processed. Process engineering has determined that, when the process is producing good product, the output can be expressed as a function of the number of trogeals loaded on the process. If we refer to the output calculated only when the process is in control as throughput, the following equation describes the functional relationship found by the engineers

$$T = X^a.$$

Here, X is the quantity loaded on the process and T is the resulting throughput during a specific period of time (one week in this case). Therefore, the throughput is what the output would have been during the specified time period if the process had stayed in control for the entire period. Furthermore, engineering has found from historical data that the value of the exponent is $a = 0.95$. Thus, it can be seen that increases in throughput occur at a decreasing rate as input increases. This is a common diminishing returns situation which occurs when the presence of related fixed factors slows down the perpetual expansion of production that might otherwise result from increased use of a factor endogenous to the model.

In order to apply the EQC approach, cost and price information is needed. To obtain this information, the quality control engineer went to the accounting department. He found that the material and labor cost invested in a widget at the point where the trogeal exits the crinkling process was $3.50 per trogeal. He also found that the cost of checking the quality of a single trogeal after crinkling (and resetting the crinkle depth adjustment when a bad crinkle is detected) amounted to $1.75 per piece checked. He found that the value of the crinkled trogeal to the next process, the mepler, was $6.00 per trogeal. This quantity was obtained by taking the sales price of the widget and subtracting all costs involved (subsequent to the crinkling process) in getting the product to market. This figure was used as the per unit product price for computing the crinkling process profit maximization raw material and QC input quantities. The final piece of information was available from his own QC engineering data. He had found from a process capability study that the crinkling depth drifts out of tolerance, on the average, 386 pieces after the adjustment has been centered. In other words, whenever the process is adjusted to bring it back into control, it will make 385 good pieces before it starts making out-of-tolerance crinkles again. The necessary information collected by the QC engineer can be summarized as

$$P_X = \$3.50, \qquad P_W = \$1.75, \qquad P = \$6.00, \text{ and } \qquad M_0 = 386.$$

The first thing the QC engineer wanted to find out was how many pieces of product should be sampled each week in order to keep the process in economic quality control given the current level of production of the process. Since the process is normally loaded at the rate of 5,000 trogeals each week, we can substitute this value into the efficiency equation along with the data above and solve for the optimum weekly QC sample size, W.

$$W = (-B + (B^2 - 4AC)^{1/2})/2A$$

Where $A = a(2M_0 + 1)$, $\qquad B = 2ax^a$, $\qquad$ and $C = -(P_X/P_W)x^{a+1}$

Plugging in the researched cost and process capability figures, he obtained the results,

$$A = 774.35, \quad B = -6205.48, \quad \text{and } C = -32{,}660{,}400,$$

which yielded

$$W = 209$$

as the optimum weekly QC sample size. Since the throughput of the machine was

$$T = X^{.95}$$

This implies equally spaced sampling intervals of:

$$X^{.95}/W = 15.6$$

or approximately once every 16 pieces of product coming off of the crinkling machine.

The plant manager was pleased that QC engineering had derived an economic sample size for quality checking of the crinkling process. However, he was also interested in adjusting production at the crinkler to maximize profits. (See note at the end of this paper.) He wanted to know the simultaneous optimum values of both the quantity of product loaded on the machine and the number of pieces checked for quality each week. Since the EQC model could be made to yield such figures, he submitted his request to the QC engineering department.

The QC engineer was happy to oblige the plant manager. He calculated the optimum crinkling process load quantity by invoking the unconstrained profit maximization equation

$$X = (aP_W(2M_0 + 1 - 2G)/G^2P_X)^{1/(1-a)}$$

where

$$G = (2M_0(P_W/P))^{1/2}$$

Substituting the researched price, cost, and process capability figures, he found that

$$X = 8{,}128.$$

Therefore, the process should be loaded with 8,128 trogeals per week in order to maximize profits. The corresponding profit maximizing sample size was obtained from

$$W = X^a/G$$

where G is recognized to be the optimum interval between samples and is calculated from

$$G = (2M_0(P_W/P))^{1/2}.$$

Using $X = 8{,}128$ pieces per week, he found that, in order to maximize profit, the plant should operate the crinkling process such that

$$G = 15 \quad \text{and} \quad W = 345.$$

In other words, 345 trogeals should be checked each week from the product coming off of the crinkling process. The samples should be equally spaced every 15 pieces or product. The process should be adjusted back into control every time a crinkle exceeding tolerance is detected during the normal check. These process control activities should take place while the machine is being loaded at a rate of 8,128 trogeals per week.

Appendix

It is postulated that the firm seeks to maximize profits where profit is defined to be the excess of revenues over costs. The general form of the optimization model for the profit maximizing firm is therefore:

Maximize: $\pi = R - C$

where π, R and C refer to the profit, revenues, and costs, respectively. If we restrict our attention to competitive firms, we know that revenues equal product price times the quantity sold; i.e.,

$$R = PQ$$

where P and Q represent price and quantity. The quantity produced is a function of the inputs, or factors of production. The equation describing how this output quantity is determined by the production inputs is called the *production function* and is described in the main body of this paper.

The costs associated with production are those that arise from using economic factors that generate that production. In our specific situation, we are concerned with the costs of raw materials and process quality checking efforts. The raw material cost is taken to be the price P_X (to the producer) per piece of product entering the process times the number of pieces, X, loaded in the given time period. Similarly, the quality checking cost is taken to be the price P_W per piece of product checked times the number of pieces, W, checked during the given time period. Hence, the relevant total expenditure on input factors can be expressed as

$$C = g(X,W) = XP_X + WP_W.$$

The specific form of the optimization model for the profit maximizing competitive firm presented above can now be developed. Recall that

$$\pi = PQ - C.$$

Thus, we can say that

$$\pi = Pf(X,W) - g(X,W)$$

which emphasizes the fact that profit is a function of the economic factor quantities. Substituting the specific equations for physical output and costs described above, we get the following model:

Maximize: $\pi = PX^a(1 - (((X^a/W) - 1)/2M_0)) - XP_X - WP_W.$

According to economic production theory, *efficiency* is reached when the marginal productivity per dollar spent is the same for one factor as it is for any other. In our model, this occurs when the marginal productivity of the quantity loaded on the process per dollar spent on load quantity equals the marginal productivity of the QC efforts per dollar spent on quality control; i.e., when

$$MP_X/P_X = MP_W/P_W$$

where MP_X and MP_W are the marginal productivities of load quantity and QC, respectively. A simple rearrangement of terms yields the equivalent observation that efficiency occurs when the ratio of the marginal productivities equals the input price ratio. The ratio of marginal productivities is sometimes referred to as the *marginal rate of technical substitution* since it defines the rate which one input factor can be substituted for another and still maintain a given level of output. The above result is derivable using calculus maximization. Derivations are readily available in the literature and will not be repeated here.

The primary conceptual notion underlying efficiency is that, for some given factor expenditure level, the inputs must be combined in such a way that the maximum possible profit obtainable is realized. The quantity of production that results from a combination of input quantities that satisfies the above equation is said to be an efficient output. The locus of all efficient outputs that results from allowing factor expenditure to vary, is known as the *expansion path*. This is the way that production should be allowed to expand in order to maintain the greatest attainable profits as input expenditures are increased.

Marginal productivities can be obtained by taking the first order partial derivatives of the production function. Hence, for our specific model, the marginal productivities are

$$MP_X = f_X = ax^a(1 - (X^a/WM_0) + (1/2M_0))/X,$$

and

$$MP_W = f_W = X^{2a}/2W^2M_0.$$

Setting the ratio of the marginal productivities equal to the input price ratio, we get the efficiency relation

$$(P_X/P_W)X^{a+1} + aW(2X^a - W(2M_0 + 1)) = 0.$$

Solving the efficiency relation for W using the quadratic formula, we get

$$W = (-B + (B^2 - 4AC)^{1/2})/2A,$$

where

$$A = a(2M_0 + 1),$$
$$B = 2aX^a, \text{ and}$$
$$C = -(P_X/P_W)X^{a+1}.$$

Note that we are only interested in the positive radical in the quadratic formula since we can never allow W to be negative. Also note that we have expressed W as a function of X. This is the expansion of the process. The expansion path equation defines the manner which increased usage of input factors should be apportioned in order to maintain production efficiency as production is expanded. A plot of the expansion path equation is called the expansion path curve.

The previous formula determines the QC sample size when a firm desires to obtain the greatest possible profit contribution from a production process when the firm's expenditure on input factors is determined by considerations other than profit maximization. This is what economists call profit maximization subject to a cost constraint, where the cost is the total expenditure on input factors. If factor expenditure is allowed to be endogenous to the profit maximization model, not only will an efficient combination of inputs be specified, but production and input costs will be allowed to assume whatever values maximize profit. Solution of the profit maximization model will be the single point on the expansion path that yields the greatest profit. That single point will specify the exact amount of each input factor required to globally maximize profit. For our specific case of the controlled process, the solution to the model will yield the amounts of X and W necessary for global (unconstrained) profit maximization.

Since efficiency is a necessary, but not sufficient, condition for profit maximization, the requirement that the marginal rate of technical substitution equal the price ratio must also hold for profit maximization. Profit maximization, however, has a stricter requirement which was not necessary for production efficiency. That requirement is that the marginal revenue products of the inputs must equal their respective factor prices. Symbolically,

$$P(MP) = P_X, \text{ and}$$
$$P(MP) = P_W,$$

where the output product price, P, times the marginal productivity, MP, of an input, X or W, is defined to be the *marginal revenue product* of that input. Solving for P in the above two equations yields

$$P = P_X/MP_X = P_W/MP_W.$$

Recall that efficiency required that

$$MP_X/P_X = MP_W/P_W.$$

Thus, profit maximization will be realized whenever an efficient combination of inputs is found such that

$$P = P_X/MP_X \text{ anD } P = P_W/MP_W.$$

We can equate P to either quotient since efficiency guarantees that the two ratios of factor prices to marginal productivities will be equal.

In order to minimize the amount of algebraic manipulation, we take the second equation (the one involving the price of W) and substitute our calculated value for the marginal productivity. We get

$$P = 2M_0 P_W (W/X^a)^2. \tag{1}$$

Solving for W, we find that

$$W = X^a/G, \tag{2}$$

where

$$G = (2M_0 P_W/P)^{1/2}. \qquad (3)$$

Plugging this profit maximizing relation back into the efficiency relation (equation 1), we can solve for the profit maximizing value of X in terms of prices and process capability. The resulting formula was found to be

$$X = (aP_W(2M_0 + 1 - 2G)/G^2 P_X)^{1/(1-a)}$$

where G is the square root value given by equation 3. Once X has been computed, it can be substituted back into equation 2 to obtain the profit maximizing value of W. Explicitly, the profit maximizing value of W can be expressed in terms of prices and process capability as

$$W = ((aP_W(2M_0 + 1 - 2G)/G^2 P_X)^{1/(1-a)})/G.$$

Thus, equations 4 and 5 describe the quantity of product which should be loaded on the process and the QC sample size, respectively, that will maximize profits. These quantities are completely determined by prices and process capability.

Conclusion

It is realized that it would rarely (if ever) be desirable in actual practice to maximize profits based on a single process in an interconnected stream of processes such as this. Instead, the output of each process would be treated as the input of the subsequent process such that an overall production function for the entire line would be generated. It would then be appropriate to maximize profits with respect to the variables of the entire production line for given parametric values of the constituent processes. In the present example, single process optimization (neglecting the effects of interacting processes) was illustrated to simplify exposition and facilitate communication of the key concepts. The author has presented the production-line maximization case in his unpublished manuscript, "Economic Optimization of a Series of Sequential Production Processes."

Bibliography

1. Chiang, A. C. *Fundamental Methods of Mathematical Economics*, 2nd ed. New York: McGraw-Hill Book Co., 1974.
2. Henderson, J. M., and R. E. Quandt. *Microeconomic Theory*, 3rd ed. New York: McGraw-Hill Book Co., 1980.
3. Silberberg, E. *The Structure of Economics*, New York: McGraw-Hill Book Co., 1978.

MANAGING FOR SUCCESS THROUGH THE QUALITY SYSTEM

Frank Caplan
Vice President, Corporate Quality Systems
Gull Inc.
Smithtown, New York

Abstract

Included in this presentation are all the basic ingredients of a management program oriented toward achieving an organization's quality and reliability objectives in a highly profitable manner. Key activities within such a system in operation are discussed in the context of a specific application but in a manner to illustrate their applicability to any kind of product. The development of a Quality System in a manufacturing enterprise is described in some detail, including the methods to be used in controlling such a system. The control devices specifically discussed are Quality Costs, System Audit, Customer-Centered Quality Audit, Yield Control, Field Problem Controls, and Customer Satisfaction Measurement. Since a system involves people to make it work, a substantial portion of the presentation is devoted to questions of organizational effect on the success of the system and the significant contributions which can be made by everyone in the enterprise. This involves programs aimed at increasing effective participation by everyone in solving problems and helping the organization to reach new heights of effectiveness. The clear relationship between such successful involvement and the resulting pride of accomplishment is carefully presented. The inspirational aspects of such programs as Zero Defects and Quality Circles are touched upon at some length — emphasizing the positive effects upon productivity and quality which can be achieved through these and similar activities. The typical outcome of the introduction of a Quality System is that of substantial reduction in quality costs. Add to that, the improved productivity resulting from both problem prevention and problem solution and the picture that emerges is one of dramatically reduced costs and significantly increased profits.

What a System Is and the Need for a Quality System

For purposes of this discussion we will describe a "system" as a planned and coordinated set of actions by people and, in some cases, equipment which is controlled by information and oriented toward producing an intended result (see Figure 1). Some systems are quite simple, requiring only a few activities to accomplish the objective.

> **System Definition**
>
> *A system is the planned interaction of people, machines, and information — controlled by information — and oriented toward meeting a common objective.*

Figure 1

Others are very complex — perhaps involving hundreds of such activities and coordination among dozens of groups of people to reach a successful conclusion.

Even in the simpler cases, but particularly as the system becomes complex, the need for coordination and, therefore, for timely interchange of information is critical for success. To achieve complete customer satisfaction with our products as a satisfactory profit to our companies, we are forced to treat questions of quality, reliability, maintainability, safety, and related subjects as the result of systematic operations. The name we give to such an approach is the quality system.

The reason that we have no choice but to adopt the "system" approach to quality is that our other approaches have failed. They have either failed to achieve the quality goals or they have managed to reach them, but at excessive cost. Let's look at what has been happening to the customer's perception of profit quality in recent years, as reported by several substantial studies (see Figure 2).

There seems to have been a substantial narrowing of the "quality gap" starting in 1980, the first such instance in many years. Thus, in the area of consumer goods, at least, the recent awakening of the American manufacturer to the survival aspects of quality has produced good results. But the profitability of American companies has, according to a recent government release, fallen below 4 percent for the first time in many years. The only conclusion I can draw is that the reduction of the quality gap was achieved through "brute force and awkwardness" rather than through the application of sound Quality System concepts of management. Should my conclusion be correct, the improvement is temporary and will erode as soon as complacency sets in and we offer the next generations of products to the marketplace.

As you might expect, as a result of the continual increases in the level of customer quality requirements, the proportion of their income manufacturing enterprises have spent on quality-related matters — quality costs — has grown substantially. Although quality costs were first looked at in the early 50s, extensive attempts to measure and report them were first mounted in the mid-50s. At that time they were estimated to be 2 to 4 percent of sales. Subsequent analyses have shown that they have grown at the rate of 5 to 7 percent of themselves each year so that today the range is from 8 percent to about 30 percent (see Figure 3). I suspect the average is well above 15 percent.

Such a radical rearrangement of spending priorities (forced, as it may be), even over 30 years, cannot help but squeeze profits unmercifully. And some companies are known to be worse than the top line of the chart. All of the quality costs are not capable of being reduced — but at least half of them are, with resulting significant profit enhancement.

Quality of Product as Seen by the Customer

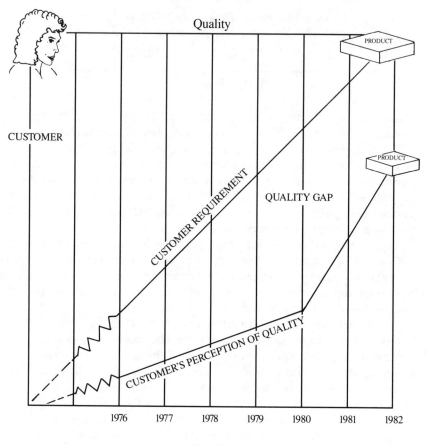

Figure 2

However, the only way to achieve the potential improvements in real and perceived quality and the reduction in associated costs of quality to maintain them permanently is through adoption of the quality system. That being the case, let's talk about that system.

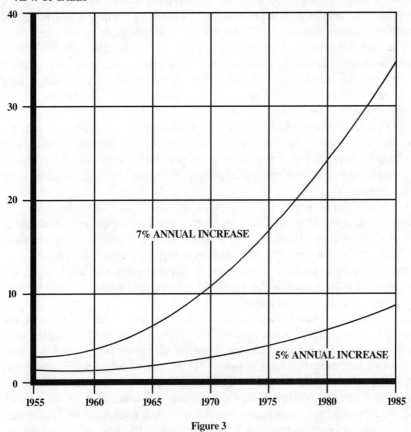

Figure 3

Development of the System

The quality system comprises nine subsystems, as follows (Figure 4):

1. Quality System Management
2. Product Development Control
3. Purchased Material Control
4. Process Development and Operation Control
5. Quality Data Programs
6. Special Studies
7. Quality Measurement and Control Equipment
8. Human Resource Involvement
9. Customer Contact

Each subsystem is made up of some number of activities, or elements, which altogether number 103, as described in "The Quality System." While some organizations may find it desirable to utilize all 103 of these elements, others (particularly, smaller ones or those with few product lines) may obtain all necessary benefit from a smaller selection. The key question to be answered is, "What are the quality system-related activities in which we now engage and which ones *should* we be performing?"

To answer this question effectively and, at the same time, to identify those elements deserving priority attention, it has been found highly useful to conduct an evaluation of the extent to which what is currently being done satisfies all the quality system requirements for each of the 103 elements. At the same time, nonrelevant elements can be eliminated from the "shopping list."

For those elements which are needed but are not currently adequately covered, it is then useful to assign a benefit rating. An example of how this might be done is given in Figure 5. The product of the need satisfaction level rating and the visible benefit rating produces a number between zero and 50 which serves as the basis for the establishment of the priority list.

In general, the high scores determine the priority elements in each subsystem, with at least one element selected from each of the nine subsystems to improve system balance. However, each element receiving an "intangible" visible benefit rating (a 1) is examined for "need" alone. If the "need" rating is high (typically, 5 or greater), the element should be put in the priority package. The thinking behind this is that such elements usually give structure to the system (quality policy, for example) or are the controls or training programs necessary to make it work.

Once the priority elements are identified, 43 of them in the example, the next step is to begin the development work required. The preferred technique for doing the work is to bring together a team of knowledgeable individuals representing all organizational functions in the enterprise which are involved in the activities of the priority elements. Typically, these include the marketing, engineering, manufacturing engineering, production, personnel, purchasing, data processing, finance, field service, quality control, and quality assurance functions — whatever their titles might be. The number of team members representing each function depend on the list of priority elements, the size of the organization, the amount of time each team member is to spend on the project each week, and the end date of the project.

In the example, we started with 23 people spending one to three days each per week (but on a staggered start schedule based on the order of attack on the priority elements) with a total project calendar time of three years. Each team member (or small group, in some cases) was charged with developing a document or family of documents for each element assigned. Documents could be procedures, manuals, instructions, computer programs, monographs, detailed training programs, or combinations of the preceding — whatever might be appropriate for the element. The documents were then submitted to design review by the complete team plus any other involved personnel, revised as indicated, and issued as the "new way" of doing business.

Element Priority Work Sheet

ORGANIZATION: ▇▇▇ DATE: 6-18 SUBSYSTEM: SPECIAL STUDIES			System Need Satisfaction Level (N)							Degree of Visible Benefit (B)						SCORE	PRIORITY
			None	Poor	Weak	Fair	Marginal	Satisfactory	Outstanding	None	Intangible	Slight	Moderate	Substantial	Very Large		
No.	Title	Comments	10	8	6	4	2	1	0	0	1	2	3	4	5	NxB	
6.01	CLASSIFICATION OF CHARACTERISTICS	S		x							x					8	
6.02	CLASSIFICATION OF NONCONFORMITIES					x					x					5	
6.03	DESIGN & ANALYSIS OF RELIABILITY & SAFETY STUDIES			x											x	35	
6.04	ENVIRONMENTAL IMPACTS			x											x	25	
6.05	MACHINE & PROCESS CAPABILITY				x										x	40	
6.06	QUALITY LEVEL IMPROVEMENT			x											x	35	
6.07	STATISTICAL TECHNIQUE APPLICATION				x										x	30	
6.08	UNIQUE DELIVERY REQUIREMENT MANAGEMENT			x											x	25	

Figure 5

The Control Aspect of the Quality System

As I mentioned earlier, the priority package normally contains some or all of the control elements of the system. The primary controls are six in number, as follows:

The first control is the one which measures the "health" of the system. It is called *Audit of Procedures, Processes, and Product* (see Figure 6). Every system begins to degrade from the moment it is installed because of lack of information or understanding on the part of the participants, uncorrected errors of omission or commission, changing circumstances which should produce a change in the system but do not, or inefficiencies in the system design. A properly conducted audit reveals examples of these conditions — providing a spur for continuous system improvement and conformance. In the example, the form of audit chosen was that of putting together small multi-disciplined teams, with rotating six-month memberships, chaired by QA personnel. When the area chosen for audit required technical proficiency, at least one member of the team had such skill. Otherwise, the primary requirements were curiosity and perseverance. Each audit took two to three days and each team audited each month, usually in different areas than previously. In addition to getting "fresh eyes" into the audit process, the technique provides excellent educational exposure for the auditors.

> **System Audit**
>
> *The planned scrutiny of product, process, and system-related activities for conformance to specification and procedural requirements.*

Figure 6

The second control is the *Quality Cost Program* (Figure 7). It is the device with which the quality system managers talk intelligently to all management at all levels in the company. In Figure 3, I showed what has happened to quality costs for industry in general over the last 30 years. The company of the case study began with an initial measure of quality costs of about 10 percent of sales. Of this amount, 68 percent was failure costs (mostly external), 28 percent was appraisal, and a charitable 4 percent was prevention. One important consideration, which heavily influences the quality costs total amount (and thus the percent of sales) for any company, is to determine what really are the contributing sources of cost. After you've listed all the legitimate sources and estimated their magnitudes, you are then in a position to decide which ones to include in *your* measure. In today's situation it is not normally cost-effective to include everything possible in the figures. What you need to do is to identify the "big ones," particularly those attributable to failure costs, because that's where the primary opportunity for profit lies; to select ones that are measurable, rather than estimates or allocations, because they will show you the effect of any changes that are made in product, quality, and shop practices; and to aggregate a big enough total (10 percent of sales) to get management attention. The measurement then becomes a rapidly responsive index which will point out areas of profit enhancement opportunity; provide the basis for sizing the reaction to that opportunity, and show the effects of that reaction.

The third control is *Customer-Centered Quality Audit* (Figure 8). It is usually conducted in the product design phase to help prevent ineffectively designed product from getting to market. In the example case, the company produced a consumer-goods product and conducted daily an evaluation of a sample from production — using highly skilled technicians. They also had enough turnover and growth that they hired a number of new assemblers to start each Monday morning. We arranged for several of the new assemblers each week to spend their first week working with the technicians on final

> **Quality Costs**
>
> The separation of accounting figures into the defined categories of prevention, appraisal, internal failure, and external failure.

Figure 7

test. Many of their comments wound up in causing redesign of existing product or in changes to new product designs. The reason is quite simple — the technicians checked the product against specification; the new employees looked at it as if they were buying and using it themselves.

> **Customer-Centered Quality Audit**
>
> The examination of the product from a user's viewpoint — using customers or simulated customers and going beyond the design criteria in the evaluation; even considering reasonable misuse of the product.

Figure 8

The *Yield Control Program* (Figure 9) is the fourth control. It provides the necessary system response to internal failure rates, similar to the quality cost response to high expenses related to quality. The technique highlights vendors, production processes, and handling practices which need attention for quality reasons. In the case being reported on, there were five production process areas identified the first month after the element procedure was implemented as needing quality improvement. Two of these also showed up in the quality cost analysis of internal failure costs. All five were attacked, using techniques developed as part of the quality level improvement element. Not only were all the target failures and costs materially reduced over the next six months, but productivity on the line rose 2 percent and two long-term nagging field problems, which nobody had been able to correct, disappeared.

The fifth control on the list is entitled *Field Problem Controls* (Figure 10). In this element we concern ourselves with the customer's experience with our product in use. Most companies producing durable goods have some sort of warranty program and obtain some data associated with failures and service during the warranty period. Given that even those data might be viewed with suspicion as to their validity, the value of post-warranty data may well be even more questionable. However, it may be practicable to get useful data — in which case, we certainly should do so. In any case, we need to examine the data for evidence of repetitive repairs for the same or a related condition performed on the same unit. The company in our case study found that 12 percent of their company-owned service facility repair personnel were involved in 85 percent of their second time and multiple repairs. The pattern could have been suggested by the

> **Yield Control**
>
> *The measurement of defect rates associated with vendors, inhouse production, and warehousing — combined with techniques to reduce those rates.*

Figure 9

Pareto distribution, but the control program identified the individuals and the particular areas of their weaknesses. A rapidly organized series of retraining sessions, coupled with some new special tools and revised emphasis in the normal training program produced a 45 percent reduction in multiple repair calls for the same condition on the same unit by the end of the year.

> **Field Problem Controls**
>
> *The measurement of product reliability and of service diagnostic and repair efficiency as experienced by the customer — combined with techniques to improve these capabilities.*

Figure 10

The last of the primary system controls is *Customer Satisfaction Measurement* (see Figure 11). Obviously, the improvements produced by the previous element will contribute to the customer's favorable impression of the company. But there is more to the customer's overall view of the company than that. First cost, delivery, product appearance and performance, cost of operation, service and repair, and ease of operation — these all affect the customer's perception of us. There are also such things as the company's community image, advertising impact, and sales representation that help form this critical reaction to us. So when we set out to measure and interpret the customer's satisfaction with our efforts, it pays to be careful about it. Even the simple step of conducting an interview with a customer can alter that person's impression of us. Therefore, obtaining a reasonably accurate picture of the company's true feelings can be extremely difficult. The case company uses four measures — warranty repair data, customer warranty registration return cards, "two years after receipt of warranty card" reliability cards, and carefully orchestrated "industry survey" of selected customers conducted by an independent survey agency. All of these data are carefully compared, with the "early warning" ones correlated with the industry survey to establish their capability as predictors. They have modified the repair data collection system and the info requested on the first return card to improve the correlation and have made some useful improvements in that capability. It is the long-term intent to reduce and perhaps eventually to eliminate the industry survey — after the first three measures come close enough to reflecting its results.

> **Customer Satisfaction Measurement**
>
> *The quantification of customer reaction to the company and its products and services.*

Figure 11

All of these controls and other measurement elements in the quality system lead to corrective action at various levels, culminating, when the others prove inadequate, in the *Corrective Action Program* element as shown in Figure 12. By this means, the enterprise can ensure the successful application of the remainder of the system elements — combining the prevention-oriented activities with those requiring and producing effective, timely, and permanent corrective action when needed.

> **Corrective Action in the System Sense**
>
> A. *Define the problem.*
> B. *Verify the existence of the problem.*
> C. *Determine the cause(s) of the problem.*
> D. *Plan and schedule the solution(s) to the problem.*
> E. *Apply the three types of corrective action.*
>   1. *Correct the item(s) — as required*
>   2. *Correct the process(es) — as useful*
>   3. *Eliminate the cause(s) — always*
> F. *Ensure permanency of solution and lack of negative side-effects.*

Figure 12

Organizational Considerations

Fundamental to the ability of any organization to produce profitably and consistently a product or service which fully satisfies its customers is a system of management which, in every respect, is dedicated to that objective. Far too frequently, top management will state a commitment to and support of quality but will fail to act in such a manner as to eliminate any doubt as to their *real* intent. Expediency decisions are made daily in most companies which are so pernicious as to belie the flowery statements in quality manuals and public declarations to which we point with pride as stating "top-level" support for quality.

While, in recent years, it has become apparent to management in some of our industries that the above practices are "hazardous to one's (corporate) health," a real understanding of what's involved in reversing the devastating effects of those practices has yet to permeate all levels of the organization even in some of the most severely affected companies. To state the requirement briefly — everything that is said and done must result, without fail, in the generation of completely satisfactory quality, reliability and safety of the products or services of the enterprise.

As a result of the recent quality shocks (perhaps I should say "earthquakes"), a lot of companies have begun to view quality as a business strategy rather than as the special province of a necessary but undesirable group of specialists buried deep in the organization. Suddenly, there are large numbers of vice presidents owning quality titles! Virtually everybody's institutional ads now bear quality messages, complete with slogans! And some of these companies are actually doing some things to improve the quality of their products, as I mentioned early in the paper. But, one of the key points to be recognized is that managing a quality system is only partially the responsibility of the functional organization assigned the quality assurance role. Every organization within the enterprise performs actions critical to the overall success of the company in satisfying both the customers and the stockholders. Therefore, when we look at organization for quality, we have to look at all the quality system responsibilities — not just those of the department with the quality label. That being the case, it is essential that we view the quality system's impact upon and operation by the people in the organization. For the system to be successful it must be strongly people-oriented and sensitive to their interpretations of what they see and hear from their bosses.

There's an old saying in this business (I know it's "old" because I've been saying it for 30 years or more) that "the product of a company looks like its president!" This really means that each manager places his own imprint on the organization reporting to him. And he causes this effect far more by what he does or does not do then by what he says. He produces an "image" which his subordinates attempt to adhere to because everyone wants to do a "good job" — usually, that devolves into what is perceived to be "what the boss wants." Years ago, I reported (as quality manager) to a manager of manufacturing with whom I had constant battles over quality. When I challenged him again one time he confided to me that, "the first time _____ (the division general manager) asks me how our quality is rather than what our monthly billings are, I'll start to pay some attention to quality!" Yet, if anyone from outside had asked the general manager what his view was of quality he would, with complete sincerity, have stated his wholehearted dedication to it.

The fundamental truth is that everybody (even the manufacturing manager mentioned above) *would* rather do a good job than a poor one. It is only after they get the negative message from above that they become "turned off" to quality. Look at what has happened when we have tried to restore the "old time craftsmanship" attitudes in this country. In the '60s we tried zero defects programs under a number of names. The technique called for each employee to commit himself or herself to producing nothing but good work, identifying for management's action those conditions not within his or her control. In a few cases the results were substantial and lasting. In all other cases there was an immediate positive reaction by virtually everyone followed sooner or later by a collapse of the program combined with a feeling of betrayal by the participants. The reason for the demise of the program here is readily apparent — management failed to follow through on its explicit or implicit promise to remove the causes for quality problems once the employees identified them. You see, it got to be a little tough for management to do this because it turned out that management had about 85 percent of the problems to solve. They had leaped into this program on the assumption that the employees were to blame for all of the problems, rather than for 15 percent of them.

The new fad is quality circles. In this program we get groups of employees to work together rather than looking to individual effort and we give them some problem-solving tools to use so as to reduce the demand for management to solve the problems it causes. But the program still requires management to take the action called for — including spending some money or stepping on some toes if need be. While, given the solution is already provided, a lot more managers take the indicated action and thus encourage the Circle to continue, there are already many cases where the Circle program has died just as zero defects died. The quality system program requires the same kind of involvement on the part of everyone to be completely successful, from the operator to the president of the company, that the zero defects and quality circle movements did — but it can still accomplish a great deal without that kind of involvement, albeit it takes longer. So there is provision for extensive training programs and for participative activities such as zero defects and quality circles in the system, with one entire subsystem, human resource involvement, devoted entirely to them.

How the System Works in Practice

Throughout much of the preceding material I have discussed details of a specific case of the application of the quality system concepts. I will add the following to my previous comments:

The company had a good quality reputation in its industry before embarking on the quality system project. Today it is considered one of the industry leaders for quality and reliability. Its market share has increased substantially even though its present pricing policies would have driven it out of the market a few years ago.

The company turns out more units of a much more sophisticated product than formerly, with fewer people and higher profits. However, this has not been accomplished overnight. From project initiation to full utilization of the quality system may take 10 to 15 years, although substantial results can be seen in as little as six months in some cases.

Like zero defects and quality circles, the quality system is not an instant panacea. However. unlike them, it does provide a long-term solid foundation for growth and increased profitability.

The Benefits of the System Approach

Since the activities of the quality system begin with the definition of requirements for a new product and end with the product working successfully in the customer's application for as long as necessary, its benefits are many. Product quality is better in the factory because the design is more readily manufactured at lowest overall cost: the processes are proven capable before start of production; and the personnel are properly trained, equipped, and supported. The customer's perception of the product is that it looks good, it does what it needs to do, and it does it when called on for at least as long as specified. The product is readily maintainable and safe to use and work on. Field documents are relevant and useful and associated tools are proper and available.

The people who work within the system contribute actively to product and process improvement and, like anyone else, when they see their ideas accepted and applied, are proud of their accomplishments. Morale is high and turnover reduced.

When new products are designed using the programs of the quality system, it takes fewer resources and less calendar time (often 20 to 30 percent less) to get them through the design cycle from concept to full, defect-free production. Since the product and process are developed together and forced to be compatible, there is much less rework, repair, and reinspection than normal. Even first time inspection and test may be safely reduced as a result. The resultant increase in productivity, however measured, is substantial.

All of the above benefits are significant and very important to the enterprise. However, they all combine to produce significant improvement in the quality costs picture (see Figure 13). What this chart shows is that the 10 percent of sales figure at which our case company started would have risen to 15 percent over the following eight years under normal circumstances. However, with a limited early investment in the quality system project, they not only avoided the increase but cut in half their quality costs as a percent of sales. With their increased sales over these years, the resultant increased profits aggregate many millions of dollars.

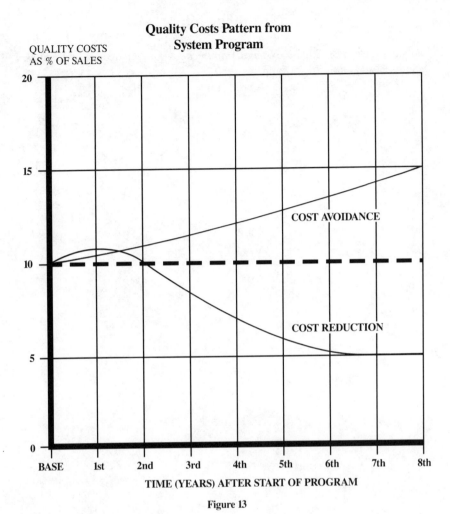

Figure 13

Conclusion

Through the application of the Quality System any manufacturing company (and, in fact, any service or governmental organization, as well) can materially improve its costs, its productivity, and the involvement of its people. Where the organization is oriented toward making a profit, it can increase its profits by an amount unavailable by any other means requiring so little investment.

Bibliography

Caplan, Frank, *The Quality System: A Sourcebook for Managers and Engineers.* Radnor, Pa.: Chilton Book Company, 1980.

PROFIT IMPROVEMENT THROUGH SCRAP REDUCTION

J. P. "Jerry" Reames
Corporate Manager — Quality Assurance
Allied Corporation
Morristown, New Jersey

Abstract

While formal cost of quality measurement and tracking programs are ideal to identify opportunities for cost reduction, many small companies or plants have neither the accounting systems nor the expertise to implement such programs. Where the major element of quality costs is in scrap, there are simplified approaches that may be utilized to quickly increase profits through scrap reduction. This paper discusses two possible approaches that may be used.

Introduction

Measuring the costs of quality is essential to a *sustained* quality improvement program to identify opportunities for cost reduction/cost avoidance and to communicate the extent of those opportunities to management in the business language of money. However, in American industry, most companies have not begun compiling costs of quality. Why? Although this is a viable, proven technique for not only identifying opportunities for increasing profits but also to measure the overall effectiveness of the quality systems, many companies (especially smaller companies or plants) have neither the quality or financial expertise or personnel to initiate and implement these programs. Even when substantial manpower resources are available in the financial organization, the quality function has some hesitancy in the expansion of their role to coordinate the cost of quality program, especially if it is to be used to measure the quality function effectiveness.

Also, for those companies or plants that experience costs of quality in the 10 to 25 percent to sales range, opportunities exist that may be identified to improve profits *without* a *detailed* cost of quality program. For these types of operations where obstacles to detailed cost of quality measurement have been encountered, there are at least two approaches that may be taken to stimulate the reduction of quality costs which can lead the organization into more formal programs. The two approaches to be discussed are:

- Plant Self-Evaluation for a Scrap-Reduction Program
- WOW (War on Waste)

These two approaches are possible interim steps to address quality losses due to scrap which is frequently the major contributor to excessive costs of quality.

Plant Self-Evaluation for a Scrap-Reduction Program

Scrap levels in a manufacturing plant frequently comprise 50 percent of the total costs of quality, especially in nonprocess oriented concerns. However, during austere times, management tends to look for cost reductions through census reductions, especially in the quality function, even though these costs comprise normally less than 15 to 20 percent of the total costs of quality. Why, you may ask, do intelligent managers ignore the opportunities for reducing costs from an area of waste? Some of the reasons are:

- Managers don't understand "total costs of quality."
- Census reductions provide almost immediate reduction in expense and increased cash flow.
- "Our scrap levels are normal for our type of business."
- "We are always working on scrap reduction."
- "Our scrap problem is related to low sales and constant changeovers."
- "We're doing good considering our poor designs and the condition of our equipment."
- "Our scrap is within budget but with sales decreases our manpower will be too high."

You could probably add many more reasons or "excuses," whichever terminology seems appropriate.

However, the real reason aggressive scrap reduction efforts are not pursued is that organizational accountability for a team effort is lacking and weaknesses in scrap reduction efforts have not even been identified, much less addressed.

If this is going to change, quality professionals must take the "lead" role in communicating with others in the organization, especially with the plant or general manager, the following:

- Current scrap levels.
- Detailed concepts of quality assurance by all."
- The need for joint efforts in identification of specific scrap reduction efforts by function.
- Plan to identify weaknesses and correct problems.

One of the ways that the quality function can initiate such a program is to meet with the plant manager using the form shown in Figure 1 *Plant Self-Evaluation for a Scrap-Reduction Program* suggesting that this be reviewed at a plant staff meeting. Each functional manager would then complete either his particular portion of the form or the complete form, depending upon the wishes of the plant/general manager. Interestingly enough, when I have requested quality managers and plant managers to fill in the complete form for their own plants independent of each other, I have found a 90%+ correlation between the two evaluations. I have also found both quality and plant managers to be objective and rather critical of plant efforts in scrap reduction.

Figure 1 *Plant Self-Evaluation for a Scrap-Reduction Program*

ACTIVITY	YES	NO	DON'T KNOW
1. Is the production function held accountable for the generation of scrap?			
A. Does each production department have a scrap budget and is each production supervisor held accountable?			
B. Does each production supervisor review scrap produced in his department daily?			
C. Does each production supervisor sign scrap tags for his department?			
D. Does each production supervisor review scrap problems with his people?			
E. Does the production supervisor take the "lead" role in training his personnel?			
F. If hourly workers are running high rejects or scrap, are communication channels established so the supervisor is aware?			
G. Are scrap levels communicated back to the hourly personnel by production supervisor?			
2. Is the manufacturing engineering/maintenance/methods function held accountable for the quality of tooling, equipment condition, and methods to assure minimum scrap generation?			
A. Is there a formal system for new tool evaluation as received?			
B. When changes to tooling are made on the floor, are drawings revised?			
C. Does manufacturing engineering use scrap cause feed-back to improve process/equipment?			
D. Does the manufacturing engineering manager review scrap?			
E. Is there a formal tool/maintenance program including preventive programs?			
F. Do manufacturing engineers perform process capability studies on new equipment?			
G. Are scrap reduction projects listed and followed-up?			
H. Are machines that are known high scrap generators taken out of service and repaired or overhauled in a timely fashion?			
I. Are standard machine rates established with minimal scrap as an operating parameter?			
J. Are CAR's sent to manufacturing engineering by production/quality supervisor on a regular basis?			
K. Do manufacturing engineering and quality management mutually document scrap problems to stimulate improved design by product engineering?			
L. Are process/machine revisions for cost reduction reviewed with the quality manager?			
M. Does manufacturing engineering take the "lead" role in equipment/tooling modifications to use substandard raw material to minimize scrap?			

Figure 1 *(cont.)*

ACTIVITY	YES	NO	DON'T KNOW
3. Is the quality function held accountable for process monitoring, first-piece inspection, analyzing causes and defects, and recommending corrective action?			
A. Are process capability studies run where appropriate to determine effectiveness of operations?			
B. Does the quality department get actively involved daily with scrap review?			
C. Are scrap levels and problems communicated to hourly inspectors?			
D. Does production consider the quality department as an important service to aid them or a nuisance with which to contend?			
E. Is there regular and continuous feedback to production and manufacturing engineering?			
F. Is scrap reduction discussed at each staff meeting?			
G. Does the quality manager stimulate a scrap review board and insist that meetings be held regularly?			
H. Are scrap reduction projects followed up regularly?			
I. Does the quality manager review scrap daily?			
J. Does the quality manager play the "lead" role in training of inspectors?			
K. Are scrap trends posted in the plant and covered with the staff?			
4. Is the accounting department held accountable for scrap reporting on a daily basis?			
A. Is there a daily scrap report listing major product scrap by part number and cause or defect?			
B. Is there a scrap report prepared that is current and discussed at staff meetings?			
C. Does the plant controller review scrap daily or serve on a scrap review board?			
5. Is the production control department held accountable for scheduling to minimize changeovers and scrap?			
A. Are formal procedures established to provide guidelines as to minimum runs agreed upon by production?			
B. Are adequate lead-times provided for receiving inspection and rework or replacement if required?			
C. Is raw material inventory rotated to reduce shelf-life problems?			
D. Is the rejected material which is to be returned to the customer returned promptly?			
TOTALS			

Overall: Would you rate your plant scrap-reduction program (circle one)

A. Outstanding — No major revisions necessary?
B. Above Average — Some revisions needed?
C. Average — Some major revisions needed?
D. Below Average — Many major revisions needed?
E. Poor — No semblance of a program?

From the completed form, the quality manager can then summarize and prioritize the organizational weaknesses using the Pareto Principle to select the five areas that could yield either the fastest savings or the greatest magnitude of savings, or both. These then could be discussed at future staff sessions and detailed corrective action projects could be assigned by the plant/general manager with tracking provided through the illustrative example shown in Figure 2 *Scrap Control Program Project Form.*

Figure 2 *Scrap Control Program Project Form* Date Issued 12/8X

ITEM NO.	DESCRIPTION OF WEAKNESSES
2.1	Lack of Manufacturing Engineering support due to capital commitment.

PROJ. NO.	PROJECT	PERSON RESP.	TARGET DATE	COST	STATUS
2.1.1	Rubber Dept. Assignment - 1 Engineer G. Racz - 1 Technician B. Schlenice	Skaggs	completed completed	-0- -0-	
2.1.2	Plastic Dept. Assignment - 1 Engineer J. Thompson - 1 Technician J. Mazure	Skaggs	Jan. 198X completed	-0- -0-	
2.1.3	Braider/Thoren Winders/String Tower/Marking/Electro Tin Plating - 1 Engineer S. Cardwell - 1 Technician B. Harris (Partial)	Skaggs	completed Mar. 198X	-0- -0-	
2.1.4	Stranding Dept. Assignment -1 Engineer J. Moore -1 Technician B. Harris (Partial)	Skaggs	Feb. 198X completed	-0- -0-	
2.1.5	Miscellaneous/Rewind/Specs - 1 Engineer/Tech J. Hudson	Wickus	Jan. 198X	-0-	
2.1.6	Hire one extrusion or drawing engineer	Forrest	Mar. 198X	XX,000	
2.1.7	Technician for string area	Cardwell	Mar. 198X	XX,000	
2.1.8	Communicate the scrap program and scrap assignments to the Engineering group	Skaggs	Feb. 198X	-0-	
2.1.9	Report relative to specific scrap problems - Item - Status - Corrective Action - Responsibility - Target Date	Skaggs	Feb. 198X	-0-	

Please note that for each weakness identified, each project has been assigned a number, and responsible individuals, target date, cost and status are shown. As projects are completed, additional projects may be added.

The elements of this program provide for the following:

- Quality provides leadership and coordination.
- Provides guide for evaluating weakness.
- Keyed to team discussions and actions.
- Leads to specific corrective action projects.
- Provides for follow-up and an ongoing effort.

Naturally, the evaluation form may be revised to fit different locations and organizational structures. The key to effective program development is not so much in the specific questions, but in the approach to team problem identification and team problem resolution through specific scrap reduction projects.

Where this approach has been used, savings of 25 percent in scrap costs in the first three to six months have not been uncommon, especially when scrap costs are 50 percent or more of total quality costs prior to program initiation. As scrap totals decline, the need for more detailed information becomes apparent to all team members so that the need for detailed scrap reporting systems are not "just another QC request" but a recognized need by all functions.

WOW "War on Waste"

While the approach just discussed works effectively in some manufacturing plants, it is generally used where some existing data is already provided for the quantity of scrap and perhaps, the part number and/or the reasons for scrapping. However, in many plants (especially the newer or smaller facilities), little or nothing is known about the quantity or nature of the scrap being generated. Frequently, "scrap" is defined as the difference between the value of the material received less the value of the material shipped, at cost. The oversimplified accounting approach does not differentiate between yield loss, inventory loss, errors in bills of material or specifications and excess material usage and normally does not factor in the other elements of in-process scrap such as labor and burden costs. Consequently, efforts at scrap reduction are not focused as they seldom are when trying to shoot at a moving target.

For these types of operations, an interim approach may be worthwhile. This approach is identified as *WOW* War on Waste. While this approach is similar to the previous one, it assumes a lesser degree of sophistication in accounting systems, especially relating to scrap costing and, as importantly, root cause definition. Additionally, this is a shorter term approach than that previously described. The key elements of the WOW approach are as follows:

- *Establish priority* for scrap reduction.
- *Define objectives* to prevent problem rather than rework or "fixes." "Root" or "primary" causes must be identified and eliminated.
- *Identify* through staff discussion the two major causes of scrap.
- *Define responsibilities* by plant manager for various scrap-reduction related activities:
    - Scrap identification and handling
    - Scrap disposition
    - Scrap information compilation
    - Scrap reporting and analysis
    - Scrap reduction projects
- *Establish deadlines* for implementation.
- *Follow-up and review* to adjust program as needed.

Figure 3 *WOW Flowchart* shows the various steps in the program and the recommended timing for the initial elements of the program.

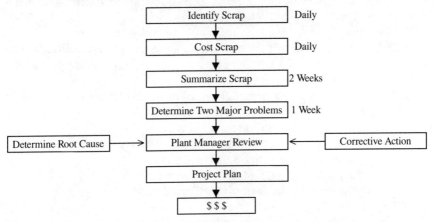

Figure 3 *WOW Flowchart*

The sequential steps of the WOW Flowchart are as follows:

A. *Identification of Scrap* - A list of parts that are scrapped should be compiled on a daily basis for a one or two week period to provide factual data for analysis. This list should include:
   1. What is it? Description, part number.
   2. Quantity? Weight, pieces, feet, etc.
   3. Who made it? Machine, operator, shift.
   4. When? Date.
   5. Why scrap? Reason, *not* cause.
B. *Costing* — Determine the cost of scrap by compiling on a daily basis the cost of each part number scrapped, including the labor and burden, if possible.
C. *Summarize* — Compile a summary sheet containing one or two weeks of the above data.
D. *Problem Identification* — From the summary sheet, determine the two most significant part numbers based upon dollars. These are the initial products on which scrap reduction efforts should be focused.
E. *Determine Root Cause* —
   1. Plant manager to assign individual or group the task of determining the *root cause* of the two major product scrap contributors.
   2. Determination of *root cause* by:
      a. Process studies
      b. Review of scrap material
      c. Knowledge of process
      d. Discussion with hourly operators and production supervisor
      e. Review of specifications

f. Hourly review of scrap segregating by defect
g. Other
3. After determination of root cause, individual or group leader will submit a written report to plant manager.

F. *Corrective Action* —
1. After receiving the root cause report, plant manager will assign an individual or group the task of determining the corrective action required to correct the *root cause*.
2. Those assigned will determine corrective action and submit written report to plant manager. Written report will include:
   a. What must be done to correct problem
   b. Costs involved (people, tooling, equipment, training, etc.)
   c. Help needed by department function
   d. Time frame required for implementation
   e. Verification plan for corrective action indicated

G. *Project Plan* —
1. Assignment of individual/group by plant manager to implement plan.
2. Establish plan for implementation.
3. Follow-up and review.

Where this approach has been used, savings of up to 25 percent of total scrap have been achieved in the first month, especially when the root cause was carefully identified and the indicated corrective action required no major capital expenditure. In one case, the major scrap problem was related to a tooling problem on one machine but the extent of the problem had just not been recognized.

The elements of this approach provide for many of the features discussed in the survey approach; however, the WOW program is unique in the following areas:

- Specifically directed at two major product scrap generators.
- Short-term approach but could expand into more continuous program.
- Establishes data to identify opportunities quickly without development of more complete systems.
- Requires the commitment and personal involvement of the plant manager.

Conclusion

The two approaches discussed, *Plant Self-Evaluation for a Scrap-Reduction Program* and *WOW* are suggested approaches for those smaller or less sophisticated companies or plants who have known scrap problems and desire to aggressively pursue scrap-reduction on a shorter-term basis. Either approach, or a combination approach, will focus organizational efforts on the reduction of waste if plant management has the proper commitment. Quality's role is to generate interest in the area of profit improvement and to coordinate the cost-reduction efforts of the various functions.

As savings are achieved and documented, future needs for more complete, detailed reporting and eventually, cost of quality reporting systems, will become more apparent to others in the organization so that the obstacles to improved systems diminish. All too often we target for utopian systems without demonstrating to others in our company the needs for these systems and subsequent savings that are achievable. My challenge to you today is to be creative on a smaller scale and let the successes in profit improvement for your company sell future systems needs.

COST OF QUALITY AND PRODUCTIVITY IMPROVEMENT

Michael P. Quinn
Assistant Vice President
Egbert F. Bhatty
Assistant Manager
Manufacturers Hanover Trust Company

Abstract

Cost of Quality is a methodology which can be applied by banks and other financial institutions to reduce their total costs while, at the same time, increasing the productivity of the operation. The methodology has been successfully applied by the Management Consulting Services department of Manufacturers Hanover Trust Company in both the domestic and international divisions.

This paper presents the theory behind the Cost of Quality program, and then describes in step-by-step fashion how the concepts of the Cost of Quality program can be applied to any clerical operation.

Introduction

The banking industry, in recent years, has become increasingly aware of the strategic importance of quality in the delivery of its products.

Heightened competition between banks has led to the use of quality as the key to differentiation of relatively standardized financial products. In addition, consumers' expectations of quality have been raised to the point where they exhibit a low tolerance for mistakes in their financial transactions. And finally, there is the historical requirement that bank transactions be processed quickly and accurately.

These three factors have led to the current emphasis on introduction of quality control as a formal discipline in the banking business.

However, many managers who have formalized the quality control process in their banks are not as yet fully aware of the impact that an effective quality control program can have on productivity as well as on operating costs.

This paper will show how a Cost of Quality program increases productivity, and reduces costs, while, at the same time, improving the quality levels of the operation.

Cost of Quality Theory

Cost of Quality theory recognizes quality as an economic consideration — it costs money to build quality into a product.

Our experience at Manufacturers Hanover Trust Company (MHT), as well as the reported experience of other financial institutions, shows that costs associated with quality control can range from 15% to 40% of a department's total operating expense.

The magnitude of these costs, and the drain that these costs represent on a bank's profits, necessitate that these costs be identified and controlled.

At MHT we have developed a Cost of Quality program that seeks to (1) identify, measure, and reduce these quality-related costs, while (2) improving quality levels, and (3) increasing the productivity of the operation.

Quality-related costs can be grouped into four categories:

(1) Prevention costs,
(2) Appraisal costs,
(3) Internal Failure costs,
(4) External Failure costs.

Prevention costs are costs incurred to reduce, eliminate, and prevent defects. Such costs typically include the costs of job training, quality report preparation, quality planning and analysis, computer/manual systems design (when geared to preclude defective output), developing and presenting quality-related seminars, and user-testing of quality control systems.

Appraisal costs are costs incurred to detect errors, and to evaluate the quality of the work done. Such costs typically include the costs for proofing, verifying, inspecting, checking, signing, balancing, and recapping.

Failure costs are costs incurred whenever a product — whether it be a simple deposit transaction, or a complex foreign exchange deal — fails to meet quality standards. Failure costs can be divided into two groups: internal failure costs and external failure costs.

Internal Failure costs are costs incurred in correcting the errors (caught at appraisal) *before* delivery of the product to the customer. In other words, internal failure costs are the costs of re-doing work a second time. Such costs typically include the labor cost of reworking items that were incorrectly processed, the cost of scrapped forms, and the cost of wasted computer and other equipment time.

External Failure costs are costs incurred in correcting errors (not caught by the appraisal process) *after* delivery of the product to the customer. Such costs typically include the cost of compensation paid, penalties, difference write-offs, interest payment on late deliveries, as well as the cost of investigations.

The costs associated with Prevention, Appraisal, Internal Failure, and External Failure can, as stated earlier, be substantial — ranging from between 15% to 40% of a department's total operating expense. However, it is the distribution of these costs that is more interesting.

Our own experience, combined with other financial industry data, shows the distribution of quality costs to be approximately as shown in Graphic 1.

Quality Category	Percent of Total Cost of Quality
Prevention costs	5%
Appraisal costs	50%
Internal Failure costs	15%
External Failure costs	30%

Graphic 1: Distribution of quality costs in financial services industry

Clearly, as the figures above show, in the financial services industry, we have all the time to do the work over, but never enough time to do it right the first time!

These figures also exemplify the typical approach to improving the quality of work — check the work, and then recheck again!

Does this increase in Appraisal costs — the expense incurred in checking and rechecking the work — lead to an increase in quality?

This relationship is explored in Graphic 2.

As more Appraisal work is done, increasing Appraisal costs (2), more defects are caught internally, increasing Internal Failure (i.e., rework) costs (3).

As more defects are caught internally (i.e., before delivery), fewer are caught externally, thus decreasing External Failure costs (4).

While the Total Cost of Quality (5) may be impacted, it often remains nearly unchanged because the decrease in External Failure costs (4) is offset by the increases in Appraisal costs (2) and Internal Failure costs (3).

Prevention costs (1) remain unaffected.

So the answer to our question — does an increase in Appraisal costs lead to an increase in quality? — is "Yes," but without any noticeable reduction in the Total Cost of Quality. All that has happened is a shifting of quality dollars from one area (External Failure) to another (Appraisal).

The Total Cost of Quality, however, is impacted by the Cost of Quality approach, which, in contrast to the typical approach, emphasizes the prevention of errors rather than their correction.

What happens when Prevention costs are increased?

This relationship is explored in Graphic 3.

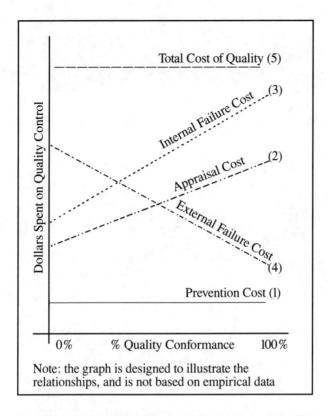

Graphic 2: Effect of increase in Appraisal costs on quality

As Prevention efforts increase (1), both Internal failures (3) and External failures (4) decrease.

Appraisal costs (2) may be reduced also — due to the overall improvement in quality.

The net result is an improvement in quality conformance, and a reduction in the Total Cost of Quality (5).

Lower quality costs translate into an increase in productivity.

This, then, is the essence of the Cost of Quality program: that dollars spent on Prevention produce a much more significant return than dollars spent on Appraisal. In other words, Prevention dollars buy more than Appraisal dollars!

COST OF QUALITY PRACTICE

At MHT we have applied the Cost of Quality methodology to clerical operations in both our domestic and international divisions.

Our results to date have been as follows:
- 37% reduction in the Total Cost of Quality,
- 10% reduction in Total Cost of Quality as % of Department Operating Expense,
- 12% increase in employee Productivity.

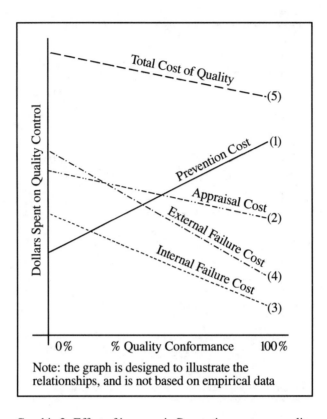

Graphic 3: Effect of increase in Prevention costs on quality

Our methodology comprises a 12-step program. The department manager/supervisors will spend 10% to 15% of their time on project activity over the investigation/installation period.

Step 1: Identify the quality problem

The analyst and the department manager together develop a PIF — a project initiation form — that defines the quality problem clearly and precisely. The PIF should also state in clear and unambiguous terms the solution required, and how the results will be quantified.

Step 2: Develop and analyze the distribution of quality costs

The distribution of quality costs for the department (Graphic 4) shows how much is being spent on each of the 4 different quality categories. The distribution of quality costs should be prepared with great care, because its analysis can yield clues to the nature of the quality problem.

The distribution of quality costs is prepared as follows:

Step 2a: <u>Draw up the department's Personnel Worksheet</u>

The analyst, in conjunction with the department manager, identifies:
(1) the different functions performed in the department — for example, data input, verification, etc.,
(2) the number of employees who perform each function, and
(3) the mid-point salary per hour for each function.

The data on employees, functions, and salaries is used in conjunction with the Activity Worksheet (see 2b below) in computing the cost of the quality-related activities of the department.

Step 2b: <u>Draw up the department's Activity Worksheet</u>

The analyst, in conjunction with the department manager and supervisors, develops
(1) the department's activity (task/operations) list,
(2) establishes how long it takes to complete each activity
(3) matches activities (for example, check customer name) with functions (verifier),
(4) identifies quality-related activities (e.g., verify amount), and
(5) assigns each quality-related activity to its appropriate quality category — Prevention, Appraisal, Internal Failure, External Failure.

Having developed data on the different activities performed in the department, the time taken to perform each activity, how much it costs per hour to perform each activity, and the quality category of each activity, the analyst is now in a position to develop the Cost of Quality Summary Worksheet.

Step 2c: <u>Develop the Cost of Quality Summary Worksheet</u>

The Cost of Quality Summary Worksheet is used to calculate the monthly quality cost for each activity in each quality category. These are then summated into the monthly Total Prevention Cost, Total Appraisal Cost, Total Internal Failure Cost, and Total External Failure Cost.

The analyst should ensure that the monthly quality cost figures reflect not just the salary costs, but should also include the fringe benefits paid by the company.

Once the total costs by quality category (Total Prevention Cost, etc.) are known, the analyst is in a position to prepare the Cost of Quality Report.

Step 2d: <u>Develop the Cost of Quality Report</u>

The total costs for each quality category (Total Prevention Cost, Total Appraisal Cost, etc.) are summated into the Total Cost of Quality. This is the amount that the department spends each month on quality-related activities.

Equipment expenses, Occupancy expenses, and Other Operating expenses are excluded from the calculation of Total Cost of Quality because (1) they constitute a small percent of the Total Cost of Quality, and also because (2) the major portion of these costs cannot be affected in the short run.

The analyst also calculates the total cost of each quality category (Total Prevention Cost, etc.) as a % of the Total Cost of Quality.

This is the distribution of quality costs (Graphic 4), and should be represented in both absolute dollar amounts, as well as percentage terms.

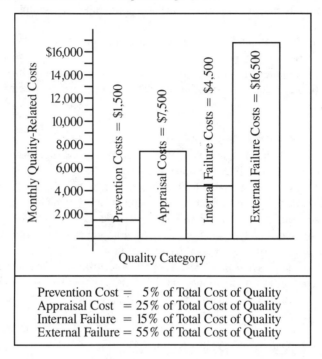

Graphic 4: Distribution of quality costs in a banking department

As can be seen from Graphic 4, the Total Cost of Quality comes to $30,000 per month. If Department Operating Expense is assumed at $100,000 per month (for purposes of illustration), then Total Cost of Quality as % of Department Operating Expense = 30%. That is, nearly one-third of the monthly operating expense of the department is spent on quality-related activities. And yet, despite this high level of expenditure, there are quality problems in the department!

What is wrong? Where is the quality control system breaking down? It is to answer questions like these that the analyst analyzes the Report and the distribution.

Step 2e: <u>Analyze the Cost of Quality Report and the distribution of quality costs</u>

What does Graphic 4 show? What conclusions can we draw from the distribution of quality costs shown?

For one, it shows that the department is heavily involved in correcting errors *after* delivery to the customer — some $16,500 being spent monthly (55% of the Total Cost of Quality) on correcting errors pointed out by customers!

Secondly, the department spends $7,500 each month on checking and verifying the work (Appraisal cost), catches and corrects some errors (Internal Failure costs), but somehow lets a large number of errors to get through to the customers (External Failure cost). Clearly, the checking and verification function is sloppy!

Thirdly, there is a tremendous emphasis on correcting errors (Internal Failure cost + External Failure cost = $21,000 per month), and almost no effort is put into preventing errors (Prevention cost = $1,500 per month).

Conclusions that an analyst can draw at this stage:
(1) the department needs to shift quality dollars away from the management of failure, and into the prevention of errors, and
(2) the checking, verification, and inspection procedures of the department need to be tightened considerably.

Step 3: Determine the different error types

The analyst, in conjunction with the manager and supervisors, draws up a list of errors being made by the department, and designs a form to record current-week errors. The same form can be used to collect data on errors from the files of immediate past weeks.

The size of the sample required can be determined by using the formula to calculate the sample size given in a statistics textbook.

The objective of this step is:
(1) to determine the total number of errors per day,
(2) to determine the different types of errors made, and
(3) to determine the frequency of the different error types.

Next, Pareto Analysis is done to identify those few errors which are responsible for generating the major portion of the Total Cost of Quality.

The analyst should also examine the large list of remaining errors, and pick out any error types which *can* lead to unexpectedly large External and Internal Failure costs.

The remaining errors can be left for the department manager to handle later.

Step 4: Determine how the errors are generated

Having determined the different error types, the next step is to determine *how* each error is generated — that is, to determine whether the error is internally generated or externally generated.

Internally generated errors are (1) errors made by the department itself — for example, input error, or verifier error, and (2) errors made by other departments within the company which lead to incorrect input by the department where the Cost of Quality study is being conducted — for example, the cable department can stamp the wrong account number on a cable that has come in for the letters of credit department.

Externally generated errors are errors made by the customer, or another bank, or another company — for example, a corporate customer may give the wrong settlement instructions in a foreign exchange deal.

It is important to classify all errors into internally generated errors and externally generated errors, because the internally generated errors are almost entirely controllable.

They can be either (1) prevented from arising altogether, or (2) reduced to the barest minimum, thus reducing the Total Cost of Quality.

Externally generated errors are more difficult to control, but measures have to be designed to effect a reduction in these also.

Step 5: Determine which function(s) are responsible for/contribute to internally generated errors

This step fixes the reponsibility for the error. It seeks to determine which function within the department (data entry, verification, etc.) is primarily responsible for making the error. It also seeks to identify all other functions within and outside the department which could contribute to the making of the error.

The function primarily responsible for making the error is, in most cases, located within the department itself — for example, a wrong credit is processed, and in most cases it is the data entry function that has made the error.

In other cases, the function primarily responsible for making the error is located in another department within the company — for example, a foreign exchange deal input and verified exactly as specified by the Trader on the deal ticket. However, it is eventually found that there was an error in the deal — the Trading function having specified the wrong receiving bank.

Today, with the increasing use of computers in clerical operations processing, it behooves the analyst to check whether the function primarily responsible for making the error is not the computer system itself. For example, the computer system installed in a letters of credit operation may not be able to uniquely identify an l/c number issued by a bank in one country from the same l/c number issued by a different bank in another country.

Examples such as the above occur, and the quality control analyst must be especially thorough in checking out the computer systems installed in the department: more and more clerical errors are the result of poor systems design and/or faulty computer programming.

At this stage of the analysis we have found it useful to draw up a matrix as shown in Graphic 5.

Frequency-rating of the function(s) responsible for making the errors allows the analyst to rank the functions by weakness — that is, which is the weakest function, which is the next weakest, and so on — and directs attention to the point(s) of greatest weakness in the department's manual and computerized workflows.

In addition to allowing the analyst to rank the functions by weakness, the matrix also allows the analyst to draw some broad and general conclusions which, in effect, is a grand strategy for reduction of errors.

For example, an analyst confronted with the situation shown in Graphic 5 would arrive at three quick conclusions:
(1) there is a general failure of both the manual and the computerized systems within the department — that is, the control mechanisms in both the manual and the computer systems have broken down,

INTERNAL INQUIRIES — Type, Frequency, and Function Responsible							
Type of Inquiry	Frequency	\multicolumn{6}{c}{Function Responsible for Error}
Charges	22 per day		DEC	VER			
Other dept errors	33 per day						OD
Wrong credit	16 per day		DEC	VER	SYS		
Duplicate payment	15 per day				SYS		
Wrong debit	14 per day	NAC	DEC	VER	SYS		
etc., etc. . . .							

Graphic 5: Matrix showing function(s) responsible for errors

(2) there is a general lack of understanding among employees of the technical aspects of their jobs, and

(3) there is a lot of inaccurate work flowing into the department from other departments within the organization.

Step 6: <u>Establish SPECIFIC reasons for errors made by the different functions</u>

This is the most delicate step in the entire methodology.

It involves sitting down with:

(1) the managers, supervisors, and, occasionally, even the workers of *other* departments,

(2) the project managers, systems analysts, and programmers in the computer department, and

(3) the manager, supervisors, *and* workers of the department itself,

and establishing the specific reasons why each function is making the kinds of errors that it does — in other words, why is each error made? what causes it?

Why is the verifier making errors? Some of the reasons that an analyst may establish as the cause of the errors are:

(1) the verifier is a part-time worker

(2) work does not come through in a smooth, even flow, but arrives in big, unexpected chunks,

(3) the computer system does not allow for "blind" verification.

Why does the data entry clerk make errors? Again, the analyst may establish the following reasons:

(1) the account number provided by the other department is wrong,

(2) the input format has not been standardized,

(3) the data entry clerk is required to pick up and also microfilm the work, and this cuts into production time.

Similarly, reasons are also established for errors made by (1) the other departments, and (2) the computer system.

Step 7: Formulate solutions for elimination and/or reduction of errors

At the end of Step 6 the analyst develops a matrix that lists:
(1) the function making the error — for example, data entry
(2) the error itself — say, wrong account number,
(3) the specific reason(s) for the error — and these may be
   (a) other department provides wrong account number,
   (b) input format not standardized,
   (c) pressure to deliver quantity,
   (d) other duties cut into production time,
   (e) low emphasis on quality,
   (f) computer system does not provide unique identifiers for 1/c numbers, etc.

In Step 7, the analyst's task is to fashion a solution for each of the specific reasons why the error was made. This, again, is done in conjunction with the relevant managers and supervisors.

The objective, of course, is to devise such solutions as will *prevent* the errors from being made.

In the case of the data entry clerk, some solutions that an analyst might formulate:

In the department itself
(1) installation of a system for prescreening the input,
(2) greater emphasis on quality, not quantity,
(3) more training in the technical aspects of the position,
(4) reorganization of workflows in the department.

In other departments
(1) stricter quality control of work going out to other departments,
(2) better understanding of the interface between the inputs and the outputs of the two departments,
(3) better training of workers.

In the computer department
(1) improved design for data entry system,
(2) system-generated rejection of incorrect input,
(3) understanding of data entry clerk's input problems.

Step 8: Establish SPECIFIC reasons for externally generated errors: as was done for internally generated errors in Step 6.

Step 9: Formulate solutions to eliminate/reduce externally generated errors: as was done for internally generated errors in Step 7.

These will range from customer education by personal contact or letters through to charging fees — for errors originating in the customer's organization — on a sliding scale.

Step 10: Develop a timetable for implementation of solutions

The timetable will list the broad range of solutions — from training employees to reorganization of workflows through modification of computer systems to educating customers — and specify start times and completion times for the implementation of the solutions.

Some of the solutions can be scheduled and completed quickly, in a matter of 1/2 weeks. Others will take time: 6/8 weeks to make major changes in a computer system. Yet others may have to be scheduled continuously: for example, the training, re-training, and updating of employees in the changing technical aspects of their jobs.

The next step is the establishment of improvement targets.

Step 11: Establish improvement targets

Given the schedule for implementation of the solutions designed to eliminate/reduce the internally generated and the externally generated errors, the analyst's next step, in conjunction with the department manager and supervisors, is to establish improvement targets. Graphic 6 shows the format we use to track targets and accomplishments.

A few things should be kept in mind during the period of implementation:
(1) externally generated errors — that is, errors made by customers — will decrease very, very slowly,
(2) internally generated errors will decline slowly at first, and then very rapidly as the cumulative effect of all the solutions designed to eliminate/reduce errors is felt,
(3) neither internally generated errors nor externally generated errors can be reduced to zero.

However, unless zero defects is aimed at, the reduction in errors will not nearly be as much as the Cost of Quality program can deliver.

Step 12: Monitor the working of the program

The analyst should monitor each and every aspect of the program during the first two weeks of implementation. Despite all the pilot testing, unexpected situations will crop up which only the analyst can handle.

Once the new work habits and new work procedures are institutionalized, the analyst should let the manager take over.

· As part of the monitoring process, the analyst should make himself available to the manager over the first quarter of implementation, directing his attention to the sequence of events that flow from the implementation of the Cost of Quality program
(1) Quality dollars are shifted from Appraisal to Prevention,
(2) As efforts to prevent errors take effect, quality improves — that is, errors decline,
(3) As both internally generated errors and externally generated errors decline
   (a) Internal Failure Costs decline,
   (b) External Failure Costs decline, and
   (c) because of the improvement in quality, even Appraisal Costs decline.

	Errors Per Day					
	Present	TARGETS				
	Dec 1983	Mar 1984	June 1984	Sept 1984	Dec 1984	Mar 1985
Internally generated errors						
Wrong bank account number	25	20	18	12	etc.	etc.
Wrong charges	21	21	19	16		
Wrong credit	17	15	10	7		
Duplicate payment	15	14	12	9		
etc., etc.						
Total	78	70	59	44	20	12
% reduction		10 %	24 %	44 %	74 %	
Externally generated errors						
Claim made in error	33	32	etc.	etc.		
Wrong instructions	28	27				
Debit wrong account	25	24				
Credit wrong account	16	15				
Cancel payment	11	10				
etc., etc.						
Total	113	108	etc.	etc.	68	
% reduction		4 %			40 %	

Graphic 6: Cost of Quality targets

(4) As a result
   (a) the number of inspectors, verifiers, and checkers can be reduced, and
   (b) the number of investigators, adjusters, compensation clerks can be reduced.
(5) This leads to a reduction in the Total Cost of Quality, while Productivity increases because the same amount of work is being done with fewer people.
(6) When the full effect of the prevention efforts is felt, a second stage of productivity increases takes place: because of better training of workers in the department, smoother manual and computerized workflows, and cleaner input from other departments, more volume can be handled by the department in the same amount of time. Or, in other words, fewer people are required to process the same volume.

    The analyst-manager meetings over the first quarter of implementation are important because they help the manager assimilate and internalize the mechanism of the Cost of Quality program how a decrease in quality costs leads to an increase in quality as well as a two-stage increase in productivity.

The analyst should, at the end of the first quarter of implementation, assist the manager in
(1) drawing up the new distribution of quality costs. It will show a moderate increase in Prevention Costs, dramatic declines in both Internal Failure and External Failure Costs, and some decline in Appraisal Costs,
(2) calculating the new, lower Total Cost of Quality as a % of the Department Operating Expense,
(3) graphing the decline in both the internally generated and externally generated errors,
(4) establishing the increase in Productivity, and
(5) quantifying the direct savings achieved as a result of
    (a) reduction in quality costs,
    (b) reduction in the workforce, and
    (c) reduction in compensation payments.

Conclusion

This paper has outlined a simple, easy-to-learn and easy-to-apply methodology called Cost of Quality.

It is a methodology which quality control analysts, internal as well as external management consultants, and managers of clerical operations can use — and effect a reduction in the total cost of quality, while at the same time increasing the productivity of their operation.

Although the Cost of Quality methodology has been tested and successfully applied in the domestic and international operations of Manufacturers Hanover Trust Company, the methodology is essentially generic in nature, and can be easily adapted to achieve similar results in any paperwork operation — in manufacturing industries as well as financial and other service organizations.

Bibliography

Duncan, Acheson J., Quality Control and Industrial Statistics, Irwin, 1974

Juran, Joseph M., Quality Control Handbook, McGraw-Hill, 1974

Crosby, Philip, Quality Is Free, McGraw-Hill, 1979

White, Bruce, "Quality Fitness Test," *Quality,* March 1981

Aubrey II, Charles A., and Eldridge, Lawrence A., "Banking on High Quality," *Quality Progress,* December 1981

Orsini, Joyce, "Three Tips for Quality Control," *Bank Administration,* April 1982

Eldridge, Lawrence A., and Aubrey II, Charles A., "Stressing Quality — The Path to Productivity," *Bank Administration,* June 1983

COSTS RELATED TO QUALITY PRESENT SITUATION IN THE FEDERAL REPUBLIC OF GERMANY (FRG) AND FUTURE ASPECTS

Rüdiger K. Vocht

Since more than 20 years costs related to quality — the so called quality costs — were collected and evaluated in the German Industry and used to direct the quality assurance activities. The suitability of quality costs as a valuable tool for quality and executive management to steer and control the efficiency of quality assurance activities and the productivity of the company is always unquestionable. Therefore there is no imperative reason to undertake fundamental changes to these. The paper deals with spread of application, practice of usage and future aspects of quality costs.

Present Situation

Starting in the sixties quality costs in German industry were systematically collected and calculated on the base of the original US-definition of the "classic" 3 cost groups prevention, appraisal and failure (1,2).

Experiences and knowledge of practice out of industry lead in 1971 to a brochure of the German Society for quality — Deutsche Gesellschaft für Qualität, DGQ — "Qualitätskosten, Rahmenempfehlungen zu ihrer Definition, Erfassung, Beurteilung" (3).

Recommendations from that time are still more or less valid in German industry. They are still part of the education program of DGQ for quality engineers and management personnel.

The usage of (comprehensive) quality cost as a supplementary management tool has been implemented and widely spread in FRG industry mainly in the electrical, machine and automotive business. This also explains the high number of German publications during the last 15 years in these areas to subject "quality cost" as well as "quality and costs."

Investigations show that 10 to 35% of all companies have implemented quality cost as an instrument for permanent control (4). In average the application rate in above mentioned industries is higher than in other industries. The same also applies for larger enterprises (from about 500 employees upwards). In other industries and branches of trade, for instance services, quality costs are rarely implemented, if at all.

The reason for the relatively sparse spread of application may be:
- insufficient familiarity and unawareness of value/advantage of quality costs by executive management
- quality costs in the defined frame require a quality assurance-system with partly organisational units and cost centres

- limited steerability of company expenses because of the relatively small quality cost-share of total company expenses
- uncertainty caused by the large range of interpretation of quality costs due to boundaries of their definition and accounting (5)
- existing cost accounting systems are not compatible and detailed enough to collect all quality cost items on a continuous basis with reasonable effort
- important quality cost items such as rework, scrap and appraisal costs of production are anyhow part of the company's cost accounting and controlling
- due to high failure-cost-potential mass and series production are more suitable than single piece or small series production.

Practice of Application

According to literature and national recommendations it is commonly known that quality costs are defined in up to 32 to 37 different cost items which are allocated to:
- failure prevention costs: 11 ... 12 cost items
- appraisal costs   : 11   cost items
- failure cost   : 11 ... 15 cost items

Both, the resolution of partitioning and the definition of individual contents of quality cost are important in order to comprehend kind and value of quality related costs, their sources and basic approaches for their reduction/avoidance. When using total quality costs as additional steering tool for optimization of quality assurance activities, which shall result in enhancement of returns for the company, then one must consider that they are composed of two main groups which have a different influenceability:
- directly occurring and attributable costs, and
- indirect costs, not directly attributable to quality assurance, for instance price reduction due to lower grade quality or warranty cost and product liability expenses.

Only directly attributable quality costs are suitable for immediate steering actions, where indirect quality costs are in most cases of value for medium range and general improvement actions.

Cost Categories

In industry usually only 12 to 16 cost categories are applied for permanent practical quality cost accounting (collection, calculation, analysis, review). These are dependent on the structure and technology of enterprises.

The selection of relevant quality cost categories for a company often depends on their individual portion in relation to the total quality costs as well as to other company expenses (or other reference figures).

Also, if cost categories are reduced or partly integrated, they still result in a duplication of cost elements if they are collected in the functional departments of the company (see figure 1). It is usual that only some selected, cost-intensive elements are collected and evaluated in the corresponding functions.

Figure 1 also gives an idea for quality cost elements which would correspond to:
Industry : Column 1 ... 6
Services : Column 1, 2 + 6
Trade business : Column 1, 3, (5), 6
Trade and industry : Column 4, 5, (6)

In larger enterprises usually cost elements are further structured into:
- processes, production lines
- products
- other cost sources (for instance vendors)

A further subdivision of quality cost categories (e.g. in process inspection and test, final inspection and test) into quality cost elements (in process inspection and test per production line, final inspection and test per product) is generally recommended when actually occurring costs are high in relation to the sum of other quality cost categories or to the corresponding shop costs.

Such subdivided quality cost elements only make sense for larger enterprises with adequate cost accounting systems. Otherwise there is a mismatch between their importance (value) and the total company expenditures.

In practice therefore is a tendency to limit the number of cost categories rather than to extend, specially in those cases where the quality cost accounting is to be based on collected values out of companies cost accounting and only related to segregated cost of quality assurance activities and/or quality assurance cost centres.

Under the consideration of a balances relation between usefulness and expenses for small and medium sized enterprises a reasonable simplification leads to a practical quality cost accounting scheme with a few cost items, for example:
Failure prevention: 1 cost items (quality control department)
Appraisal : 2 cost items (inspection and test, test equipment)
Failure : 3 cost items (rework, scrap, warranty)

Techniques

For using quality costs to control quality assurance activities optimally and for improvement of return of enterprises some techniques are implemented as standard practice today. Generally these are:
- limitation of quality cost categories to be recorded in respect to a balanced advantage/expenses relation and adaptation of quality cost contents to changing needs
- product and/or process related recording, monthly (or quarterly) accounting and reporting through controller department
- periodic reviews and analyses by quality management mainly in view of:
  • expenditure split up by sources, in order to identify short term improvement measures
  • comparison of relative quality cost against previous figures (in respect to various reference expenses and/or to quality cost groups/categories)
  • analyses of deviations compared to planned quality costs and flexible budget respectively
- short and medium term quality cost planning which reflects results to specific measures directed to quality and quality system improvement activities.

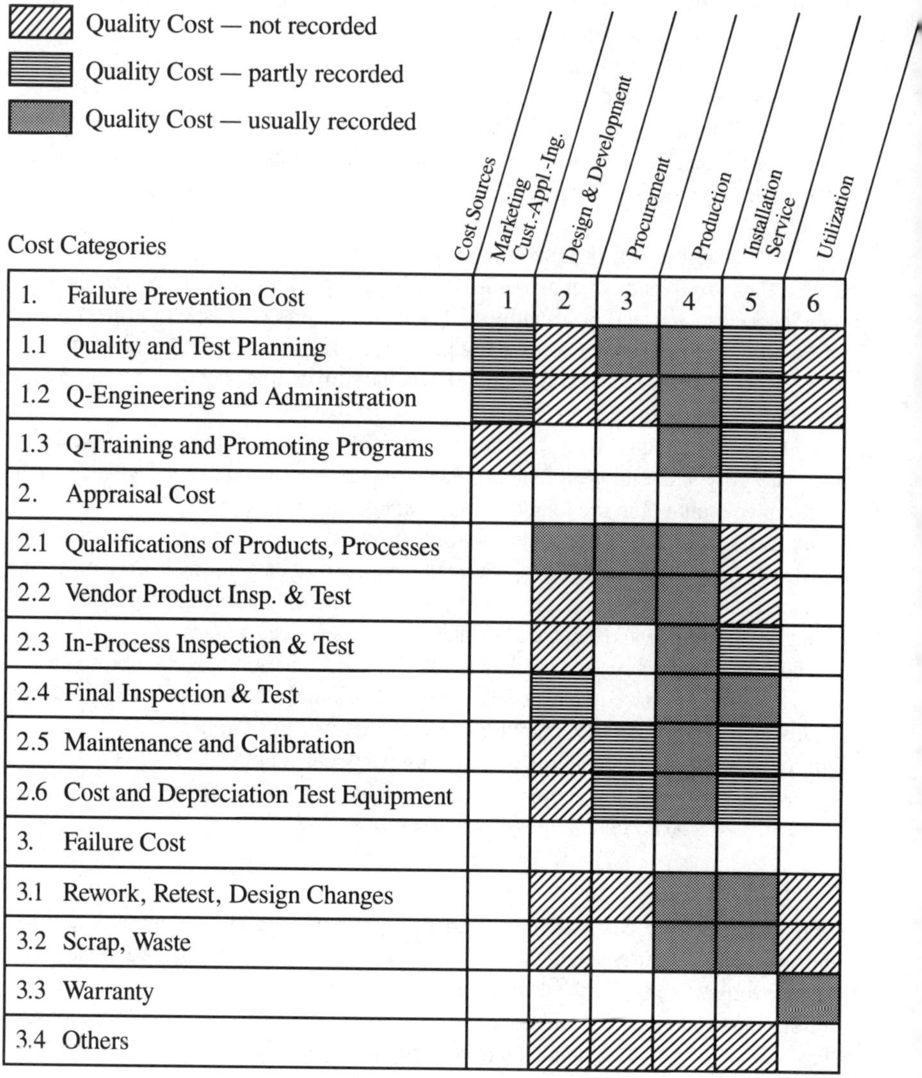

Fig. 1: Quality Cost Categories and Sources

Legend:
- ▨ Quality Cost — not recorded
- ▤ Quality Cost — partly recorded
- ▦ Quality Cost — usually recorded

Cost Categories	Cost Sources					
	1 Marketing / CustAppl.-Ing.	2 Design & Development	3 Procurement	4 Production	5 Installation / Service	6 Utilization
1. Failure Prevention Cost						
1.1 Quality and Test Planning	▤	▨	▦	▦	▤	▨
1.2 Q-Engineering and Administration	▤	▨	▨	▦	▤	▨
1.3 Q-Training and Promoting Programs	▨			▦		
2. Appraisal Cost						
2.1 Qualifications of Products, Processes		▦	▦	▦	▨	
2.2 Vendor Product Insp. & Test		▨	▦	▦	▨	
2.3 In-Process Inspection & Test		▨	▦	▦	▦	
2.4 Final Inspection & Test		▤	▦	▦	▦	
2.5 Maintenance and Calibration		▨	▤	▦	▦	
2.6 Cost and Depreciation Test Equipment		▨	▤	▤	▤	
3. Failure Cost						
3.1 Rework, Retest, Design Changes		▨	▨	▦	▦	▨
3.2 Scrap, Waste		▨	▨	▦	▦	▨
3.3 Warranty						▤
3.4 Others		▨	▨	▨	▨	

Quality Cost Budget

Planning of quality cost is not an act on its own, however, it must always be the result of strategy planning for quality and productivity improvement.

It is not the aim of quality cost planning to set up quality cost as a budget value against which actual figures could be measured in order to obtain positive variances. Actual quality costs and their development rather serve through comprehensive analyses possibilities for improvement. Measures for quality improvement and rationalization which result from quality cost analysis in respect of expenditure and deviation then must lead to quality cost changes which are to be reflected in planning of the quality cost budget.

Therefore in reality quality cost planning is planning of measures of technical, technical-organizational and/or personnel nature which result in changing of quality cost.

Useful and beneficial quality cost planning demands the following steps:
1. Analysis of cost-intensive expenditures in quality assurance/control, in processes, methods, organizational procedures, etc., in order to determine the causes and potential corrective measures
2. Establishment of measures and actions which are realizable (action program)
3. Determination of their cost effects to quality cost and to individual cost categories/ elements respectively.
4. Incorporation of agreed measures and their resulting cost effects into the company's budget (for instance shop cost standards, rework and scrap charges)

Only such proceedings, which take into consideration all potential and realizable improvements as well as intensified quality control efforts for new technologies and/or procedures and which incorporate all cost effects in the company's budget and all corresponding calculations, can be called real "quality cost planning." Otherwise "quality cost planning" is only a quality cost forecast with an ineffective passive prognosis of a quality cost development.

Real quality cost planning does not allow hiding displacement of cost elements. Improvements or worsenings through taken measures at cost sources should become transparent which are for instance material consumption, personnel employment, qualification of personal, capital investment etc. (see figure 2). Quality cost planning which is valuable for the company must always consider these aspects.

Therefore in practice the following topics are part of strategic quality planning and require elaboration:
- analysis of situation and evaluation/consideration of company's business plan
- set-up of medium term improvements in respect to problem areas (technically, economically)
- listing of measures and actions
- quality cost budget with corresponding changes to cost sources

Experience in SEL

In Standard Elektrik Lorenz AG, Stuttgart, a German company of ITT with about 1.665 Mio $ sales which I represent, quality costs are used uniformly in all divisions

Kind of changes \ Nature of changes	Planned Quality Costs	Changes + (−) in Personnel Cost (Compensat.)	Personnel Employment (Manpower)	Material (Costs)	Burden, Charges etc. (Costs)	T. Equipment Depreciations (Costs)
Quality Cost current year (per year end)		▓▓▓				
+ Effect of Content − and for recording changes						
+ Effect of Improvement − measures — current year						
+ Effect of Volume − (e.g. quality of shop output)						
+ Effect of structure changes − (change of production)						
+ Effect of changed − charging standards			▓▓▓			
+ Effect of improvement − measures — plan year						
= Quality Cost Budget (per year end)						

Fig. 2: Nature and kind of cost changes in the Quality Cost Budget

by products and processes as a management tool since 15 years. Over the years no significant changes to quality cost definitions did take place. The tendency in defining quality cost categories/items is leading towards simplification rather than to an extension/sophistication.

The degree of attention which quality cost did cause was extraordinary high at the executive management level in the years after implementing quality cost reporting. The reason is very simple. In the first years cost improvements and/or cost avoidances of millions of DM could be achieved by strong and effective measures in quality assurance/control and other affected cost centres supported by proper quality cost accounting and comprehensive analyses (6, 7). The quality cost in percent of sales could be reduced by 25% (6.5 to 5.0% of sales, see figure 3).

Later on quality cost came down to a stabilized value (4, 85% of sales), although all techniques of planning, recording, analysis, search for corrective and improvement measures as well as permanent review and control of quality costs were strongly applied. This lead to reduced attention at the top management level in comparison to the initial years.

Today quality costs are permanently used as before by the quality management as an indispensable and valuable steering and promoting tool for quality. However, they are not ranked as high as they used to be.

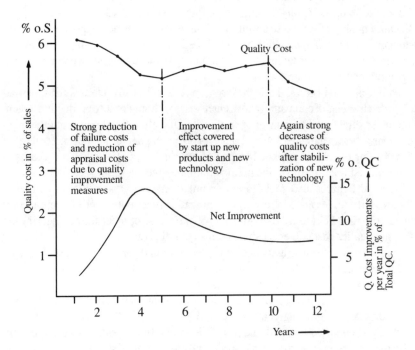

Fig. 3: Quality Cost and Quality Cost Improvement

Trends and Tendencies

The capability of quality cost for short term (failure costs) and medium term (prevention and appraisal costs) steering directions on quality related results in industry is recognized by users since years and by potential interested parties. Over that tremendous advantages were again and again proven and demonstrated by many examples in public presentations. Therefore quality costs in their "classical" structure will further be maintained as a valuable management tool without fundamental changes.

What is the future trend?

A more extensive spread of use of quality cost and medium size companies as well as in other parts of economy will depend either on implementation of quality assurance systems and related quality organizations or the profit and loss responsible management wants to know the quality costs.

Otherwise one would always be satisfied with recording of parts of quality costs which mostly are part of the company's cost accounting and controlling system. Also structural changes in industry as a consequence of new technologies and techniques will hardly effect the "classical" quality cost structure and contents. Perhaps there may be a shift in portions and relative amounts of quality cost categories.

If somebody reports that quality cost accounting is not implemented as a formal procedure, that may not mean that parts of quality costs like rework, scrap, test and retest are not used in the company to control and improve productivity.

Because quality cost require a well functioning and independent quality assurance system their spread of application has developed. In this respect quality cost must be considered as a second order instrument for quality management when compared with other more technical control tools.

This also is valid in particular in the large area of services in trade as well as internal service departments of companies and in enterprises with single piece production, where in the latter direct quality control operations are mostly a small portion of the main operations. The expenditures for segregation of operations and/or for the exact collection and recording of costs of attributable quality assurance activities are in most cases in an unbalanced relationship to the predicted improvements.

Similar difficulties also apply to development and design activities of computer software. Difficulties arise in respect to segregation of activities and collection costs related to quality. For the area of computer software development and engineering the following quality cost contents can be defined, for example:

Appraisal costs are those associated with detection and correction of defects during development and such related to tests as are required for user acceptance. They also include the cost of reviews and their preparation.

Failure costs are costs for repairs, after user acceptance and shall include all costs occurred under warranty agreements.

Failure prevention costs include the cost of training and the apportioned cost of tooling, where these costs have been occurred specifically for the purpose of improving quality. The cost of vendor service could be included under this category.

Also the intention to define quality costs mainly customer oriented does not lead to a practical and useful approach for the producer. In most cases there is only a little chance to collect deficiencies, late consequences or product life cycle behavior etc. in order to have a timely, proper and quality cost related feedback.

Therefore every user and industrial customer should define and account his own quality cost as a measure for reasonable product assurance and maintenance in order to achieve an optimum in planning and appraisal efforts versus avoidable failure cost and consequential effects. The aspect of economic optimization in the past obviously was taken into account for the definition and structure of quality cost. As long as present quality cost and their elements are indicators for the quality capability of products, processes, performance, design and can be used as an economic control means they fulfill their purpose.

Also a new proposal of the DGQ containing updated quality cost definitions, recommendations for calculation and evaluation does not deviate from this basic principle, even considering actual experiences and knowledge of industry.

For future quality costs discussions the following questions remain open:
- Can the present quality cost accounting system be replaced by a simplified system in the future when considering that quality assurance remain an interdisciplinary function with varying extension?
- Can quality cost become a system element of a standardized quality assurance system?
- Is it desirable and reasonable to unify or even standardize quality cost internationally?
- Does a threshold value for quality cost (e.g. in % of sales) exist at which they become totally uninteresting for company management? Is it therefore necessary to artificially increase this cost item by addition of new cost elements, to be defined?
- At which grade of automatization do today's inspection and test cost of manufacturing disappear as a main item of appraisal costs?
- Will there ever be a chance to come to trade/product specific quality costs standards without subsequent discussions about contents and their details, accuracy of recording and accounting, etc.?

In spite of the (above) mentioned questions we see quality costs today and in the near future — even considering the inherent weaknesses — as a valuable control tool for the quality control department and to a smaller degree for the executive management. It is suitable to reaffirm implemented and planned quality assurance measures and to improve the company's return through reduction of actual failure and other cost potentials and through influence on the product design from the quality point of view.

Literature

(1) Masser, W. J.: Quality Control Engineering, Industrial Quality Control, 12 (1956)

(2) Masser, W. J.: The Quality Manager and Quality Costs, Industrial Quality Control, 14 (1957)

(3) DGQ (Hrsg.) : Qualitätskosten, DGQ 17, 4. Auflage, Berlin, Köln, Benth Verlag 1978

(4) Hahner, A. : Qualitätskostenrechnung als Informations system zur Qualitätslenkung, Forschungsbericht für die Praxis, Produktionstechnik Berlin, 23 (1981)

(5) Vocht, R. K. : Quality Costs — Value and Limitation of their application, Proceedings of the 27th EOQC-Conference, 1983

(6) Vocht, R. K. : Handbuch Qualitätsförderung, Verlag Moderne Industrie, 1974

(7) Vocht, R. K. : Returns Planning by Quality Costs, Proceedings of the 23rd EOQC Conference, 1979

HOW TO INSTITUTIONALIZE QUALITY COST IMPROVEMENT

J. M. Groocock

Although quality costs have been a principal interest of the quality profession in the USA for over 20 years, the level of successful use of quality cost improvement systems has been disappointingly small. One reason is that the approach has usually been one of limited application in special areas. A more ambitious approach is to "institutionalize" a quality cost improvement system throughout all of the divisions of a company. The paper shows that this is a practical approach, and defines the key steps by describing how it has been done in two major multifunctional companies.

The Americans love to coin new business jargon. Perhaps it's because they find it easier than sorting out the correct English words. Or perhaps it's because fame, and even immortality, can be the reward of the creator of a particularly colorful piece of jargon.

Two current pieces of American business jargon are "high-leverage" and "institutionalize." A high-leverage project or activity is one which, when compared with other projects or activities, requires relatively little effort to produce a relatively large effect. It is analogous to the old 80-20 or Pareto rule. To institutionalize some activity is to repeat it over and over and over again in a standardized way so that everyone knows how to do his piece of it, the activity becomes an ingrained habit, and (hopefully) no special effort is required any longer to make it happen.

At the present time American business is much involved on the one hand in searching for high-leverage situations and on the other in trying to institutionalize good activities. The logical impossibility of institutionalizing high-leverage activities is causing great concern.

In the USA quality cost improvement has been a major preoccupation of the quality professionals since Feigenbaum discussed it in his classical book, "Total Quality Control," published in 1961. In 1984 the ASQC issued a booklet compiling 91 papers on quality costs which had been published in the ASQC's various journals between 1970 and 1982. Interest did not end in 1982 — since then the ASQC has published several more papers on quality costs. Unfortunately, few if any of these papers describe successful systems in action or give actual statistics of real achievements. The April 1983 issue of "Quality Progress," of which 35,000 copies were mailed, contained a one-page survey designed to gather information about the implementation of quality cost systems. There were only 40 replies from the US and Canada, and five from overseas, giving further evidence of a lack of real achievement.

One reason for this lack of success in implementation of quality cost systems is because a "high-leverage" approach has usually been used. The quality people have identified some special situation in which quality cost improvement might be particularly advantageous. The situation might concern a particular division of a multi-division company, or a certain product line or plant. With backing from senior management, the quality people persuade the accountants, as a special exercise, to measure the quality costs over a prescribed period of time. Presuming the special situation has been selected with good judgment, the quality cost to sales ratio will be high and some good quality cost improvement projects will be identified and addressed energetically. However, after some months — or at most a year — the special situation has ended, the project teams have completed their main actions, prime members send deputies to meetings which become infrequent, and the accountants stop measuring the quality costs. The hope that other divisions or plans will pick up the system by example is not realized, and soon the whole exercise is over.

In principle, a much more effective approach is to "institutionalize" the quality cost improvement system. This however is a task of daunting difficulty. It is hard enough to get quality cost improvement into one or more model divisions for a temporary period, but to install it as a permanent activity in every division of a major company seems to verge on the impossible. However, having succeeded in institutionalizing quality cost improvement in a major company not once but twice, I affirm that it really is possible, and I think I can describe the way of doing it.

The first example of an institutionalized quality cost improvement system concerns Standard Telephones and Cables, which at that time was a subsidiary of the American multi-national corporation, ITT. STC is one of the UK's principal designers and manufacturers of telecommunications and other electronic equipment. Quality cost reporting on a monthly basis commenced in all of STC's divisions in 1967. A sister company, Standard Electric Lorenz, in Germany, started at the same time, and over the next year or two the system was introduced into all of the companies of ITT Europe, which was then a $10 billion a year company manufacturing in every country of Western Europe. For the next ten years some hundreds of ITT Europe divisions reported quality costs every month. In the later years these costs exceeded $400 million and quality cost improvement projects saved $30 million to $40 a year.

The second example concerns TRW, another multi-national company, with sales of about $6 billion per annum. Quality cost reporting was started in all of TRW's 80 divisions in 1981, totaling about $400 million a year. Quality cost improvements of $30 to $40 million were achieved in 1982 and 1983, and are forecast for 1984.

How was this done? There were only four essential requirements:
- A brave champion;
- A special event;
- Some simple know-how; and
- A reasonable level of top-management support.

The book, "In Search of Excellence," which headed the USA's non-fiction best seller list for many months in 1983 and 1984, describes the characteristics of a selected group of excellent companies. For successful innovation — introduction of new products, start-up of new processes, etc. — an essential ingredient was a "champion." The book states, "All the activity and apparent confusion we were observing resolves around fired-up champions. . .was possessed by his idea, a fanatic on the subject." (I am not sure whether the use of champion in this context is a natural extension of its normal meaning of, "militant advocate or defender," or whether this is another example of colorful American business jargon.) As for these other innovations, the drive of a champion seems essential for the institutionalization of quality cost improvement.

I will now briefly describe how it happened in STC and TRW. I was appointed Company Quality Manager of STC in October, 1966. The start of the UK's first Quality and Reliability Year was imminent and STC had no plans for it. Apart from being a missed opportunity, lack of any action was likely to cause the company embarrassment with its major customers. That was the special event. My new boss said to me, "What do you suggest we do for Q and R Year?"

I replied, "Introduce quality cost reporting and start Crosby's 14 Step Quality Improvement Program."

(I knew my answer would cause me great trouble and my hands shook under the edge of the desk as I spoke, but what the heck.) Crosby had just given me a copy of a first draft of his now-famous program, and the recently constituted ITT Europe Quality Council had defined a quality cost reporting procedure, but no one had actually used it. This procedure provided much of the needed know-how.

My boss agreed to my suggestion and said he would get approval from the managing director — the reasonable level of top-management support. Later it was agreed that I would present the ideas at the forthcoming company management conference. I was quite young and not a practiced public speaker, and I was scared, but nothing terrible happened.

Then I sought out the right accountant. He was the man responsible for accounting systems throughout the company, and reported to the Comptroller. He showed me how to write a Comptroller's Procedure and together we did so. He suggested we should make quality costs a quarterly report instead of a monthly report. That seemed rational. Quality costs don't change all that rapidly, and he said that it would save the divisional comptrollers, who were already greatly overworked, a substantial amount of effort. I had never asked a senior accountant for anything before in my life and was surprised that he was taking my request so seriously. However, I felt intuitively that as most of the financial system was on monthly reporting, a quarterly measure would always be a low-prestige bother (it would never become institutionalized, in the term I heard for the first time many years later). So, I kept my nerve, dug in my heels and got away with monthly reporting.

I used the Q & R Year and the annual accounting cycle to create an aura of urgency. We circulated the draft to the division quality managers and controllers and quickly incorporated their comments. Most of them felt it wasn't serious and that it wouldn't really happen.

Early in December the Comptroller's procedure was signed-off. People had been surprised to find that they could not think of any good reason for disapproving it. No one had stood up and shouted "Nonsense!" during my presentation at the conference. The division people had had their chance to table rational objections, and these had been taken account of. My colleague in accounting had devoted so much personal effort that he was getting to be a champion himself.

Suddenly, it was no longer an academic exercise, and amazingly at the end of February most of the required reports for January, 1967 came in (January is a quiet month for accounting). The delinquents were a minority and were exposed to the pressures that minorities suffer, and soon came into line.

I nursed the fledging system for three years, and after I went to Brussels as Director Quality ITT Europe in 1969, I championed the system throughout Europe for the next 11 years.

Towards the end of 1980 I joined TRW in the newly-established position of Vice President, Quality. TRW has a strong tradition of decentralization, and the decision to give a new level of centralized leadership in the quality area was one expression of a major policy decision to make quality one of "four priority projects" for the company in the decade of the 80s. For a foreigner to emigrate to the USA is a complicated business of long duration, so I had to spend my first few months working with TRW's European operations, and I did not reach the company headquarters in Cleveland until March, 1981.

The already-existing company Quality Steering Committee had started an investigation of quality costs, and there was a certain level of interest. The combination of this, the four priority projects, and my own appointment created the special event, and once again I became the champion. Contact had already been established with the appropriate accountant. His name is Bill Hamilton and I immediately felt comfortable with him, and his successor Tom Connell. I suggested we should draft a "Standard Practice Instruction" for the accountants, and in a few days this was done.

I tabled the draft at the Quality Steering Committee and immediately there were strong objections. The representatives of the sectors did not want a centralized procedure — every division should be free to have its own system. Some of them were already doing quality cost reporting in their own way. Why should they change? For a company whose product range extends from spacecraft (in 1983 TRW's Pioneer 10 passed Neptune and left the solar system still transmitting data 11 years after its launch) to car engine valves and electronic resistors, this seemed very reasonable. However, I was no longer a young man. I expected to retire in 10 or 15 years, and I felt it would take at least that long for 80 divisions each to develop and implement its own system. So, through many battles I persisted with the centralized system.

We circulated the draft to the divisional quality and accounting people, and instantaneously incorporated their comments (11 drafts in two months). As with ITT, I think that most of them felt it wasn't serious and that it wouldn't really happen.

Then I had a stroke of luck. Bill Hamilton told me he would be conducting a series of training programs for divisional controllers on inflation-adjusted accounting. Would I like to tack-on a session on quality cost reporting? I jumped at the chance. In the next couple of months I taught some three hundred accountants the principles of quality cost reporting in sessions in Cleveland, Los Angeles, Chicago, London and Frankfurt. I exploited the "priority-project" theme without shame, and once again built up an aura of urgency. If we could install the system in the fourth quarter of 1981, then 1981 would be the shake-down year and we would get a good lead-in to 1982. If we delayed we would lose a whole year.

By July the Standard Practice Instruction was ready for approval. I signed it and asked the Chief Financial Officer to do so also. Once again, by now there was no rational reason no to. All of the divisional comments had been taken account of. The accountants had been trained and no one had said the system wasn't practical. Bill Hamilton had become a champion. The Chief Financial Officer felt the President's opinion should be determined. Of course, he gave full support. If he had felt the approach was wrong he would have reined me in long before. In fact, within TRW in 1981 far more top management support was available than the minimum provided by STC in 1966.

The instruction went out. To many people's surprise it was really happening, and in the middle of November, 1981 the quality cost results for October came in from 68 TRW divisions.

Since then, my Director Quality, Tom Hughes and I have nursed the system and together with the Quality Steering Committee have progressively strengthened it. It is essential to keep doing this. In 1981 the reporting started. In 1982 every division had to forecast its quality costs for the year and total its quality cost improvement projects. Then in 1983 we refined the forecast to include an actual measure of 1982 results. Also in 1983 every Group Vice President presented the quality costs of each of his divisions to the company's Large-Management-Meetings. In 1984 most divisions will formally identify and describe their quality cost improvement projects as part of the forecast package, and the Internal Auditors will audit the system.

Figure 1 shows the quality cost report for the company for 1982. It is the sum of the year-end reports for 80 divisions. Figure 2 shows the 1983 forecast, which includes the 1982 forecast and actual.

It was stated before that one of the four necessary conditions for institutionalization of quality cost improvement is simple know-how. There is not space here to cover this requirement fully but some key points are as follows:

- Monthly reporting by controllers against a formal procedure is essential (computer reporting is very desirable);
- The procedure must be simple with a small number of categories each of which contains a big piece of money (TRW's nine categories each contain more than $14 million);
- The reporting system should concentrate on one area. (Manufacturing is the simplest. All of TRW's categories are concerned with either manufacturing, or things just before and after manufacturing);
- Quality cost reporting works best in operations with substantial amounts of inspection and test;
- The emphasis should be on appraisal and failure costs (prevention costs are small and tricky);
- The best two bases for comparing quality costs are sales and manufacturing added cost;
- Requirements to forecast or budget the coming year and measure the previous year are essential;
- Specific quality cost improvement projects are the key (in 1983 TRW had 400).

MR-5 Quality Cost Report

(1) Report required from all Operations
(2) Machine transmit
(3) Do not mail hard copy

EXHIBIT 5.1

Division or Subsidiary	TRW Company
Date	3-30-83
Prepared by	Groocock
Phone	216-383-2992

(In Thousands)

Description	Current Month			Year To Date December, 1982		
	Quality Costs	As A Percent Of Sales	MAC*	Quality Costs	As A Percent Of Sales	MAC*
1. Prevention Costs						
1.1 Test & inspection planning costs 096-201				27,319		
1.2 Quality Department prevention costs 096-202				25,825		
1.3 Qualification test costs 096-203				19,580		
1.4 Other prevention costs 096-209						
Total Prevention Costs				72,723	1.39	3.55
2. Appraisal Costs						
2.1 Incoming & source inspection & test costs 096-211				20,954		
2.2 In-process & final inspection & test costs 096-212				129,396		
2.3 Test and inspection equipment costs 096-213				30,791		
2.4 Other appraisal costs 096-219						
Total Appraisal Costs				181,141	3.47	8.83
3. Failure Costs						
3.1 Rework costs 096-221				37,452		
3.2 Scrap costs — 3.2.1 Gross 096-222				95,478		
3.2.2 Salvage Value 096-223				(9,521)		
3.3 External failure costs 096-224				14,054		
3.4 Other failure costs 096-229						
Total Failure Costs				137,463	2.63	6.70
Total Quality Costs				391,327	7.49	19.1
Forecast Quality Costs 096-231						

Memo Data	Current Month		Year To Date	
	Forecast	Actual	Forecast	Actual
Net Sales *Manufacturing Added Cost (Labor and Overhead) Quality manpower (average equivalent full time)			096-241 096-243	(1) 096-242 096-244

Figure 1

ASQC Quality Costs Committee

EXHIBIT 5.2

QCF Quality Cost Forecast

Division	TRW Company
Date	3-29-83
Prepared by	Groocock
Phone	216-383-2992

In Thousands of Dollars

	Quality Costs			Quality Manpower		
Quality Manpower — Average Equivalent Full-Time	1982 Fcst.	1982 Actl.	1983 Fcst.	1982 Fcst.	1982 Actl.	1982 Fcst.
Initial Year Quality Cost and Quality Manpower	399	400	391	8056	8117	7577
Effect of production volume changes	(2.62)	(13.3)	4.79	41	(318)	176
Effect of compensation rate changes	18.1	15.5	9.98	—	—	—
Effect of other changes (define below) (Note 1)	5.90	12.8	49.4	34	(9)	265
Expected Quality Cost and Quality Manpower Without Quality Cost Improvement	420	415	456	8131	7790	8018
Effect of carry-over quality improvement projects reducing quality manpower	(2.57)	(2.97)	(1.55)	(58)	(55)	(37)
Effect of carry-over quality improvement projects not involving quality manpower	(2.36)	(3.54)	(4.64)	—	—	—
Effect of new quality improvement projects reducing quality manpower	(4.69)	(4.17)	(9.79)	(159)	(158)	(356)
Effect of new quality improvement projects not involving quality manpower	(12.0)	(12.7)	(13.0)	—	—	—
Planned Cost and Manpower Improvements	(21.6)	(23.4)	(29.0)	(217)	(213)	(393)
Percentage of expected (without improvement)	5.14	5.63	6.36	2.67	2.73	4.90
Final Year Quality Cost and Quality Manpower	399	391	427	7914	7577	7625

Memo Data **(In Millions of Dollars)**				Quality Cost As % Of		
	1982 Fcst.	1982 Actl.	1983 Fcst.	1982 Fcst.	1982 Actl.	1983 Fcst.
Net Sales	5722	5224	5749	6.96	7.49	7.42
Manufacturing Added Cost (Note 2)	2224	2051	2265	17.9	19.1	18.8

Commentary

1. 1983 Forecast adds $34 million of DSG's software-development, failure costs, which were not forecast or reported in 1982.

2. Includes my estimate of MAC for ESG.

Figure 2

HOW TO SUCCEED IN ACTIONS TO REDUCE QUALITY COSTS. THEORY AND REALITY.

K. Ullberg
M. Karnebjer
Ericsson Radio Systems AB
Quality Assurance Department
Mölndal, Sweden

The literature concerning quality costs often deals with how to define, collect and report these costs. It is seldom described how to proceed to reduce the costs. This work has to be organized and special procedures have to be implemented, especially for cost reducing actions concerning several departments, where responsibilities are distributed. This paper describes our experiences of quality cost follow-up and a possible way to reduce quality costs, exemplified with a case.

Introduction

Ericsson Radio Systems AB, with headquarters located in Stockholm, Sweden, is a subsidiary company within the Ericsson Group. Its activities are divided into a number of divisions, acting as profit centres. This paper describes quality cost activities within those divisions that are located in Mölndal, on the West Coast of Sweden. The divisions at this site develop, design and produce high technology military equipment, often with short series of production. It can be as few as 1 to 5 systems of the same kind to be delivered during one year. Every system contains many different subsystems and sub-subsystems, meaning that the flow of units are high, but the number of similar units is normally low. This fact has to be considered when studying this paper.

When implementing a quality cost system, it is natural that cost reducing actions are handled with each responsible organizational unit and that the actions will be tied to its products. The Quality Assurance Department is implementing a quality cost system with the aim to measure, report and reduce quality costs. The goal is high and the first time schedule has been modified. The first planned quality cost system did not turn out well. It is now being modified. The experiences during these first two years are brought to your attention in this paper, with the emphasis on procedures for cost reducing actions.

Quality Cost System. Principles and Experiences

The principle with the first planned quality cost system was that all units in the company should follow up the costs in the same way. Due to the problems that occurred (described below) we are now developing and implementing a quality cost system, according to figure 1. The principle is that each organizational unit shall follow up the quality costs in its own way, but controlled by general rules given by the Quality Assurance Department.

The quality cost system at company level will primarily be used for highlighting areas of problems. The purpose is to initiate actions, that in a rather long term affect the quality costs. This system will be used as an instrument to achieve "break-through," not as an instrument for "control."

During the period of developing and implementing the first planned system, we worked with problems that can be grouped into:
- definitions
- measurement
- motivation

Definitions

The definitions we use, do not principally differ from the traditional ones. The main categories of quality costs are:
- prevention costs
- appraisal costs
- failure costs

Furthermore we split the failure costs into three groups:
- internal failure costs
- external failure costs
- correction costs

We define the correction costs as: "Costs for activities in order to avoid a recurrence of a failure."

The correction costs must not be mixed up with other failure costs or with prevention costs. We consider the correction cost as something in between, like a positive (constructive) failure cost.

The internal and external failure costs are primarily costs for repair, rework, scrap, etc. These activities are labelled recovery actions.

Measurement

The measuring problems we met are:
- The accounting system is insufficient for quality cost measuring. Most of the quality cost elements cannot be found separated in the accounting system. The costs that can be found are often insufficiently specified, e.g. failure costs are not shown per cause of failure.
- Problems to find new measuring methods. These problems occur when you try to find general measuring methods applicable for all organizational units in the company.
- Problems to find bases for comparison.

Motivation

A quality cost system is only operating effectively when persons concerned are motivated and interested. If the quality costs shall be reduced, it is important that persons

concerned are analysing the costs and are working with cost reducing actions. There are problems to motivate and convince all persons concerned that quality cost follow-up is something profitable that ought to be implemented. The arguments against quality cost follow-up run as follows:
- It is too difficult and too expensive to measure the quality costs, especially per project (product).
- Current cost accounting systems and failure reporting systems are sufficient instruments in order to reduce the costs.
- It is too difficult to evaluate quality cost changes from time to time, depending on types of projects that are in production, production methods used, present stage of manufacturing, etc. The results of planned quality cost reducing actions cannot be separated from the results of other circumstances affecting quality costs.

Other possible reasons for the lack of motivation and interest are:
- Fear to show high quality costs.
- Lack of knowledge.
- Quality cost is not expressly requested from top and middle management.

Solutions and Plans for the Future

In order to motivate and interest persons concerned with quality cost follow-up and in order to find the right way to continuously follow up the quality costs, we are working with the following activities:
- Education and information for all persons concerned, from top management to foremen.
- Evaluation of quality cost follow-up by the use of "pilot-projects." A product, a process or a department are examples of "pilot-projects."
- Special quality cost analyses of problem areas. These analyses are detailed analyses of a specific product, process, etc, where problems have occurred, but where sufficient corrective action has not been taken. We calculate all quality costs the problem is causing, in order to find out where the significant quality costs arise. The purpose is to convince persons responsible, that corrective actions are necessary and profitable. The case study shows an example of such an analysis.
- Implementation of procedures to handle quality cost reducing actions, effectively. These procedures are described later.

As recommended in the literature we have started with follow-up of failure costs. Mainly, they will be studied and evaluated by using existing failure reports, in which time for repair, rework, etc will be added. By using the failure reports, the costs can be measured per cause of failure, per causing department, per type of failure and per component. This information is necessary to initiate corrective actions. Thus the priority for corrective actions will then depend on the costs for failures, instead of failure rates only.

With exception of the use of the failure reporting system the follow-up of quality costs will not be based on measuring methods common to all organizational units in the company. Each unit is allowed to use the methods they consider being most suitable. But they have to follow general principles given the Quality Assurance Department, i.e., definitions to be used, types of costs to be included, etc. The purpose is to motivate persons concerned to follow up the quality costs as well as possible, to make them conscious of costs and not to give them figures, calculated by measuring methods which they consider as incorrect.

Each organizational unit shall:
- measure, analyse and follow up the quality costs
- suggest quality cost reducing actions
- implement decided actions
- distribute quality cost reports to their management and to the Quality Assurance Department

The Quality Assurance Department shall:
- define the costs to be included
- suggest measuring methods
- develop and inform about methods, procedures, etc, which can be used in order to handle quality cost reducing actions effectively
- summarize and analyse the total quality costs
- distribute total quality cost reports to top management
- initiate quality cost reducing actions
- perform special quality cost analyses
- be a consulting function regarding quality cost matters
- educate persons concerned

Our experience is that it is difficult to implement a quality cost system applicable to all organizational units in the company. Such a system is very sensitive for changes of organisation, accounting systems, etc.

Procedures to Handle Cost Reducing Actions

In spite of all problems with definitions, measurement and motivation described above, the most important problem remains. That is to handle the quality cost information effectively so that these costs will be reduced. It is not enough to have a measuring system for quality costs. Resources have to be spent on quality cost reducing activities, the work has to be organized and special procedures and methods have to be used. Otherwise there is a risk that the quality costs remain unchanged from time to time. Persons might be interested in quality cost matters, but are fully occupied with other matters with higher priority. Bearing this in mind, one possible solution of the organization and work with quality cost reducing actions is as follows:
- Assign a person to be responsible for quality cost reducing actions. Reduce his field of responsibility to an organizational unit, working with only one kind of product. This philosophy creates a number of quality cost responsible persons in a big company, but will undoubtedly be more effective than assigning one person in the top quality organization.

- Find problem areas by listening, looking, studying reports (accounting, failure rates, scrap, inspection, etc) and by implementing a measuring system for quality cost, although it is not absolutely necessary to have a measuring system. You can achieve acceptable results without it, but you might not be able to reduce the costs as effectively as with a measuring system.
- Make detailed cost analyses. The analyses have to be performed to the depth required, to see singular costs that can be reduced by a corrective action. With detailed analyses you can prove to hesitators that you are right. If you only guess, the decision makers will not take the decision. Use the Pareto principle for prioritizing.
- Make cause analyses. Use known aids as brainstorming, fishbone diagrams, etc. Cause analyses do not have to be performed by the quality cost responsible.
- Prepare alternative solutions. If you need to convince people, calculate the cost for each solution. (Gross "investments" contrary to estimated reductions of quality and production costs.)
- Gather a corrective action board and present the case. A corrective action board with participants from marketing, design, production and quality has such a broad knowledge, that their common decision has all the possibilities to be the right decision. Also the motivation for the participants is high when they implement the corrective action, since they will have a lot of facts instead of guesses.
- Follow up is a very important activity. This is the best way to learn from your mistakes and to improve the system. It makes persons aware of your attention.

As we see it, the most important thing in order to handle quality cost reducing actions effectively, is to assign persons who work with these activities and to implement functions (e.g., corrective action boards) which have the authority to decide corrective actions.

Case Study

Background

The theory of a quality cost system is to measure, analyse the costs, analyse the cause, take corrective action and follow up. In this case study we did not follow the theory. At a meeting we heard that a special type of modulator was very difficult and expensive to produce. We decided to measure the quality cost as a separate test to find out how expensive the production of the modulator was and how detailed the cost could be measured with existing systems. No special reporting activities were used. The design department had already taken the decision not to implement any more design changes. We had the feeling that they were wrong in their decision and hoped that we could prove it. Another reason for choosing the modulator was that high voltage technique is difficult and that we might gain some valuable experience for the future.

The Hardware

The modulator is an electronic black box, producing high voltage pulses in a radar. The modulator contains transformers, capacitors, resistors, high voltage cables, pc-boards, spark gaps, is filled with oil and is hermetically sealed. The design is very compact and the unit is expensive to repair.

The Quality Costs for the Modulator

Basic Data. The quality cost investigation covers 24 units, produced in 7 lots. The precalculated production labour cost is in this report 100 per unit and the total 2.400. The number of design changes implemented are 12. The costs reported emanate from assembly and test of the modulator, mounting of the modulator in the radar and warranty repairs. The costs for receiving inspection, inhouse production of components and material are excluded.

Quality Costs Included. In this case, the quality costs contain labour costs for repair, rework, reinspection, retest, design changes, implementation of design changes at the work shop, cost for standstills at radar level and loss of interest due to late deliveries. Costs for scrapped material and loss of interest due to late invoicing are excluded. These costs are not significant.

Measuring. Data from the existing accounting system have been used. Some data are estimated. The cost for the measuring, analysing and reporting activities is 3% of the failure costs for 24 units.

Result. The result is shown in table 1 and figure 2.

TABLE 1 - Total Quality costs for 24 units.
The precalculated Production labour cost is 2400

1	"Failure costs" as difference between measured and precalculated production cost. (Note. Part of difference might be error of precalculation)	5.670
2	Failure costs as measured at assembly	250
3	Failure costs as measured at radar level	250
4	Design changes and implementation of design changes in work shop	3.170
5	Warranty costs	250
6	Costs for standstills at radar level	830
7	Loss of interest due to late deliveries	750
	Total	11.170

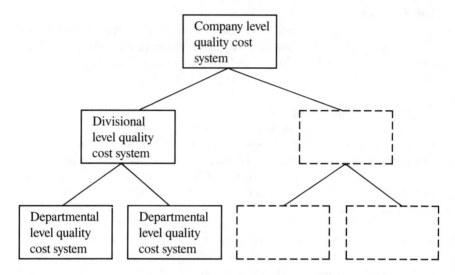

Figure 1 Levels of quality cost systems

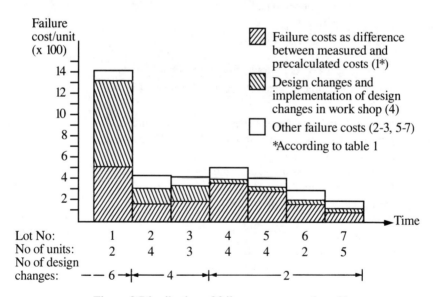

Figure 2 Distribution of failure costs per unit and lot

Corrective Action Board

When the quality cost report was distributed, a corrective action represented board was gathered. Participants represented marketing, design, production and quality. The presentation of the quality cost figures gave the participants some minutes of deep thinking. Naturally, criticism of the figures came out. Some participants had no idea at all that the costs were so high and questioned the figures. Some said "Here is the proof of my suspicions" and others "The cost for the design changes are higher than presented."

After one hour of discussion, the corrective action board decided:
1. No more design changes shall take place. They are not cost effective to the limited future production.
2. A study of all failure reports shall take place and the result shall be incorporated in "Design rules for high voltage equipment."
3. Measuring of failure costs for the next type of modulator shall be performed.
4. A study of costs for keeping extra modulators in stock in order to be able to reduce costs for standstills and loss of interest shall be performed.
5. Checklists for design reviews shall be modified in order to include the latest experience.
6. The failure reporting system at modulator production shall be improved.

QUALITY COSTING IN THE UK

Peter A. Daisley
Daisley Associates Limited
J. J. Plunkett
B. G. Dale
University of Manchester
Institute of Science and Technology

Introduction

The present status of quality costing in the UK is neither easily recognisable nor easily defined. UK has of course a great many companies with many differences. Whilst there are firms with well established uses of quality costing most businesses have yet to take advantage and many are still unaware of the subject. The growth of interest in quality as a contributor to competitiveness and marketability has stimulated concern for the economic effects of quality but attitudes are still dominantly defensive. Despite the initiative of institutions, government and practitioners, there are still far more questions about the cost of quality effort than about the overshadowing expenditure of quality failures.

Some advances are being made. With the help of national and regional collaboration through Colleges and Universities, Professional Bodies and consultancies, and particularly through the energies of quality specialists in businesses, there is progress. UK companies are beginning to assess the true costs of quality. We believe we need to accelerate. Thus we may have to break with some of the practices and assumptions of earlier costing to develop more readily acceptable methods. It may need new approaches and vigorous effort to lead the bulk of manufacturing industry towards finance-based planning for quality.

The Present Status

There are some firms with long standing and well established Quality Costing. Some use their cost evidence to guide and direct their work and others just maintain a measure of quality performance in financial terms. The state of development of quality practices is no doubt closely associated with the costing of quality but it is likely that the lead comes from other than financial arguments.

Recent experience suggests that more firms are getting partial evidence on quality related costs including data on waste, downtime and warranty costs. It is encouraging to find such facts being used even if they are not assembled as formalised Quality Costs.

Many firms in manufacture have no form of quality costing. Indeed there are still numerous industrial managers who have no concept of costs of failures being associated with investment costs on quality. Even among the quality mature businesses few as yet regard their quality losses as the drainage of lifeblood from the business.

Constraints

There is in UK industry a widespread attitude of quality as a necessary evil, as a consumer of resources contrary to company interests. This does not mean that there is less concern for quality but only that the concern is defensive. Realistic costing of quality causes and effects would help to modify such attitudes but getting started is difficult in such a climate. Someone in authority has to do the wanting. A few companies do view quality with enlightened self-interest, assessing what is right for their business and managing their quality to meet company objectives. These companies are the ones that use quality costing.

UK companies do tend to compartmentalise their functions. A harsh comment perhaps with an element of truth is that the main competitor of a UK manager is the head of the next department! With strong divisions the desirable concensus on quality is still unusual. The cross function cooperation, particularly between technical and finance, necessary for quality costing is difficult to establish. The initiative of the Quality Department in seeking to expose quality costs has been counter-productive in some cases: Proclaiming the sins of others is not guaranteed to win cooperation!

Executives do have to delegate responsibilities for action, of course, but on quality issues there is too often delegation of decision making. This might be regarded as an abdication of responsibility. While ensuring that the subject of quality is given attention by a specialist department, as is common in most companies, the executive may be contributing to continued separation. Another generation of managers may grow up thinking departmentally. The alternative to such separation is widespread understanding and quality competence. It takes long term investment in education and training to effect changes in attitudes and competence to that extent. In UK some firms are actively moving in that direction. Some will be demanding a measure of the effects.

The need for decision data, in financial terms, is evident for most aspects of the business but not for "quality" where a measure of investment is easy but measures of returns are obscure. Until there is more evident and accountable benefit from attention to quality, it is unlikely that company executives will demand answers. With a lack of conscious planning of quality the question of VALUE of quality effort arises only in budgeting for the department.

National Initiative

It is encouraging that a lot of effort is being made in UK to encourage companies to examine their quality costs. Many organisations including the Institute of Quality Assurance now run short courses on quality costing. Consultancy and Training businesses are also playing a part in extending quality costing.

At a national level too there have been developments. The British Standards Institute produced their standard on quality related costs (ref BS 6143) which has become a necessary reference on the subject. With the added impetus of the National Quality Campaign there has been a surge of interest in the subject of quality with corresponding growth in concern for its economics. By concentrating attention on system approval and compliance to regulations as the first need, there is a danger though that economic implications may be neglected. There are however several study and development projects with official backing aimed specifically at quality costing.

Practicalities

Further evidence of the interest in quality costing is the initiative of the SERC(1) in funding a research project by Plunkett and Dale at UMIST(2) to investigate the collection and use of quality-related costs. The research includes collaboration with six manufacturing companies, mostly involved in machinery manufacture, with annual turnovers from £5 million to £50 million.

As yet incomplete, the research yields some useful guidance rather than quotable cost figures at this stage. Experience gained in this work shows that costs need to be suitably qualified before they became meaningful and comparable with other published data. This suggests that the value of much of the published data is questionable in view of the lack of qualification. This paper warns the unwary of possible pitfalls, suggests a rational approach to cost collection and the need for qualification of cost data.

A literature review covering about a hundred papers and books. A dominant feature is the Prevention - Appraisal - Failure categorisation of quality, as recommended in BS 6143 (3) and an ASQC paper (4). Whilst this is a useful way of looking at quality-related costs, it is important to remember that the task is primarily a cost-collection exercise and is subject to other different criteria.

(1) Science and Engineering Research Council.
(2) University of Manchester Institute of Science and Technology.
(3) BS 6143 — The determination and use of quality-related costs.
(4) ASQC — American Society for Quality Control.

It is suggested that an appropriate set of criteria for a cost collection exercise are:
Purpose
Relevance
Ease of Collection
Size
Accuracy
Completeness
Potential for change
Recording and presentation
Use

The collection of quality-related cost data is discussed against each of the criteria.

Purpose

There is little point in collecting data just to see what they may reveal. Many managers who have successfully resisted pressure to co-operate in the collection of costs would not reveal anything of which they were not already aware.

Getting the purpose of the exercise clear can go a long way towards avoiding such problems and can also influence the strategy of the exercise. For example, if the main purpose of the exercise is to identify high-cost problems, coarse scale costs in known problem areas will suffice. If, on the other hand, the purpose is to set a percentage cost-reduction target on the companies' total quality-related costs, it will be necessary to identify and measure all the contributing cost elements in order to be sure that costs are reduced and not simply transferred elsewhere.

In a wider context, knowledge of quality-related-costs may be required for comparisons with other divisions or other companies. An important aim may be to impose financial stewardship on quality as applied elsewhere: As such it also enhances the role of quality as a business parameter comparable with production, marketing, R&D, etc.

Relevance

In any cost collection exercise the costs must be relevant to the topic. This may seem to be obvious, but deciding on which activities to include is by no means straight forward. There are many "grey areas" and even such eminent authoritative sources as the ASQC and BSI publications differ in their categorisation. In the present study four main areas of difficulty have been highlighted.

Firstly, for certain activities such as testing or running-in, it is unclear whether it is a quality task or an integral and essential part of the manufacturing operation. The costs can often be substantial and can alter quite markedly the proportions in the cost categories prevention, appraisal and failure. Because each case is different there is no general solution. However, where it is unlikely to be amendable to change by quality management influences the cost should not be defined as quality-related.

Secondly, similar problems of categorisation arise for costs generated by functions other than quality and production. Notable examples are the contributions of the Purchasing to supply quality assurance and the fitness for purpose of purchased goods, the activities of engineering design departments involved with concessions and design modifications prompted by quality considerations. Quantifying, classifying and costing such inputs is very difficult but they are believed to be significant proportions of prevention and failure expenditure.

Thirdly, there are those factors which clearly influence the attainment and preservation of the desired quality. Such factors serve to ensure the basic utility of the product, guard against errors, and protect and preserve product quality. Examples are use of design codes, preparation of systems and procedures, capital premiums on machinery, document and drawing controls, and handling and storage practices. Whether or not such factors give rise to costs which may be regarded as being quality related is a matter for judgment in individual cases.

Fourthly, there is the matter of whether quality related costs should include a portion of the costs of site services such as catering, security, etc. Obviously such problems need to be discussed with purchasing, engineering, production, and accountancy personnel. It is unlikely, however, that there will be a uniform view, and it is improbable that the accounting practices in use will yield satisfactory solutions.

Ease of Collection

Before attempting to collect quality-related costs it is worth looking briefly at the history and evolution of the company accounting practices. Most companies have materials and direct labour costs analysed in considerable detail for the purpose of measuring production efficiencies. However, indirect labour including "staff" costs are seldom analysed in detail even though they may total four or five times the direct labour costs. In those cases where indirect personnel are engaged on single activities or a narrow range of closely related activities full time (e.g. inspectors) the cost is easy to collect. In practice, however, many personnel are involved in a number of disparate activities, sometimes wholly quality related, sometimes not. Hence it is the lack of data about how people spend their time which makes the collection of cost data under the different quality categories and elements difficult.

For personnel within the quality department it is presumably within the quality manager's authority to require routine accounting of staff time. Weekly returns in half-day increments would suffice. The quality topics against which to book time, and personnel costs to be used in conjunction with such bookings are discussed later. Within the production function some costs can be readily obtained from routine monthly accounts, e.g. rework, scrap material. Analysis of failure costs by functional causes is not usually readily available. Other quality activities such as sorting and inspection work by production personnel are not easy to identify and cost.

An appreciable "failure" cost often not collected and very difficult to estimate is the cost of non-quality-department personnel where part of their work is generated as a result of quality failures. Examples are the involvement of purchasing and account personnel in dealing with supplier rejects and production control personnel rescheduling rework.

Finally, there is the matter of labelling of account codes. It sometimes happens that accounts departments wish to keep separate accounts of some non routine matter e.g. a product recall, and give it a title from which it is not immediately obvious that it may be a quality cost. Hence it is prudent to work very closely with accounts departments when investigating quality costs.

Size of Costs

Size is often regarded as being synonymous with importance, though it is size coupled with relevance and potential for reduction which determines the real importance of costs. Clearly it may be much more advantageous to pursue a small percentage reduction in a large cost than a large reduction in a small cost, depending on the ease of achievement.

Deciding what magnitude of cost may be regarded as being insignificant in the cost collection exercise is another problem. Unfortunately there are no useful guidelines. In the end it becomes a subjective decision about what sum of money is significant in the context of quality costs or company profit. It is suggested that all costs which are readily available should be collected, but that cost elements which are unlikely to be less than £1,000 per annum are not worth pursuing. Having decided on a figure it is important to recognise that it is much more likely that that magnitude of cost may fail to be picked up as part of a large cost element than of a small one. The strategy must therefore be to concentrate on the large cost items.

It is important to remember that sizes of costs, both absolute and relative, can become grossly distorted under some accounting practices. An example met with is the practice of including full overheads in direct labour charges to quality related costs. Thus rework and scrap costs become grossly inflated compared to prevention and appraisal costs which are incurred via salaried and indirect workers. Such a system is not tenable of course, in that there is some double counting inevitably taking place.

Accuracy and Precision

It is the mark of an instructed mind to rest content with that degree of precision which the nature of the subject permits and not to seek exactness where only approximation of the truth is possible.

Aristotle

Against the background of difficulty in definition, selection of inclusions and problems of fact-gathering there seems little point in pursuing precision even where it is known to be obtainable. On the other hand, the costs must be precise enough to be credible even to those whose efficiency of performance is perhaps impugned by the resulting report. A cost report which does not have credibility is a waste of time, and unfortunately, the credibility of a whole report can easily be undermined by the exposure of a single weakness. Hence, only costs endorsed by accounts department should be used in the report. Costs produced by the accounts department have greater acceptability and are more likely to be compatible and consistent with other cost efficiency measures.

Accuracy of costs is dependent on the quality of the underlying data. Knowledge of how that data is obtained, and the purposes it is used for, may give good indications of the levels of accuracy expected but care to avoid inaccuracy is always necessary. For example, where calculation of production efficiencies deducts "non-productive hours" for allowable peripheral activities, this may encourage false recording and "loss" of quality-related work such as extra sorting.

Poor control of paperwork may lead to inaccuracies. With process rejection notes for example, each note carries a unique number but there may be no check that all notes issued are accounted for when it comes to costing the various disposal outlets, scrap, rework, etc. It may be possible that a significant number of rejection notes which ought to go to the quality and accounts departments for costing purposes, never get there.

When trying to get a feel for accuracy of cost information accumulated in accounts departments it is sometimes useful to look for independent measures as corroboration. In one case study there appeared to be inconsistencies between provision made for warranty claims, actual warranty charges accumulated, product failure rate, and the level of effort devoted to customer rejected products and warranty.

In apportioning personnel time the use of actual costs, which may vary from month to month, may be be warranted. Annual figures for employment cost may be sufficiently accurate. Nor is it necessary to have an employment cost for every individual. Two or three levels of costs could give sufficient accuracy. These figures could also be useful in making interdivisional or intersite comparisons.

Specifying the elements and categories against which to apportion time can lead to inaccuracies if not thought out carefully. If the individual concerned does not think that the elements are appropriate to his work he will not bother to make accurate entries. So, we may get participants to specify their own element definitions or cooperate to provide a very comprehensive list of elements at the start.

A final point of accuracy is that guesses are useless! Experience has shown that where people are involved in disparate activities estimates by individuals, their colleagues and their supervisors vary widely. The danger is that guesses may be ascribed the same credence as other figures in the report. They may be identified as a weakness through which the report may be discredited. They will certainly not be kept in their proper perspective.

Completeness

Once it has been decided which costs are relevant, which are insignificant, it is important to collect and display all those costs which are available, and also to indicate the existence of relevant costs which cannot yet be quantified. It is important because (a) reporting only part of the costs can be very misleading, and (b) reporting the existence of unquantified costs keeps them in view and encourages attempts to measure them.

Potential for Change

The criterion "potential for change" is derived from the principle that there is little point in gathering and presenting on a routine basis, costs which do not change. Hence, before setting up a cost collection system, it is advisable to examine the potential for change of a cost element in both absolute and relative terms. The inclusion of fixed or immutable costs also has the effect of reducing the sensitivity of costs to performance changes.

However, it is also important to know the total quality-related costs, including those which do not change, so that the effects of changes in elemental costs on total costs may be seen. The classic and often-quoted example of failure to do this is that of reducing inspection costs only to increase failure costs by a far greater amount.

Recording and Presentation

The format for the collection of costs should make provision right at the outset for all those elements and costs sources which are deemed to be worth collecting. The creation of a quality-related-cost file, integrated with existing costing systems but perhaps with some additional expense codes, should not present many problems.

Collecting the data will probably be much more difficult. Cost data from within the quality department may be easy to obtain but that from other departments may be difficult. Any suggestion of attack may well deter cooperation. Nevertheless provision should be made in the file for collecting such data, even though it may take a long time to obtain satisfactory returns on a routine basis.

Presentation of costs under prevention, appraisal and internal and external failure categories as advocated by the BSI and ASQC publications, is the most popular approach, albeit with different cost elements appropriate to different industries.

This format is favoured by quality managers perhaps, because it forms a kind of balance sheet for the quality function: Prevention is equivalent to investment; Appraisal to operating costs and Failure corresponds to (negative) profit. However, in some industries the time lags arising between action and effect are such that concurrent expenditures on, say prevention and warranty, bear no relation whatever to each other. Looking for other, better ways of presenting and using cost data is a part of the research programme yet to start.

One view among quality practitioners is that quality costs reports should indicate the department of origin of failure costs, e.g. production, engineering, marketing, etc. It may be hoped that such adverse publicity will provoke remedial action. Unfortunately, it may also antagonise departmental managers so that they become uncooperative in providing information. It may even result in the deliberate obscuring of quality performance evidence. Emphasis on "improvement opportunities" rather than "attributable failures" may help to overcome the difficulty.

The need of the recipient is an important consideration in the presentation of quality-related-costs. It may be worth presenting information in several different formats for company top managers and for middle managers separately according to particular interests and likely use of the information.

Use of Cost Data

Clearly this is one of the most important criteria in setting up a cost collection system. The principal uses may be defined as:
  (a) to display the importance of quality-related activities to the company management in meaningful terms, i.e. costs
  (b) to show the impact of quality related activities on important business criteria, e.g. prime cost, and profit and loss accounts
  (c) to assist in identifying projects for improvement by departments or operations
  (d) to enable comparisons of performance with other divisions or between companies
  (e) to establish bases for budgets with a view to exercising budgetary control over the whole quality operation

It should be clear that the purpose of a quality related cost report is not to promote the interests of the quality department. Nor is it to display quality cost contributions of other departments. Such analyses are separate issues and should not be allowed to distract attention from the stated objectives.

Lessons from the UMIST Project

1. Most of the published quality-related-cost data needs to be qualified if it is to be meaningful.
2. Manufacturing industry literature concerned with quality-related-costs is almost exclusively prevention/appraisal/failure orientated.
3. Cost collection exercises must satisfy other different criteria than those imposed by the cost topic.
4. The definition of quality-related-costs or the relevance of cost elements to quality is not always obvious.
5. Standard accounting practices are unlikely to provide satisfactory solutions to the problems of relevance accuracy, etc.
6. A major problem in estimating costs is finding out how indirect personnel spend their time.
7. Only costs produced or endorsed by accounts departments should be used in quality cost reports.
8. The quality of the underlying data is an important factor in the credibility of quality cost reports.
9. There is difficulty in relating costs across the time lapse between activating a quality programme and the expiration of the warranty period.
10. Wherever it is possible, independent corroboration of quality costs should be sought.
11. Never use guessed costs or costs based on guessed data, not even from informed guesses.
12. Indicate all existing cost sources in the report including those which may not be currently quantifiable.

Future

The use of quality costing has been limited and after 20 years of publicising only a minority of companies have formal quality costing as part of their management system. Surely we need to rethink. Recognising the difficulties referred to perhaps we should be seeking another approach. We may need to sacrifice some of the formulae we have held on to despite their low acceptability. We may have to move away from the costing models of the pioneers. At least we should perhaps change emphasis, with less play on omnibus costing and more attention to particular costing of projects, controls and quality improvements.

In UK we may need to give more attention to the effects of quality on revenue and earnings an area largely neglected. Perhaps our concern should be directed to the commercial implications of quality, of marketability and growth. The challenge for the quality specialist of today may still be competitiveness in costs and conformance but for the next generation of advisers it may be the business-winning aspects of quality that dominate.

Acknowledgement: The support of the Science and Engineering Research Council for the work reported here is gratefully acknowledged.

INCREASED PROFITS THROUGH COMPANY-WIDE COMMITMENT

William J. Ortwein
United Technologies Corporation
Pratt and Whitney Manufacturing Division
East Hartford, Connecticut

Abstract

Quality performance and productivity improvement requires the assistance and cooperation of every discipline within an organization. The active participation of areas such as production, purchasing, materials, personnel, engineering, and the like are necessary to initiate and maintain an effective, comprehensive quality system.

A quality cost program is one of the few measurement parameters sufficiently comprehensive to adequately satisfy overall trend data and at the same time provide detail for the application of analytical techniques.

Most systems and performance evaluation techniques encourage the worker to accomplish his task as quickly as possible, depend on the "system" to "catch" errors, and feel not in the least responsible for deviant material. We must reward for good work and encourage the identification of defects. Consider some of the reasons for error — machine capability, operator error, faulty stock, engineering drawing error, process planning error. Teamwork requires not only the assistance and cooperation of fellow employees but also a commitment and environment established by upper management which allows a job to be made right the first time.

Text

Every viable concern in a business environment which is attempting to generate a profit is affected by some form of competition. This competition may be in the form of similar products offered by other organizations or the availability of substitute products. An organization does not exist which has exclusive rights to perfection. Everyone experiences problems throughout the spectrum of design to field application and service. Why does one concern succeed and another fail?

Economic theory explains scarcity of product has an impact on price. Monopolies and shortages allow for higher prices and pardon or at least soften the impact of inefficiencies, waste, and mismanagement. However, most of us cannot afford to tolerate any effort which does not maximize the utilization of available resources. One of the most significant concepts which can contribute to the performance of an organization is the integration of all disciplines into systems which result in a saleable product. In particular, quality assurance needs to be an integral part of the engineering, planning, and production systems as opposed to filling the role of a policeman. A team effort is required to assist manufacturing and suppliers in performing their jobs better.

Teamwork Approach

Changes in skills and trends will require extensive training and continuous reinforcement and updating of personal skills. The quality professional, like any other professional, must adapt to the changing environment. Those among us who are the most perceptive about what must be done and the most realistic about what can be done will be in business long after others drop off. The success of organizations will be measured by the extent to which they can manage costs.

Product cost control has improved within the last three years with an emphasis on early in-process detection as well as final inspection. A shift away from end item inspection toward the control of process variables, a closer working relationship with engineering, increased application of computers and automated equipment, and expanded statistical applications are some of the methods being employed to realize productivity improvements.

Managers need to develop the skills of communication, listening, and clarifying in order to specify requirements, delegate assignments, and enable employees to find the source of problems rather than continually "put out fires." Understanding motivational theory and the management skills of delegation and positive reinforcement is necessary for the manager interested in involving people in the identification and permanent resolution of quality and productivity problems. A comprehensive system links employee participation and human resource development activities directly to the financial goals. Employee attitudes and behaviors are the very basis of quality and productivity goals.

Quality is the key to company survival. Production of high-quality, first-time yields with low scrap, rework, and repair costs will significantly improve profits while maintaining a competitive price.

Creating the Proper Environment

A teamwork approach for the implementation of an effective quality system is not restricted to a group of employees on the shop floor attempting to resolve a problem. The chief executive officer and all other members of management must identify with the concept, and more importantly must provide an environment which allows for the implementation of their own programs.

Everyone recognizes the complexities of contract compliance, which include delivery of the product in a timely manner within specifications. The teamwork approach demands quality first, then schedule. To state that philosophy satisfies the public image requirements. Do we have a "make-it-right" philosophy or a "say-it-right" philosophy? The latter too often prevails. How does shop supervision get the real message as to what is important to management?

What are the measurement parameters for determining the foreman's performance? Why is a supervisor given a raise in salary?

Is the question "how many pieces has his department shipped?" or "how many pieces has his department made correctly the first time?"

Measurements reflect the specific indicators most significant within an organization. Heavy emphasis on an industrial engineering indicator of performance with quality measures being nonexistent broadcasts to supervision and workers an extremely clear message.

Establish goals and measures in various departments which compliment rather than oppose each other. Instructing the purchasing department to expand the supplier base and procure from a network of suppliers in a widely disbursed area while at the same time instructing the materials department to develop and install a just-in-time inventory system is a tall order.

Teamwork requires management to study the strategic plan in depth in order to determine no conflicting plans are employed.

Implementation of improvement programs such as process and systems improvement create an environment which allows for continued confidence in product output. Consistent compliance with specifications permits the selection of a few critical characteristics for review. Validating control of materials and processes through programs such as Operator Certification, permits implementation of McGregor's Management Theory Y. It recognizes the human being will not only accept but seek responsibility; he will exercise self direction and self control in the service of objectives to which he is committed. The machine operator is responsible for inspecting and reporting his own work — good or bad. Do not penalize the operator if he tells the truth, even if he produces a nonconforming part. Take action if he misrepresents his work.

Companies are turning around quality control concepts. Those performing the work are becoming more aware they are responsible for the quality of their work.

People want to be proud of the company they work for. They want to take pride in what they do and in what the company does. They do not want to hold back, or be held back. They want to work and they want to give a full day's effort every day, for an appreciative employer.

It is the responsibility of management to stimulate and reinforce the work ethic through leadership.

Providing the proper tools, gages, and equipment to the operator of a piece of equipment along with good instructions makes the entire work experience more convenient. Communication between operators establishes a sensitivity to what other workers are doing and what problems they might be having. Working as a team eliminates "finger pointing."

Integration of manufacturing, quality, financial, and other departments through establishing long-term quality objectives as key elements in corporate strategy are essential. Quality support must be maintained when significant trade-off decisions are made. Encourage behavior that seeks to improve rather than maintain the existing business. Avoid giving conflicting messages to employees.

The return from paying for high quality exceeds the incremental costs of achieving it. Develop employee potential through education, training, delegation, and positive reinforcement.

Consumer orientation is essential to long-term success. The entire system of operations must place primary importance on meeting customer requirements, whether these requirements occur inside the company or outside. Accepting the definition of a customer as someone who receives the product or service means various functions must work together to establish requirements and clarify them. When changes occur, internal customers determine whether requirements need to be changed. No assumptions can be made. Ongoing and extensive communication and the development of measurements, documents, and procedures which are reliable and concise result in the development in an operation that people can trust.

Traditional Financial Measures

For a quality management system to be effective, communication of problems and issues must take place across functional boundaries. A commonality of terms and measurement basis are necessary. Productivity will suffer from a variety of statistics and data generating reports employing constantly changing measurement basis. Traditional cost accounting does not attempt to quantify quality. Accounting structures do not easily provide for the collection of cost segregated in such a way so as to satisfy the requirements of a quality cost system.

Accounting has been charged with the responsibilities of collection, accumulation, retention, and publication of financial data pertinent to a business, with heavy emphasis toward fulfilling legal disclosure requirements. Their many output statements have displays intended to clearly present the interests of many different parties; such as, owners/stockholders, managers, creditors, prospective investors, governmental agencies, employees, and citizens.

A financial structure is generally developed as follows:

1. Cost of Goods Manufactured
   A. Overhead Costs (Overhead - That Part of Operation Costs Which Cannot Be Traced Directly to Items Produced)
      • Factory Overhead
      • Material Overhead
   B. Prime Costs
      • Direct Labor
      • Direct (Product) Material

2. General and Administrative
   A. Division Counsel
   B. Financial
   C. Personnel and Industrial Relations

3. Selling Costs

4. Profit

Financial measures of performance developed in a company on a standard cost program usually are calculated as follows:

MEASUREMENT PARAMETERS
Direct labor

- Standard Labor — The labor value, at standard cost, of a saleable product placed into an inventory account.
- Labor Variations — Labors costs of a saleable product not included in the standard.

Typical Labor Variations
(Calculated as a Percent of Standard Labor)

Performance
Setup
Inspection
Rework (By Production and Inspection)
Substitution — Surface Treat
Labor Savings

MEASUREMENT PARAMETERS
Overhead rates

$$\text{Factory Overhead Rate} = \frac{\text{Total Factory Overhead Costs}}{\text{Total Direct Labor Costs}}$$

$$\text{Material Overhead Rate} = \frac{\text{Total Material Overhead Costs}}{\text{Total Direct Material at Standard Costs}}$$

$$\text{QA Factory Overhead Rate} = \frac{\text{QA Factory Overhead Costs}}{\text{Total Direct Labor Costs}}$$

$$\text{QA Material Overhead Rate} = \frac{\text{QA Material Overhead Costs}}{\text{Total Direct Material at Standard Costs}}$$

Industrial Engineering Studies

Productivity is generally measured through an industrial engineering department on the basis of performance. Comparing actual hours to a predetermined standard for that operation establishes performance.

Direct labor variances consist of those direct labor operations which are difficult to measure. They include heat or surface treat operations, inspection, rework, etc., and performance which is hours expended in excess of standard hours. Standard direct labor plus direct labor variances added together equal total direct labor.

Total direct labor is then employed as a base for calculating overhead costs which, when added together, equal cost of goods manufactured.

Selling expenses and general and administrative costs are then calculated as a percent of the cost of goods manufactured and added to cost of goods manufactured to determine the cost of goods sold.

Within accounting, cost accounting attempts to measure and assign costs of operation with respect to processes, products, departments, territories, and other centers of managerial interest. This includes elements entering into the cost of producing and distributing goods or rendering services, in order that unit costs may be determined and that management may be supplied with much more detailed information for evaluating operating performance. The principal cost accounting statements are functional operating budgets for the various cost centers, and periodically updated lists of standard unit cost data and variances. It is in the area of cost accounting, therefore, that the needs of quality cost analysis can best be satisfied.

Because of the primary goals of cost accounting (producing functional operating budgets and standard unit cost data), various elements vital to complete quality cost analysis are hidden.

Quality Cost Program

Industry needs to develop quality cost accounting as a significant new specialty within the corporate structure, so that (1) it identifies what is really being spent on quality, and (2) it creates a baseline from which to measure and target precisely the needed effectiveness of preventive systems installed to reduce costs while improving customer quality satisfaction.

Our ultimate goal must be to achieve 100 percent conformance, and we must actively strive to achieve that goal. Strategically, how do we approach the problems?

Quality's high price tag is the result of quality expenditures being spent the wrong way — on trying to sort bad products from good and on rectifying mistakes after they have been made rather than on preventive measures that might have made products conforming initially. Consider the following:

- Validating control of materials and manufacturing processes as opposed to the product.
- Statistical process control techniques, operator certification, and total process control.
- Integration of systems for calibration, inspection methods, and tool design.
- Automated measuring systems which identify the actual dimension and immediately display to engineering for review and disposition.
- On-line recording of inspection results.

To determine areas for application and also establish a measurement base for productivity improvement, Quality Assurance should provide the initiative to review with the Accounting Department their entire existing chart of accounts, and help identify those segments which relate to accurate quality cost analysis program needs.

Quality Costs

Quality costs are costs which would disappear if all possible quality deficiencies disappeared, and if perfect control of material, people, and processes were possible.

The primary areas of cost segregation in a quality cost program are:

1. Prevention: — Costs associated with personnel engaged in designing, implementing, and maintaining the quality system.
2. Appraisal — Costs associated with measuring, evaluating, and auditing products, components, and purchased materials to assure conformance with quality standards and performance requirements.
3. Manufacturing losses: — *Internal* costs associated with defective products, components, and materials that fail to meet quality requirements and cause manufacturing losses.

External costs generated by defective products being shipped to customers.

A quality cost program concentrates attention in areas which traditionally have consumed resources. The program highlights those areas where costs may be reduced or avoided without decreasing the value of the product to the consumer.

A quality cost program simply identifies a portion of the financial structure in a slightly different manner than have traditional financial methods. This program concentrates in areas where expenses are not incurred when the organization operates at 100 percent efficiency.

When implementing a quality cost program, use the same nomenclature as present financial documents. Supervision and management will understand the program more readily if terms are familiar to them. Acceptance of the concept is enhanced when those who will use the program feel comfortable with the terminology.

Employ account descriptions from the chart of accounts, unit or department names, product line nomenclature, and any other source of terminology which will lend itself well to analysis of your operation.

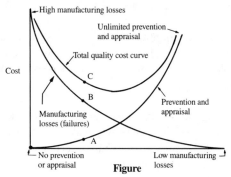

Figure

A plot of the cost categories along with a total quality cost curve identifies the concept of maximizing total quality costs. A model can then be developed to explain the concept which is displayed in the figure above.

The application of this quality cost model identifies those resources which are consumed to assure fitness for use of the product.

Manufacturing losses are expected to be extremely high when little or no effort is devoted to prevention and appraisal activities. As ever increasing resources are expended on prevention and appraisal activities, manufacturing losses begin to diminish. They eventually reach a point where unlimited application of prevention and appraisal resources yield a minimal reduction in the cost of losses. Adding the two cost curves together at points A and B all along the curves generates a total quality cost curve, from a series of points at C.

The central portion of the curve represents the optimum application of quality resources. When a company's program matures and the optimum or indifference zone is reached, the primary function of a quality cost program is control. The program must continue in order to identify elements in the model which appear to display an unfavorable variation from that which is considered to be acceptable.

Implementing a variety of improvement programs could yield savings in all of the cost areas identified and in most cases requires an outlay of funds in one or more specific accounts. Quality improvement program benefits yield results in the same areas. However, these savings are realized in expense areas not required for final delivery of the product. By that I mean, these costs do not generate a characteristic or assemble a product into usable form. Quality costs accumulate the value of resources expended to establish systems which prevent the shipment of nonconforming material, appraise the product to determine compliance with specifications, and dispose of deficient material. All costs are those which could be eliminated if the product were always correctly made in the first place.

Conclusion

The further we advance in producing a quality product initially, the fewer the resources (quality costs) required. Up front detection of variances must be resolved through process improvements and operator education which required the assistance of all departments.

The quality function must be an integral part of all cost reduction and productivity improvement programs. Excuses such as the following cannot be employed to absolve the quality department:

- Marginal processes may present potential field or schedule problems — additional inspection is required.
- State of the art processing requires a high rate of testing with an accompanying 10 percent cost increase.
- Inspection is mandatory to ensure adequate performance.

Everyone, especially quality, is responsible for reducing costs. The most lucrative area for savings lies within the boundaries of a quality cost program. How do we address this "gold mine" or "hidden factory"?

Prevent the occurrence of defects by attacking problems in the system as soon as possible.

How does anyone determine if the efforts of a program to prevent errors are effective? A quality cost program is one of the few measurement parameters sufficiently comprehensive to adequately provide overall trend data. Traditional financial measures are of course essential to the operation of a company. The impact will be evident in the profit picture as well, but a quality cost program will segregate expenses not required for final delivery of the product.

Quality's strategic weapon — prevention through teamwork — will increase worker satisfaction, customer loyalty, market penetration, and the value of the company's stock. Everyone but competition benefits from prevention through teamwork.

Acknowledgment

For permission to publish:

Mr. J. J. Robinson
Quality Assurance Vice President
United Technologies Corporation
Pratt & Whitney Group
Manufacturing Division

Bibliography

1. Figenbaum, A. V. *Total Quality Control*. New York: McGraw-Hill Book Company, 1983.

2. Juran, J. M. *Quality Control Handbook*. New York: McGraw-Hill Book Company, 1974.

3. Rehoer, R. and F. Ralston. "Total Quality Management: A Revolutionary Management Philosophy," *S.A.M. Advanced Management Journal*, Summer 1984.

ON THE ROAD TO QUALITY SAVINGS

William D. Goeller
LTV Aerospace & Defense Company
Vought Aero Products Division
Dallas, Texas

Abstract

The theory of quality cost has been discussed and publicized many times. The best known is "Quality Cost — What and How." The question of "how to," combined with the results of a quality cost system, need to be addressed in-depth in order to give an understanding of the methods and data necessary to convince corporate management on the advantages of a quality cost system. That is the purpose of "On the Road to Quality Savings." This paper briefly reviews the theory of quality cost analysis, recommends a corrective system that is based on quality cost, and discusses implementation methodology including a recommended time phasing. The icing on the cake comes in the results section. Here, two companies are studied over a seven and five year period, respectively. The first case study is a quality cost system that is used as a means of evaluating suppliers and purchasing agent performance. This system has been in effect at the company studied since 1976. The contrast between the traditional "bean count" system and the quality cost system is dramatic. The second case study begins with the initiation of a quality cost system and tracks the results over a five year period. The cost savings documented by the corporate Controller were in the multimillion dollar range, and are exceptionally useful in convincing management of the necessity of a quality cost system. Combining the "how to" with the results of in-place systems is one of the most effective approaches to convincing an unenlightened corporate management team of the benefits of a fully integrated quality cost reporting system.

Introduction

Good work by the Quality Cost Technical Committee has enabled the quality community to understand the advantages of using quality cost as a measurement of product quality in communication with upper management. Unfortunately, most of today's executives are not trained in the quality discipline, making it difficult to get their attention long enough to explain the workings of a quality cost system. Too often in American industry, the chief executive will only give lip service to product quality — except when in trouble with the customer. Getting a real understanding that quality levels are measurable in dollars, and that corrective action can be administered most effectively based on cost, is difficult. They often believe this to be an over simplification, or an "it's too good to be true." The purpose of this paper is to outline methodology and schedule guidelines, show how quality cost is a proven criteria for corrective action and a measurement of supplier effectiveness, and then show specific results from a company that has a quality cost system. For definition purposes, the cost of quality in this paper is the cost of doing things wrong. It includes all redo costs from both functional and service departments.

Background Research

Market share can be increased by providing a higher quality product at a competitive cost. Intuitively, executives are aware of this, but you should research specific cases in your industry and plot sales versus quality. Profit Impact of Market Strategy (PIMS) has completed a broad base research on the subject of quality versus sales. There have been a series of articles published in *Harvard Business Review* using the PIMS data as a base. There is also depth of data publicly available from both Japan and America, especially in manufacturing areas of electronics and automotives. Figure 1 is a typical curve.

Figure 1

One of the things that every executive understands is the cost of goods produced. But, a comparison seldom seen in the board rooms is the cost of goods produced versus quality. Using the data from your own company product lines, plot the cost of goods produced versus your quality index. From this curve, you can develop the quality level expected at lowest cost. Figure 2 is a typical curve. But, neither of these examples produce optimum quality level. Optimum quality is in a range where long term profit is at a maximum. It includes all disciplines of the company associated with product design, production and sales.

Supply and demand curves need to be developed for quality similar to pricing supply and demand curves. There is a level of quality so low that your product will not survive in the market place. Here, demand goes to zero. There is also a level of perfection that is so high, production becomes limited, and in the zero tolerance case, become zero. We are all aware of the supply and demand limits from experience. From these extremes, we have the beginning points for the supply and demand curves. Cost of production drops off sharply from either extreme and demand becomes more price driven. For

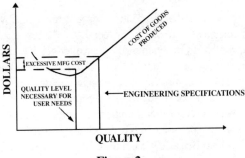

Figure 2

years, the consumer was more price conscious because quality levels between companies were similar. But, technology and foreign competition have changed the shape of the demand curve, moving it toward higher quality. The effect is to narrow the range of optimum quality, because the demand curve has moved to the right in Figure 3. As you look at Figure 3, optimum quality becomes a range to the right of the maximum sales curve where profit margin is at maximum and demand exceeds supply.

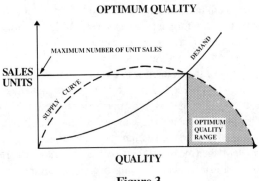

Figure 3

In completing your research, look for the data available within your own factory. Look for such items as scrap cost, rework cost, shop utilization, off-standard hours, quality department cost, and engineering liaison cost. Nearly every company, in one way or another, has these costs in their cost accounting system. They are all quality cost items. If you need help in the constitution of quality cost, reread, "Quality Cost — What and How."

Obtaining Your Controller's Assistance

It is absolutely mandatory that the controller's assistance be obtained in presenting the case of quality cost measurement to the executives. The research data obtained must be presented to the controller. In obtaining the controller's cooperation, the most graphic attention-getter is to plot quality cost (your best estimate numbers) as a percent of profit. Whatever you do from the controller up, always talk in dollars. Do not substitute. In fact, you will achieve much more if you make the controller the team leader in this effort. If you are unable to get the controller's support, do not proceed. Almost every failure of a quality cost measurement system has been because a quality professional attempted to present the dollars lost. Because it is not the controller's system, he will be defensive and most probably an adversary. At best, it comes off as another quality system. If you did not get his support, the best solution is to get outside help in presenting your case.

After you have obtained the controller's ear and he has obtained the chief executive's attention, the next step is to establish a council to run the program. The head of this council should be a senior executive in the corporation. Each function should be represented on the council. The controller and quality vice president are the energy behind this council, supplying the data necessary to keep it moving — but are not necessarily the head(s) of the council. The council should be responsible for establishing the measurement system and implementation schedules.

The measurement system must be fair above all other considerations. An example of such a measurement system for suppliers is developed later in this report that has proven extremely effective. In fact, one of the responsibilities of the council should be to assure that a fair and effective measurement system is maintained.

The second major duty of this council is to establish an implementation schedule. The total schedule will probably require at least five years to implement. The first year should be a preparation, planning, and study period for the council. The second year should begin the initial measurement. These measurements should be at the individual level for items that are easily understood within the entity being measured. For example, if rework is a familiar item, then measure the cost of that rework, starting at the individual and summing up the organizational ladder. Improvement incentives are mandatory. If the system is without them, it will slowly die; and if it is used for punishment, it won't even get off the ground.

In the third and fourth years, begin measuring the other operating departments. Only after you are proficient in measurement of the operating departments should you begin measuring the service departments in your organization. These will prove more difficult in most instances, because a definitive product that can be easily costed is generally not associated with them.

Corrective Action Based on Cost

Several years ago, the government released MIL-STD-1520, which allowed the separation of significant defects and the insignificant. Only significant defects require one-on-one corrective action. The specification definition of a significant defect is: high cost, or recurrence probable. Both are quality cost drivers. Studies show that 30 percent of the defects equate to 85 percent of the cost. This Pareto Lorenz Distribution becomes one of the most effective tools in controlling quality cost at the working level. It can be used in developing a corrective action data management system for the line quality engineer. Such a system is necessary for the quality engineer to make decisions based upon cost. Figure 4 is a pictorial of the system used at the Vought Aero Products Division.

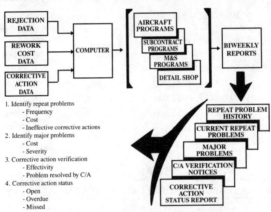

Figure 4

The system relies heavily on the computer to establish the routine reports. This piece of the quality cost system does not need the controller's blessing to work effectively, since it is a lower working level tool that can easily be based on standard hours as man-hours. In fact, industrial engineers can estimate the repair, rework, and scrap cost in hours if desired. This will allow you to keep away from tracking actuals if that is a problem in your plant.

Results Section

A supplier performance improvement program was begun at Vought in the mid-70s based upon the cost of doing business with a particular supplier. The system was begun because it had become obvious that the old defect counting system was not producing the improvement desired. The design of the system had two basic constraints. These were: it must produce report cards, and it must be cost effective. A standard of quality cost, $5 cost of quality per $1,000 of supplier purchase order value was established as a Level I standard. A Level II standard of zero to $2 for each $1,000 of business was also established. The supplier had to maintain these levels for a period of one year to qualify for an award at either level. A third level of excellence was also developed which stated Level II must be maintained for an additional year. To assure the purchasing agents were vested, the same awards criteria was established for them. See Chart I.

PERFORMANCE TARGETS AND AWARDS

SUPPLIER

LEVEL	TARGET	MEASUREMENT INTERVAL	AWARD
I	$2.01 - $5.00/$1,000	1 year	Letter of congratulations
II	$0.00 - $2.00/$1,000	1 year	Letter of commendation
III	$0.00 - $2.00/$1,000	2 years	Certificate of excellence

PURCHASING AGENT

I	$0.00 - $5.00/$1,000	1 year	Certificate
II	$0.00 - $2.00/$1,000	1 year	Unit plaque
III	$0.00 - $2.00/$1,000	2 years	Personalized aircraft model

Chart I

SPIP RESULTS

YEAR	LOST DOLLARS	LOST $/1,000$'s
1976	$151,497	$2.09
1977	89,666	1.64
1978	56,990	0.74
1979	73,862	0.55
1980	74,037	0.35
1981	121,898	0.95
1982	116,850	0.98
1983	115,778	0.98

Chart II

Chart II shows the results were dramatic the initial five years, with the average cost dropping from $2.09 per thousand to 35 cents. Major new products were begun the next three years and the cost crept up to 98 cents per thousand and despite this, has never returned to the former levels.

Finally, the system changed our attitude toward many suppliers. Many of the thought-to-be-the worst were not; and some of those considered the best were the most expensive to do business with. Chart III shows three examples of how bean counting can produce an erroneous supplier rating.

SYSTEM DESCRIPTIONS
EXAMPLES

VOUGHT SUPPLIER CODE NO.	EQUIPMENT FURNISHED	"BEAN COUNT" DEMERIT SYSTEM	LOST $ PER $1,000 PURCHASED
966781	Electronics	10.20	$103.25
983806	Machined Parts	12.40	$ 1.65
992854	Forgings	45.50	$ 0.18

Chart III

The payoff for Vought was a standardization for comparison of suppliers and a motivation for pride in workmanship. The executives of the corporation easily understood the system and quickly worked to get their costs down. In-house, we also had a powerful new measuring device for evaluating new quotes. The truth is simply the lowest cost proposal is not always the cheapest.

Results of a Corporate Quality Cost System

The company studied incorporated the quality cost system five years prior to the study. The system incorporated the external and internal failure costs, appraisal cost and preventive cost in this analysis of the cost of quality. In other words, it basically followed the guidelines of "Quality Cost — What and How." The system was begun in 1978 and tracked through 1982. Revenue for the company increased from $273 million to $665 million during the period.

The cost reports shown include the cost for the supporting departments to redo their jobs, as well as the primary (causing) department cost. For example, a workmanship rework cost includes the Manufacturing Department's cost to perform the rework, the quality department's cost for material review board activity and the reinspection cost, engineering cost to disposition the rework, any new material cost involved in the rework, production control rerouting, and allocations for facilities (maintenance and utilities). The study did not attempt to analyze gain in sales due to improvement in quality, although sales did increase dramatically in the period. The gain was definitely greater than economic improvement during the period. In included a gain in market share.

The company is a basic manufacturing company. It designs, machines, processes, and assembles goods that are used in oil well drilling. An analysis of the quality assurance department prior to 1978 shows it was basically an inspection department charged with the responsibility of appraising product acceptability. The prevention and corrective systems were nearly nonexistent. This required that resources be committed in the beginning to establish the prevention leg of the quality cost system. Fortunately, the top management of the company was in agreement with this initial release of funds. They had just absorbed several million dollars in warranty costs and had several extremely irate customers because of field failures of a $50 item.

During the years of 1978 and 1979, a modern quality assurance department was created from the inspection department. It included closed loop corrective action management system, planning functions, auditors, design reviews, statistical analysis, and all other functions that made it a balanced quality assurance department. Chart IV shows how the cost of this department initially rose as a percent of revenue, then began to reduce in a steady manner as prevention and corrective action took effect. Interestingly, the scrap and rework cost followed similar courses (Charts V and VI), but were delayed approximately a year as the effects of the new system took hold in the company. The number of malfunctions per million dollars of revenue on the other hand did not wait the year, but began an almost immediate drop as seen in Chart VII. The number of malfunctions per million dollars of revenue appeared as a leading indicator of the scrap and rework cost. A one-time happening, obviously, does not make it a positive leading indicator, but it is something worth watching in future studies. A leading indicator early in the program is a blessing to the hero of quality when you've spent corporate resources on a new system.

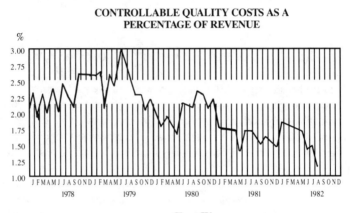

Chart IV

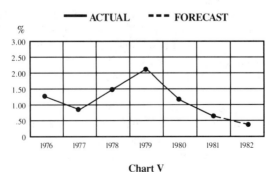

Chart V

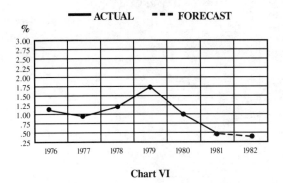

Chart VI

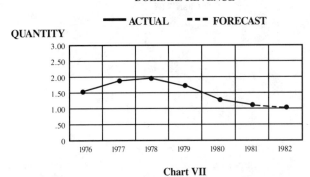

Chart VII

The most dramatic depiction of the results can be seen on Chart VIII, where the revenue curve overlays the cost of quality curve. Here we see a dynamic, growing company in which the cost of quality actually decreases during four of the five years studied.

Finally, the documented savings by the corporate controller for the period was $54,171,409. This included an actual negative cash flow the first year of the system. The savings in the failure area, both internal and external, was $40,779,819. The remaining $13,391,590 was in the appraisal area, since the actual size of the Quality Department shrunk when compared to revenue.

The quality cost system of the company studied included only those functional departments associated with delivery of hardware. It did not include a quality cost system for service departments such as the computer department, sales, travel, etc., or the results would have been much higher. Even so, the nearly $25 million contributed to profit in 1982 had a significant impact on the operating results of the company.

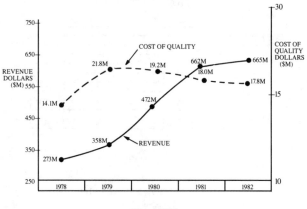

Chart VIII

QUALITY COSTS II
THE ECONOMICS OF QUALITY IMPROVEMENT

John T. Hagan
ITT Corporation
New York, New York

Abstract

The original intent of Quality Cost Systems was twofold: (1) to create a management awareness of quality costs as an opportunity for improvement; and (2) to provide a useful tool for the quality manager to justify and support specific quality improvement actions. To a large degree, this objective has been accomplished — but its potential usefulness has not been realized. It is now clear that the ultimate business value of a Quality Cost System cannot be achieved without significant change.

From the experience of industry in general there are several areas which literally cry out for change:

1) As presently conceived, quality cost systems should be limited to costs associated with "conformance quality" in manufacturing operations. They are not capable of handling costs associated with service or administrative quality, design quality or the quality of product requirement.

2) Today's appraisal costs are clearly affected by many things beyond the control of the Quality System. Therefore, they need to be more clearly defined within the confines of Quality System objectives.

3) Most obvious is the need for the quality cost system to stand alone, to become untangled from the other cost systems and programs which it now overlaps. It should include those costs over which some control can be exercised through the normal execution of a Quality Management program.

With these thoughts in mind and coupled with the basic concept of a simple, practical, stand-alone system, a new approach has been formulated. This new approach focuses on clearly defined failure and appraisal costs, both of which are targets for systems and cost improvement.

These cost elements relate directly to the need for assuring customer satisfaction, the need for minimizing inspection and test costs, and the need for eliminating or minimizing the cost of defects. The article contains a detailed description of the new system and recommendations for effective use.

Text

The business of company wide quality management is Quality Improvement — improvement in the performance of every company activity involved in the life cycle of a product. This means improvement in Marketing, Product Design, Manufacturing and every other function that supports these key elements of product life — through the integration of quality control disciplines into the daily operations of each of these functions. Quality Cost systems were created many years ago to highlight the cost of poor performance — the cost of doing things wrong. As conceived, however, Quality Cost systems never reached their full potential. While many uses have been genuinely productive on a broad range of products, their non-use clearly indicates an opportunity for system improvement.

There is much to be learned from many years of experience in quality cost applications and this learning cannot yet be found in the text books of business management. Quality Costs II attempts to crystallize this learning and apply it in a new way — while still maintaining the original intent of highlighting the cost of poor performance as a basis for corrective action and performance improvement. The benefits continue to be that extra costs related to performance, especially costs hidden from normal purview, can be exposed in each area where quality improvement is being promoted. Individual applications are limited only by imagination and the ability to accumulate identified costs.

History of Quality Cost Usage

To recap how this new concept was arrived at and to describe it in some detail, let's start by exploring some of the history of quality cost development and use. The ITT Quality Cost system was started in 1967. Its original purpose was twofold:
1. To create a management awareness of the "cost of quality" as an opportunity area for cost improvement.
2. To provide a useful tool for the quality manager to justify and support specific quality improvement actions.

(See Attachment #1 for a brief description of the original quality cost concept.)

After more than fifteen years experience within ITT, it can safely be said that considerable management awareness has been achieved — but use of quality costs as a management tool has not been universally accepted. Some managements fail to pursue its profitable use because they accept some level of quality costs as inevitable — and uneconomical to reduce. That is, they view something less than 100% perfect performance as an industrial fact of life. And there is no pressure to change this view so long as the cost of quality, or some other measure of quality performance, is *not* an operating performance issue in senior management meetings.

This is not only true in ITT, it has also been experienced by the Quality Costs Committee of the American Society for Quality Control (ASQC). For many more years than ITT the ASQC has been promoting the use of Quality Cost systems. In spite of many publications, seminars, tutorials and a constant flow of new papers and presentations on the subject, the outcry of the membership continues to be, "Help me to convince my management to use this important tool of Quality Management."

It has been estimated that less than 15% of the opportunities for quality cost application in manufacturing companies in the U.S. are actually being pursued in a profitable manner. In fact, this failure to effectively develop and use quality costs in direct support of quality management programs has inadvertently contributed to a certain lack of credibility for the present concept. The time is ripe to learn from a wealth of varied experience and to organize systems improvements that will, in fact, allow the full potential of quality cost usage to be achieved.

Development of New Concept

Based on the current history of Quality cost system usage, it is clear that the ultimate business value of quality cost systems will not be achieved without some changes to the original concept, coupled with an effective marketing plan to attract top management to its cause. To determine those changes that will have the most value, a survey of current quality cost usage was conducted on a broad scale of manufacturing companies, involving the general manager, the quality manager and the comptroller. Key observations from this survey are as follows:

1. Where quality costs are being effectively utilized by management, the prime focus is on failure costs. In these cases, management fully understands the negative impact of product defect costs, which can be prevented, and forthrightly demands improvement actions.
2. Better than 90% of all prevention costs being reported are quality administration and quality engineering costs — in other words, Quality organization/function costs. Measuring the cost of prevention activities in other functional areas proved to be very difficult. For the most part these costs are considered to be minimal and are normally either estimated or ignored. Even when clearly understood, prevention activities, or disciplines, are most effective as a fully integrated part of each function.
3. The appraisal costs of quality contain the most inherent variation due primarily to variations in the complexity of the product and thus the amount of inspection and test deemed necessary. The trends of manufacturing technology are toward more sophisticated, automated and expensive inspection/test equipment. In many cases the demands of the marketplace and an efficient manufacturing operation seem to be contrary to the theory of striving for continuous appraisal cost reductions. Efficient test and inspection also has a tendency to obscure the real need for process improvement. The appraisal costs of quality, as currently measured, are impacted by things other than quality performance.
4. Quality Cost summary reports are often being used as stand-alone management reports when, in fact, they are an after-the-fact measurement of the effect of quality management actions. Elements of quality costs can only be used to support management actions. Without a quality management system to support, quality costs have *no* practical value.
5. Quality cost systems are being misapplied to a broad area of company operations. It is now clear that quality cost systems, as presently conceived, are *limited* to the

improvement of costs associated with the "conformance quality" of manufactured products. All current definitions directly relate to "conformance to requirements" (except for that portion of Quality Function costs associated with overall company quality improvement). In their present form, quality cost systems cannot directly apply to, or have an effect on, costs related to the quality of design, the quality of the marketing specifications, or the quality of service functions.

Based on the message of these observations, any new concept of quality costs should consider the following objectives:

1. It should be clearly limited in application to costs associated with the conformance quality of manufactured product.
2. It should continue to focus on failure costs as a prime target for reduction.
3. It should maximize the definition of appraisal costs and clearly establish these costs as another prime target for reduction (through improved quality control of the entire manufacturing process).
4. It should eliminate prevention costs as an element of quality costs, thereby allowing total quality costs to become a target for total reduction.
5. Use of quality costs should start as part of a strategic plan for Quality Improvement and continue as a monitor of progress against that plan.

In addition to the objectives based on survey results, there is an underlying need for any new quality cost system to become untangled from other cost control or improvement systems with which they often become intermingled, such as Manufacturing Cost Controls, Inventory Controls and formal Cost Reduction programs. The Cost of Quality needs to stand alone and be easily recognized by management as an entity with obvious merit.

From an overall analysis of these objectives, two clear pictures emerge:

1. The opportunity for significant improvement in the total failure costs of quality relative to the total cost of the manufactured product. This picture is represented by failure costs essentially as they are now defined, offset by an investment in Quality function prevention activities.
2. The growing cost of seemingly necessary product inspections and tests compared to the total cost of manufacturing is a picture of increasing importance. This picture is represented by modified appraisal costs, also offset by prevention activities aimed at their reduction.

As these two pictures emerge, it also becomes clear that there could develop a picture of product design failure costs compared to total design costs and offset by an investment in their prevention; and even perhaps a similar picture in marketing, product support and administrative areas. Unfortunately, data does not now exist to develop such views but *should be undertaken* in the future. These thoughts lead to the conclusion that the new concept of quality costs can be a modular concept, allowing for the continual development of new segments as cost data and opportunities permit.

As a result of this study and analysis, a new quality cost system for manufactured product has been developed. This new approach satisfies the objectives listed and has been entitled, "Quality Costs II," to clearly distinguish it from the original system. Its subtitle, "The Economics of Quality Improvement," more clearly defines the goal of the system. That is, to achieve the payback in quality performance that is there to be gained from an investment in prevention activities for all areas of the operation. The payback will be reflected in revenues and profitability and, in addition, it will have a positive impact on customer satisfaction. Therein lies the economic justification for companywide Quality Improvement.

The most significant change in Quality Costs II from the old system is the elimination of *Prevention Costs* as an element within the system. This accomplishes two important goals:
1. It allows the total cost of quality (appraisal and failure costs) to become a target for cost improvement — without compromise or confusion.
2. It allows the quality control theory of "prevention" to be directly applied to all company functions as an integrated investment concept. In reality, the prevention of errors is the personal responsibility of each company employee, achieved through the application and integration of quality control disciplines into individual work. Investment in these self-controls will have a payback in performance — not unlike an investment in modern equipment.

Appraisal Costs are being expanded (or clarified) to include all costs associated with the inspection and test of supplies, tools, product and new operations. Altogether, these are the true total cost of product appraisal — a growing cost which needs to become a prime focus for reduction opportunities. Integral with this thrust is the need to discourage "automation to reduce inspection and test labor costs" when process improvement is the answer. Statistically controlled processes require the absolute minimum in effort and equipment to verify the results.

Failure Costs, with the exception of a few additions and deletions, are essentially the same as before — and they continue to be a prime focus for cost reduction.

In *general,* design engineering costs and other functional costs are eliminated except as they may be a direct cost of Appraisal or Failure as herein defined.

Quality Costs II is applicable to all manufacturing companies. Its intent is, as with the original quality cost system, to supply economic awareness and support for the company's quality improvement strategies and goals — by bringing attention to costs that can be eliminated as quality performance is improved. That is, it strongly supports performance improvement through the use of quality control disciplines in all operations that can effect the integrity of the product. It describes elements of cost that should be compiled by the Comptroller for use by the Quality Manager. The frequency of reports, as well as the necessary support data and the detailed breakdown needed (product line, production area, etc.), must be determined jointly by the Comptroller and the Quality Manager based on the current status and needs of the company Quality improvement program.

Description of Cost Elements

The discussion of Quality Costs II will start with Failure Costs, continue with Appraisal Costs, and then deal with implementation and use. The Failure Costs of Quality include all the tangible accumulated costs of product defects and they should be clearly defined and accounted for as individual elements of the Company Cost Accounting System. That is, the quality cost system definition of key failure cost items and the Comptroller's definition should be one and the same. These items are either important enough to be accounted for separately or we should re-examine their value for our purpose. If they are to be a prime focus for cost reduction they cannot be buried somewhere within other cost accounts. They must be visible.

From a management viewpoint there is a major difference between failures discovered in-house and those discovered by the customer. Internal failure costs are what one renowned author refers to as the "hidden plant" — an unexposed opportunity for cost improvement. External failures are individually more significant, affecting both cost and reputation. For these reasons the failure costs of quality are separated into internal failure costs and external failure costs.

Essentially, Internal Failure Costs are the cost of all product failures to meet requirements discovered in-house prior to installation or delivery. It includes all rework and scrap due to nonconforming product, losses from substandard product and extra manufacturing operations (such as buffing, deburring, sorting, etc.) caused by the lack of capability to produce 100% conforming product.

Rework is the total cost (labor, material and burden) of all rework or repair work done to manufactured product, regardless of whether done by specific work order, personal assignment or as a planned part of the manufacturing operation.

Scrap is the total cost (material, labor and burden) of nonconforming material or product that is discarded (scrapped) because of irreparability or uneconomical rework cost, regardless of whether it is accounted for on a scrap ticket, counted as planned or unplanned yield losses, or simply discarded on the factory floor.

Scrap and rework allowances are sometimes built into the standard cost of a product. Whether or not this is the case, all scrap and rework is to be reported. Inventory write-downs for excess and obsolete product are specifically not included because all such losses should be addressed and controlled as an integral part of the Inventory Control System. If the salvage value of specific scrapped product is significant, reported scrap can be a net figure representing actual losses incurred.

Substandard Product Losses are the total cost of all losses from full sales value incurred in the sale of product that failed to achieve planned requirements.

Extra Manufacturing Operations are the total cost (labor and burden) of manufacturing operations (planned as either a standard cost or a temporary operation) added to the primary operation as an in-line sort/rework activity — to compensate for a primary process deficiency which does not relate to a state-of-the art limitation. (Example — a deburring operation due to inadequate tooling.) This is a new item in quality cost definitions. The idea of labeling and treating such extra work for what it is, a failure cost, is to prevent these actual losses from being overlooked and ultimately accepted as a necessary cost of doing business.

External Failure Costs are represented by the cost of all product deficiencies discovered during an installation phase or after delivery to the customer. They consist of Warranty Costs and Installation Failure Cost, which is the total cost of all rework and scrap incurred during the installation process.

Warranty costs are the total cost of customer complaints, returned goods, field repairs, recalls and warranty claims due to faulty product or service. Not included in these costs are distributor stock adjustments. Also excluded are the cost of liability claims and other contractual penalties which occur infrequently and are not directly related to day to day conformance quality levels.

The Appraisal Costs of Quality are costs directly associated with the appraisal of products being manufactured or installed for use — to determine their degree of compliance to requirements.

In the world of high technology and complex products the challenge and the cost of adequate inspection and test programs is growing in significance each passing year. Throughout the entire spectrum of manufactured products, approaches to manufacturing inspection and test proficiency are as varied as the products themselves. There is no magic formula for achieving the ultimate in this dynamic, and fast-changing area.

Regardless of current approach (technically and organizationally), the intent of this portion of quality costs is to help a company to develop and justify a gradual transition from today's status to the ultimate appraisal proficiency — total operator self-control; all operations and supplies well within required performance capabilities; effective use of automated, computer-controlled inspections and tests; all integrated into basic manufacturing operations to the extent possible; and altogether having a minimum impact on actual product cost. Simply stated, this objective means to eliminate or minimize inspection and test costs (the Appraisal Cost of Quality) through their elimination or complete integration into individual manufacturing operations, while at the same time allowing the Quality Management function to evolve from a hardware evaluation phase into a true operations performance measurement phase.

Before proceeding with a discussion of the cost elements for this section of quality costs, there are two aspects of costs within this realm that need to be understood:

1. In the ultimate proficiency, the cost of inspection and test incorporated into operator self-control activities should be considered as part of the standard cost of product. These costs are desirable. They will normally replace higher independent appraisal costs and are not to be estimated or calculated for inclusion in these measures.
2. Inspection and test costs normally comprise two elements — initial product inspection and test, and extra inspection and test due to product defects. The latter element should really be classified as a failure cost but, since it would be too costly to discriminate between the two elements, total inspection and test costs are viewed as a single target. Existence of the failure element, however, is one of the reasons for these costs being high.

The Appraisal Costs of Quality are subdivided into Supplier Appraisal Costs, Tool/Process Appraisal Costs, Manufacturing Appraisal Costs and Installation Appraisal Costs. Supplier Appraisal Costs are the total costs associated with the quality control of purchased materials, product and service. These costs include but are not limited to the cost of supplier quality evaluations, purchase order quality reviews, first article inspections, source inspections, inspection and test planning, incoming inspections and tests, inspection and test equipment (expense or depreciation costs), supplier rating systems and supplier corrective action.

Tool/Process Appraisal Costs are the total costs associated with the quality control of production tool and process developments. These costs include but are not limited to the costs of tool/process design reviews, tool inspections, tool/process capability studies and trial or proofing runs.

While supplier quality controls and manufacturing inspection and test are normally budgeted and accounted for separately, quality costs for the control of tooling and the development of new manufacturing operations are usually buried in some larger account. Allowances for such efforts are often very meager or nonexistent — probably at the expense of some unnecessary failure cost. For this reason, and for increased emphasis, this appraisal cost is identified separately.

Manufacturing Appraisal Costs are subdivided into labor costs and equipment costs. Manufacturing Appraisal Labor Costs are the total costs incurred for all in-process and final inspection and acceptance tests of manufactured product, including the cost of inspection and test planning. Primarily, this includes the fully burdened total cost of all in-process and final inspections and tests of manufactured product, whether they are planned as a normal part of the manufacturing operation or are needed because of some special situation.

Manufacturing Appraisal Equipment Costs include the total calibration, maintenance and depreciation or expense costs for all inspection and test equipment.

Installation Appraisal Costs are the total cost of all inspections and tests, and the calibration, maintenance and depreciation or expense costs of inspection and test equipment — as applicable to the installation process.

Implementation and Use

In the concept of quality costs, products are considered to be any deliverable item, including hardware, computer software and product support documents. As a Quality Management support system, Quality Costs II is equally applicable to large and small companies, the only difference being the amount of detailed data required by the quality manager. It is clearly intended that with this approach the amount of data generated, and its specific use, be at the discretion of the quality manager, who should be using only those quality costs needed to support the company's current objectives for quality improvement.

The quality manager will need quality cost details by product line, by major manufacturing area or by any other breakdown that supports ongoing quality management strategies. Also needed will be appropriate baseline support data against which quality cost trends can be calculated. The data will likely be needed in different time increments — daily, weekly, monthly, quarterly. For these reasons, implementation of this system requires a detailed plan and schedules to be developed, between the comptroller and the quality manager, for each application. This individual plan and schedule will provide maximum flexibility to the quality manager in using quality costs as a tool in support of the quality management program.

It is hoped that the simplicity and flexibility built into this new system will promote creativity on the part of the practicing quality manager to develop productive uses based on the realities of the unit or company's overall Quality situation. It's also up to the individual quality manager to sell the values of this system to appropriate levels of management, and to produce whatever reports are needed to support its objectives — such as summary reports for general management and individual product line or area reports for each functional manager's review and action.

Long term trend measures that support the continuous need for quality improvement and highlight current available opportunities and progress, should be used as summary reports for general management and background support for individual action reports. For this purpose a basic Quality Cost Report (see Attachment #2) is proposed. This report requires the following support data to be included:
1. Investment Costs — total Quality Function costs less those accounted for in Appraisal Costs (costs devoted to overall quality improvement).
2. Net Sales
3. Value Added Production Costs — the part of the cost of a manufactured product attributable to work performed on constituent raw materials.
4. Purchased Material Costs
5. Tool/Process Development Costs

This support data will allow the use of ratios like:
- Internal Failure Costs as a percent of Production Costs
- Warranty Costs as an average percent of Net Sales
- Supplier Appraisal Costs as a percent of Purchased Material Costs
- Manufacturing Appraisal Costs as a a percent of Production Costs
- Quality Function investment costs as a percent of the total Quality Cost opportunity

Measures such as these will no doubt take time to evolve into accepted standard business measures like backlog, turnover or overhead rates — but they can, in time, become just as important.

At this point it may be useful to review two of the basic purposes of a company's investment in the Quality function:
1. Prevention of defects in delivered product and of excess manufacturing costs — through the measurement and enforcement of quality disciplines in manufacturing operations.
2. Elimination of failure costs through the discovery and elimination of defect causes (Formal Corrective Action).

Quality Costs II will help achieve these objectives. Normal use of quality costs starts with an annual plan and budget for overall cost improvement and the necessary Quality Function budget to support its achievement. This plan will take into consideration all external sources of change, such as product mix, volume, inflation, cost of labor and materials, pricing, etc. When the net effect of these external sources is calculated, quality cost reduction planning can begin. Planning for incremental annual reductions in quality costs is one of the most important steps to be taken in reinforcing the basic issue of quality improvement. Taking this step will cause the analysis of the next levels of quality cost details and lead to identification of the specific quality cost data needs (unit quality cost plan) to support its achievement. This is the essence of quality cost usage.

Whenever there is room for overall performance improvement in a company, Quality Costs II can be invaluable — but *not* by itself. It is not a substitute for a quality management program. It is a practical assist tool for the Quality Manager and it can provide another important yardstick of company performance for the General Manager in an era when Quality Performance more often is making the difference between a first and second rate company.

Attachment #1

ORIGINAL QUALITY COST SYSTEM CONCEPT

Simply stated, Quality costs are a measure of all costs specifically associated with the achievement of product Quality (defined as conformance to requirements). More specifically, Quality costs are a total of:
  a) the costs of *appraising* product for conformance to requirements.
  b) the costs incurred by *failure* to meet product requirements.
  c) the investment costs of *preventing* product failures.

The strategy for using Quality costs is quite simple:
  1) take direct attack on failure costs and try to drive them to zero.
  2) reduce appraisal costs according to results achieved.
  3) invest in prevention activities to the extent necessary to maintain and improve your accomplishments.

This strategy is based on the premises that:
  1) for each failure there is an *assignable cause*.
  2) causes are *discoverable* and *preventable*.
  3) prevention is *always cheaper*.

QUALITY COSTS — GENERAL DESCRIPTION

PREVENTION COSTS The Cost of all activities specifically designed to prevent defects in design, purchased materials and deliverable products. Includes activities accomplished prior to and during design, purchasing and manufacturing to assure that acceptable results will be achieved in each case. Examples are design reviews, vendor capability surveys and process capability studies.

APPRAISAL COSTS Costs incurred in the conduct of appraisals (inspections and tests) of design, purchased materials and manufactured products to determine compliance with established requirements. Requirements include marketing specifications, product and process specifications, engineering drawings, company procedure, operating instructions, professional or industry standards, government regulations and any other document that can affect the definition of product.

FAILURE COSTS Costs required to evaluate, disposition and either correct or replace defective product, tools, design and associated product documents. Includes both material and labor costs, with full burden for all direct labor involved. Compiled as internal or external costs.

Attachment #2

QUALITY COST REPORT

Company _____ For Month Ending _____ Currency _____ (Local/US)

1. FAILURE COSTS	Current Month	Year-To-Date		2. APPRAISAL COSTS	Current Month	Year-To-Date	
		Current	Prior Yr.			Current	Prior Yr.
1.1 Internal Failure Costs				2.1 Supplier Appraisal Costs (g) Purchased Material Costs (h)			
1.1.1 Rework Costs							
1.1.2 Scrap Costs				Ratio (as a %) — g/h			
1.1.3 Extra Manufacturing Operations				2.2 Tool/Process Appraisal Costs (i) Tool/Process Develop. Costs (j)			
1.1.4 Substandard Product Losses							
Total — Internal Failure Costs (a) Value Added Production Costs (b)				Ratio (as a %) — i/j			
				2.3 Manufacturing Appraisal Costs			
Ratio (as a running average) a/b				2.3.1 Manufacturing Appraisal Labor Costs			
				2.3.2 Manufacturing Appraisal Equip. Costs			
1.2 External Failure Costs				Total — Manufacturing Appraisal Costs (k) Value Added Production Costs (b)			
1.2.1 Warranty Costs (c) Net Sales (d)							
				Ratio (as a %) — k/b			
Ratio (as a %) — c/d				2.4 Installation Appraisal Costs (l) Installation Costs (m)			
1.2.2 Installation Failure Costs (e) Installation Costs (f)							
				Ratio (as a %) — l/m			
Ratio (as a %) — e/f							

	Current Month	Year-To-Date		Full Year	
		Current	Prior Yr.	Budget	Forecast
Total Failure Costs					
Total Appraisal Costs					
Total Quality Costs (n) Investment Costs — Quality Function					
Ratio (as a %) — Investment/Opportunity (n)					

256

RELATIONSHIP OF FINANCIAL INFORMATION AND QUALITY COSTS A TUTORIAL

Earl T. Szymanski
Pharmaceutical Products Division
Abbott Laboratories
North Chicago, Illinois

Abstract

Quality costs have been discussed and editorialized for at least a decade. Many of the people responsible for managing and analyzing the data may not realize that all of the data is contained in the financial statements. They are:

Balance Sheet — A statement of the assets, liabilities, and capital as of a particular date.

Income Statement — A review of operations (profit & loss) during a specific time period.

Net Working Capital — Statement of source and use of funds.

Manufacturing Costs & Variances — Statement of actual costs compared to standards.

Each of the financial statements are reviewed and key items are selected for examples. Several levels of detail are analyzed for quality cost content. It is not until the second and third level of detail that certain quality costs become evident. For example, scrap costs that appear as a single line item in the Manufacturing Costs Report may be a consolidation of:

Scrap which is inherently part of the process, destructions because of externally caused damage, obsolete materials, an accrued reserve, and non-conforming material.

Balance sheets that state the *balance* of assets and liabilities can be useful in investigating opportunities to reduce costs. Reserves that are set up for expected rework, destruction, or warranty services can be looked upon as opportunities to reduce costs. Consider the item that is called warranty services owed. In the case of articles that are sold with a warranty for 1 year, a certain $ amount is set aside as a liability for services owed when a unit is sold. If the reliability can be improved it may not be necessary to use all of that assigned reserve. Hence the savings or potential savings can justify efforts dedicated to reducing expenses.

The purpose of this paper is to demonstrate the sources and opportunities of quality cost control by use of typical financial information.

Introduction

Financial information has a very long history, which has its beginnings around the same time the human race began to move to communal life and property ownership. The earliest forms of writing other than art can be traced around 5,000 B.C. It represented financial information. Impressions of property symbols were made in soft clay. Those impressions represented a net *balance* of property owned.

Financial Information

Today, some seventy centuries later, financial data has become complex. The data that is analyzed and reported is integral to the function of business. It is the main measurement base that describes the "health" of the enterprise. Quality costs whose source is financial data are equally important to the Quality and general management functions.

Quality costs become apparent from a different view of accounting data. Figure 1 displays how financial information generated from the various functions is used for operational and cost control management. Quality costs emanate from the same source but the data is used for strategic analysis and resource management.

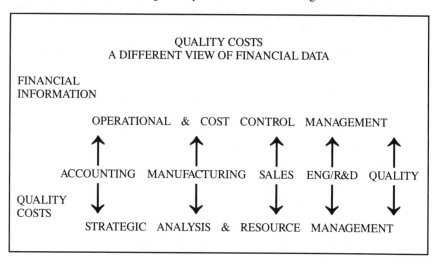

FIGURE 1

Accounting Information Descriptions

In order to understand the sources of quality costs a review of the basic financial statement is necessary.

Balance Sheet — A statement of the assets, liabilities, and capital *as of a particular date.*

> The asset side consists of:
> Current assets — cash, inventories, prepaid expenses and others.
> Fixed assets — property (land), plant (buildings), and equipment.
>
> Balancing the asset side are the liabilities:
> Current liabilities — notes payable, accrued expenses, warranty service owed.
> Long-term liabilities — bonds, leases.
> Shareholders equity — stock and retained earnings.

The Balance Sheet describes the "balance" of total assets being equal to total liabilities.

Income Statement — A review of operations *during a specific time period.* This report is intended to display revenues of which costs are subtracted resulting in a profit or loss, hence the "P & L Statement."

Cash Flow — A review of the sources of cash matched to the uses of cash for assets & liabilities *during a specific time period.* This report is used to identify the sources of funds; net income provided from operations, proceeds from sale of stock, and increases in long-term debt.

> The use of the funds examples are: additions to fixed assets, stock dividends, increase in cash, increase in inventories.

The financial reports previously mentioned are the sources for quality costs. When the accountants assemble the financial statements, there may be many levels of detail for each item appearing on the statement. Figure 2 is an example of a balance sheet. Under Current Assets, Inventories are listed as finished products, in-process products, materials and parts, and service parts. Each item has a separate accumulation of data from warehouse, departments, manufacturing areas and service areas. The accumulation of defective material is a direct effect on inventory. Consider a manufacturing area that produces an average of 5% non-conformance. Converting this rate to inventory dollars means that to deliver 100 finished units, manufacturing has to make 105 units. It also reflects that 105 units have to be carried in materials and in-process inventory.

ABBREVIATED BALANCE SHEET

Assets	1978
Current Assets	
Cash	$ 929,753
Receivables Less Bad Debts	7,879,350
Inventories:	
Finished Products	2,595,568
In-Process Products	2,697,264
Materials and Parts	1,922,327
Service Parts	1,274,258
Deferred Taxes	155,200
Prepaid Expenses	219,111
Total Current Assets	$17,672,831
Fixed Assets	
Total Plant, Property & Equipment	$ 8,468,500
Less Depreciation	3,091,849
Net	$ 5,376,651
Other Assets	$ 193,984
TOTAL ASSETS	$23,243,466

Liabilities & Shareholders Equity	1978
Current Liabilities:	
Notes Payable	$ 2,399,000
Long-Term Debt	173,200
Accounts Payable	1,360,771
Accrued Expenses	2,491,497
Product and Service Warranties	283,000
Unearned Service Contract Revenue	718,000
Total Current Liabilities	$ 7,425,468
Long-Term Debt	$ 1,089,200
Deferred Income Taxes	238,700
Shareholders Equity	14,490,098
TOTAL LIABILITIES	$23,243,466

FIGURE 2

In Figure 2 there is a liability called Product and Service Warranties. This item reflects a reserve that is set aside for each unit that is sold to cover Warranty Services. This item is a direct measure of product reliability. If the reliability can be improved it may not be necessary to use all of the assigned reserve. If the product has more failures than expected the reserve would have to be increased. The source of funds that covers this increase comes directly from earnings.

The line item directly under Product and Service Warranties lists Unearned Service Contract Revenue. This item covers the cost of service that is owed to contract service customers. If the product is more reliable than the planned service frequency the favorable balance is reflected in the bottom line.

ABBREVIATED INCOME STATEMENT

		1978
Net Sales		$27,099,551
Operating Costs and Expenses:		
Cost of Sales	13,328,644	
Selling and Administration	7,594,768	
Research & Development	2,062,301	
Total	22,985,713	
Operating Income		4,113,838
Other Income		436,553
Income before Taxes		4,550,391
Provision for Taxes		2,003,000
Net Income		2,547,391
Net Income per Share		1.57

FIGURE 3

Figure 3 represents an abbreviated Income Statement. The two top lines deal with Net Sales and Operating Costs and Expenses. Let us look at the several levels of detail for Operating Costs and Expenses in Figure 4.

DETAILED OPERATING COSTS & EXPENSES

Operating Costs and Expenses:

Cost of Sales:
   Total Manufacturing Costs:
      Standard Manufacturing Cost of Goods Sold
      Controllable Manufacturing Variances
      Other Variances from Standard:
         Purchase Price
         Activity
         Capacity
         Spending
      Costs not included in Standard:
         Obsolescence, damage
         Scrap/Rework
         Returned Goods
         Production Testing/Process Development

Selling and Administrative Expense:
   Selling:
      Royalties Net
      Freight
      Warehousing
      Marketing/Advertising
      Sales Force
   Administrative Expense:
      Production Administration
      Quality Assurance
      Corporate Charges
   Research & Development Expense:
      New Product Research
      Process Improvement

FIGURE 4

Each line item listed in Figure 4 represents the sum of accounting ledger entries. There may be hundreds of entries that add to or subtract from an expense item. Quality costs have their beginning at the "Ledger entry" stage.

Quality Costs as a Part of Manufacturing Standards

Many of the quality cost segments originate from the Income Statement, however the manufacturing standards can be another source. Figure 5 lists a brief outline of product standards and probable quality cost sources.

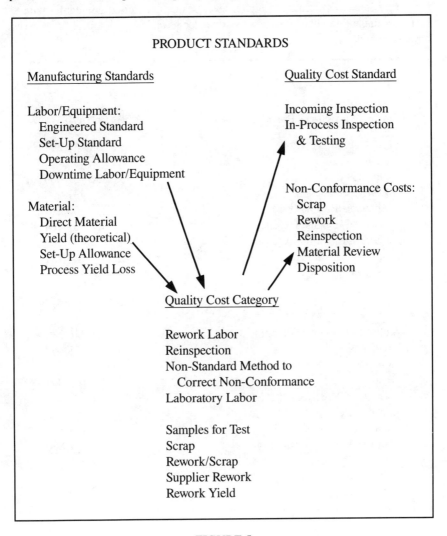

FIGURE 5

Product Cost Standards can be a valuable source for identifying sources of appraisal and non-conformance costs.

Cash Flow Statement

Figure 6 displays a typical Cash Flow Statement. This report is generated from changes in assets/liabilities from one year to the next. The notation (QC) indicates a possible impact of quality costs. Increases in inventory may indicate longer lead times and larger stock pile because of higher level of supplier caused non-conformance. Increases in liabilities can indicate warranty reserves and non-conformance reserves have increased.

CASH FLOW STATEMENT

Financial Resources Provided	1978
From Operations	$3,471,557
Increases in:	
Notes Payable	767,000
Long-Term Debt	437,200
Accounts Payable (QC)	500,215
Accrued Expenses (QC)	567,464
Current Liabilities (QC)	453,000
Proceeds from Sale of Stock	177,157
Total	$6,373,593

Financial Resources Used	
Additions to Plant & Equipment	$1,731,230
Dividends	455,157
Increases in:	
Cash	100,005
Accounts Receivable (QC)	2,098,324
Inventories (QC)	1,650,702
Other Assets (QC)	163,375
Decrease in Long-Term Debt	174,800
Total	$6,373,593

(QC) — Possible impact of Quality Costs.

FIGURE 6

Quality Cost System

A Quality Cost report consists of selected quality-related costs from the *total* cost accounting system which are organized and combined in a meaningful way to measure the cost of quality. Each company that has accounting systems that collect and report financial data have the capability to collect and report quality costs. For the most part the Income Statement is a primary source for quality cost information. In many cases quality cost information does not become noticeable until the second or third level of the Income Statement are reviewed. Figure 7 displays how the Income Statement, Quality Assurance Departmental Budgets and the Quality Cost Report are related.

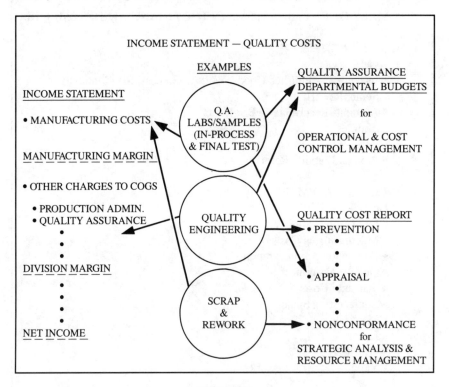

FIGURE 7

Conclusion

Quality cost reports are a management tool for strategic analysis and resource management. Management of Quality functions by using only the quality cost reports is not complete. A good understanding of their source and accounting method is necessary for the *total* management of the Quality Function. This awareness is useful for making judgements in the grey areas of non-conformance costs versus operating variance. Figure 8 summarizes the financial statements and the key line items that are entirely quality costs or a portion is considered a quality cost.

SUMMARY OF KEY LINE ITEMS IN THE FINANCIAL STATEMENTS

A. BALANCE SHEET
  1. Assets
     - Inventory — Finished Goods, Work-in-Process, Raw Materials, Scrap, Rework
     - Accounts Receivable — Bad Debts
  2. Liabilities
     - Reserves
     - Product Liability Reserves

B. INCOME STATEMENT
  1. Net Sales
  2. Cost of Goods Sold
     - Manufacturing Costs
     - Production Administration
     - Quality Assurance
     - Warranty Costs
  3. Manufacturing Standards
     - Materials
     - Labor
     - Overhead
     - Yield — Scrap & Rework

C. CASH FLOW
  1. Source of Funds
     - Changes in Assets
     - Changes in Liabilities
  2. Use of Funds
     - Salaries
     - Purchase of Assets
     - Reserves

FIGURE 8

Quality costs are a tool that displays trends for management to take action. In the case of unfavorable trends there are specific actions/decisions associated with their change. Besides being a key measure of quality performance, quality cost reports can be used to track cost improvement opportunities, resource allocation and strategic investment decisions.

Bibliography

1. ASQC; *Quality Costs — What and How;* Second Edition; 1971.

2. *1978 Annual Report;* Gilford Instruments.

3. Quality Costs Technical Committee, ASQC, *Guide for Reducing Quality Costs,* American Society for Quality Control, 1977.

4. Weston, J. F., *Managerial Finance,* Fifth Edition, The Dryden Press, 1975.

EUROPEAN ORGANIZATION FOR QUALITY CONTROL
ANNUAL CONFERENCE PROCEEDINGS

QUALITY COSTS — FAILURES AND POTENTIALS

Frank M. Gryna
Juran Institute
Wilton, Connecticut

Experience with the quantification of quality costs applied in manufacturing industries shows that the results have been mixed.

This paper presents five examples of failures of quality costs by discussing the symptom, cause, and remedy for each example.

Manufacturing activities represent a narrow scope for the quality costs concept. Application to other activities of the manufacturer and the user represent a potential.

Introduction

In the 1940's, a methodology was developed to estimate the costs associated with defects ("quality costs"). Many companies determined these costs and found them surprisingly high. Making the number known to management often gained support for a quality program. But there were some problems. In the USA, although the desirability of calculating quality costs have had much publicity, only about 40% of the companies have actually measured their quality costs. Furthermore, some quality cost programs have failed.

This paper discusses quality costs from two viewpoints:

1. Reasons for failures of quality cost programs. These reasons are presented in the form of scenarios. Each scenario describes an actual situation in terms of symptom, cause, and remedy. "Symptom" means the outward evidence of a problem in a quality cost program. "Cause" means the reason for the symptoms. "Remedy" is my suggestion on how to prevent such causes in the future.

2. The potential for quality costs in areas other than manufacturing. The paper proposes some applications to achieve this potential.

Now, on to the scenarios of failures.

Scenario No. 1 — Hide the Scrap

Symptom

At a paint manufacturing facility, a worker trips on a hole in the ground outside a building. He recovers and notices paint all over his trousers. A subsequent investigation reveals a serious problem. This company measures quality costs in order to stimulate manufacturing supervisors to reduce spoilage. Unfortunately, the supervisors are told that it is their responsibility to control the amount of spoilage but are given no assistance in how to do it. The supervisors soon learn that unless they achieve some reductions in spoilage, it will seriously affect their performance evaluations. Their efforts do not meet with success. Some of them decide that there is only one way to get an acceptable scorecard on spoilage — hide the spoiled paint. They tell their workers to dig a hole in the ground, pour the spoiled paint into it, and cover the hole. (In inclement weather the worker trips and uncovers the hole.)

This case is not an isolated example. A manufacturer of surgical needles places production supervisors in a position of hiding the scrap. The supervisors play a game and delay reporting scrap that exceeds a goal of 3%. If the scrap is 4% they turn in 3% and hold back 1%. The 1% is turned in when a "good month" occurs. There is sometimes a long wait for that month, if it comes at all!

If you believe that these situations only occur at the level of production floor supervisors, hold your seats. A tire manufacturer has a quality improvement program consisting of much cheerleading but no tools. This pressure from the corporate office causes a plant manager to hide defective tires. He issues orders to place the tires in railroad cars which are to be kept moving between plants so that the scrap is not recorded. (When this is discovered, he is fired.)

Before we criticize the actions of these people we should ask ourselves how we would act with the same pressures on us.

Cause

Upper management failed to realize that to improve quality they must (1) set up machinery to identify specific problems and (2) provide the resources and skills to attack those problems. Some people believe that a scoreboard and the subsequent reporting of quality costs to management is sufficient to stimulate people to achieve an improvement. In the writer's experience, this simply is not so.

Remedy

The steps required include:
- Identify specific projects for improvement.
- Form a management steering arm for improvement.
- Form individual project teams.
- Plan and execute a training program for project team members.
- Provide people with the time required to pursue projects.
- Establish a new priority for quality by changing performance evaluations and taking other measures to emphasize quality.
- Have upper management personally lead the improvement program.

Coupled with this machinery must be a successful case example of improvement to illustrate that the machinery does work. It was hate at first sight in a chemical plant when production supervisors read a quality cost study. They felt that it simply dramatized the scrap and rework in their departments. Later, however, when the study led to projects that solved problems, one line manager said: "The cost study identified a project in my department and the result was that management authorized the purchase of new equipment. I have been requesting this equipment for several years, but could not get approval. The quality cost study showed that the new equipment would have a short pay-back period due to lower quality costs. I see now how quality cost can be useful to me."

Scenario No. 2 — Language to Fit the Territory

Symptom

A manufacturer of pharmaceutical products learns that quality costs are $52,000,000 per year, mostly in the failure cost categories. The information is presented to management but does not have the strong impact expected. Management listens politely but comments that, although the number is large, many financial numbers of the company are equally large. The company has yet to act on the quality problem.

Cause

The language of money alone was not sufficient to hit the right nerve with force. Some examples below make the point luminously clear.

Remedy

In a classic example, a manufacturer of mechanical assemblies calculated the quality cost at $76,000,000 per year. This number was translated into a dramatic statement. Quality Control, working with Accounting, determined that $76,000,000 was equivalent to 2900 extra personnel, 1.1 million square feet of extra floor space to store the scrap and rework, and $6,000,000 of extra inventory to keep the production lines going. In effect, the company had one entire plant making 100% defective work every working day of the year.

In an example from a chemical company, the quality cost per year in dollars was divided by the number of shares of common stock outstanding. The resulting quality cost was 13¢ per share per year. Upper management immediately compared this number to the earnings per share. Action took place.

In another case, an accountant noticed that most of the quality costs were due to rework of 40% of the batches. The goal was to reduce the rejected batches to 10%. The accountant determined that the savings would make additional capacity available and eliminate capital investment for plant expansion for a period of two years. (Expansion was necessary each year to handle a steady growth in sales.) This presentation won approval for the quality improvement program.

The break-even point can also be useful in clarifying the significance of quality cost. In one organization, the quality cost was estimated at 10% of sales income. At that level, the overall break-even point for the plant was 66% of capacity. If a quality improvement program reduced the quality costs from 10% to 5%, then the break-even point would drop to 53%.

The final example involves the use of the DuPont Financial Model to calculate return on investment (ROI) using asset turnover. A reduction in the costs of poor quality can have a dramatic effect on return on investment as calculated from the relationship, ROI = Profit Margin × Asset Turnover. An example of this approach as applied to a company making health care products is shown in Figure 1. Note that the recognition of asset turnover highlights the effect of quality improvement on the return on investment.

Scenario No. 3 —
Comparisons, Comparisons, Comparisons

Symptom

A large electronics manufacturer institutes a quality cost system for all plants. The corporate office develops the system with participation from the plants, a pilot run is held at one of the plants, some changes are made, the system is implemented by all plants. Quality costs are summarized monthly and reviewed quarterly by upper management. At the quarterly reviews, management uses the reports as part of a comparison of plants. The meetings at which the comparisons of plants on quality cost are made become heated and plant managers burn with a slow steady fever. Those plants whose quality cost is high complain that comparisons should not be made between plants because conditions are too different. Some in upper management perceive this as an excuse. The arguments continue through several meetings.

Cause

First, upper managers are not recognizing that there are important uses for facts on quality costs. Second, they are overlooking the fact that differences among plants can make a simple comparison of numbers inappropriate.

Remedy

In a multi-plant company, comparisons of quality costs (and many other parameters) will commonly be made. However, differences among plants *can* make valid comparisons difficult. Plants differ in product lines, age of facilities, degree of experience of the work force, and a host of other respects.

Quality costs are measured in order to achieve quality improvement and subsequent cost reduction. What each plant needs is a structured improvement program that identifies projects, and sets up the organization and other machinery to pursue these projects. Comparisons should evaluate the degree to which each plant has identified projects and set up the improvement approach. The number and nature of projects and the implementation machinery can vary widely by plant and are best decided by each plant. Upper management should insist that the size of the quality problem be determined, projects be identified, and the machinery set up. With plans explicitly defined and resources assigned, reduction in costs will follow.

Scenario No. 4 — Quality Costs Over a Dinner Table

Symptom

A multi-division corporation embarks on a quality cost program. The corporate quality office explains the objectives of the system, develops broad guidelines and recommends that each division develop its own system. One division chooses not to act.

Cause

At dinner on the evening before a quality seminar the vice president of the division told a consultant that he understood there was a methodology on quality costs, but that he did not believe it was as important to pursue as other issues in his division. Here was an executive who did not know the size of the quality problem, and therefore, he chose not to act.

Remedy

Let's listen in at the dinner table.
Consultant: Do you think it would be worthwhile to estimate these quality costs to see what the total size of the quality problem is?
Vice President: Well, I suppose it's a good idea, but we haven't had time to do it.
Consultant: I've heard that you have a formal cost reduction program. How much did it save last year?
Vice President: About 1.2 million dollars were saved. We were quite satisfied.
Consultant: That saving sounds reasonable to me but I wonder how large is the cost associated with poor quality. Why don't we try to come up with an educated guess right now at this table?
(The consultant then mentions some of the major categories of quality cost and, for each category, the Vice President is asked to provide a conservative estimate.)
Consultant: Well, let's add it up.
Vice President: The total comes to 11.5 million dollars per year.
(There was an oaken silence like that in a library.)
Vice President: Good heavens, my cost reduction program saved 1.2 million last year, but the quality loss is almost ten times higher. We should be able to do something about that 11.5 in quality. It could be a greater saving than the cost reduction program produces.
Consultant: It's not unusual to set a goal to reduce the quality costs by half in a four-year period.

Vice President: I'm stunned about two aspects of this. First, the quality cost is far higher than I expected. Second, each year I personally conduct a cost reduction meeting. We review our progress on cost reduction projects, propose new projects, and discuss the resources needed. I decide which projects we will pursue, allocate resources, and assign someone to oversee each of the projects. That form of planning is routine for cost reduction but I don't have a similar meeting to plan quality improvement. I'm beginning to understand what you mean about the importance of estimating the quality costs and identifying specific projects.

Scenario No. 5 — The Plant Manager Takes Over

Symptom

At a large chemical company, management emphasizes the importance of accuracy in the quality cost study. As the present accounts are not able to provide all of the data required, a decision is made to change the accounting system. Many months pass and still there is no quality cost report. Upper management is now uneasy about the entire program.

After some discussion, the task is turned over to the corporate quality office. The quality manager assigns several people to define the categories and collect the necessary data by making a pilot study at one plant.

The results are greeted with an avalanche of argument concerning the categories included in the study. Upper management finally directs that the study be put aside until there is more time to consider it carefully, thereby burying it in the grave of a file cabinet.

Cause

The original stalemate is due to the insistence on perfectionism in the data. To determine the size of the quality problem and identify opportunities for improvement, it is sufficient to have an estimate that is within plus or minus 20% of the true answer. This initial study can be completed in several weeks without changing the accounting system. (Where quality costs are used for comparisons or other purposes, a better precision is necessary.)

The rejection of the report is due to the lack of agreement on definition of quality costs. Experience shows that there are recurring examples of controversial cost categories. Agreement is essential if there is to be useful application of the study.

Remedy

Here's an approach used at a small steel company when a consultant was asked to evaluate the quality program. To start, the consultant suggested a quality cost study and the plant manager immediately agreed. The plant manager expected the consultant to make the study but the consultant suggested that the plant manager create and chair a task force for the study. Here is the parade of events. The consultant worked with the quality manager of the plant, reviewed the literature on quality costs, and together they drafted a list of the categories of quality cost for the plant. The plant manager called a meeting of the line managers. At the meeting, the categories of quality cost were presented and discussed, resulting in useful clarifications and additions. Toward the end of the meeting, the plant manager observed that if numbers were attached to each of these categories, the total could be large. This told him that the study was important and must be made. The plant manager had set the tone for the future.

At the second meeting, again chaired by the plant manager, additional refinements were made to the definition and agreement was reached. The accountant lamented that it would take considerable time to collect the data because "the books of the company were not kept that way." Before either the consultant or the quality control manager could respond, the plant manager interceded. Only an estimate was needed, he said, and extreme precision was not necessary. Further, the study was important and the accountant would have to change priorities to assure that the study was completed soon. Indeed, the study was completed, and the results were accepted. Adept leadership by the plant manager was the key.

The Potentials

From these failures of quality cost programs, we proceed to a discussion of some potentials. These include application of quality cost concepts to non-manufacturing activities.

Product Development

As product complexity increases, the evidence grows that the potential for improvement of the development process is large. About 40% of the problems encountered by the ultimate user are due to design and development. (Another 30% are due to manufacturing, and 30% due to operation, maintenance, spare parts, and other phases of the field use.) The numbers come from my experience and apply particularly to electronic, mechanical, and chemical products having at least a moderate level of technology.

One large electronics manufacturer reports that the cost associated with design changes is about ten digits per year. A manufacturer of mechanical products has measured the costs of identifying, analyzing, and correcting design deficiencies as 30% of the total engineering activity.

An aerospace manufacturer tracks the number of changes to each *production* drawing issued. The average is 5.6.

At one chemical plant, 50% of the lots shipped failed to meet specifications. Costly analyses revealed that every lot was completely suitable for customer needs. The specifications proved to be woefully obsolete.

In the cases cited, the design changes necessary were due to errors in the original design and not due to requirements changed by a customer.

The activities required to correct product development problems are really a form of "rework." The costs, however, are often not separately measured and highlighted; the extra activity is considered a normal part of the evolution of the product. Many products do go through stages of development that involve a learning process but the cases cited above show results after specifications have been released.

The approach to improvement involves three steps:

1. Review the entire *process* of product development from the collection of information on user needs to the issuance of specifications for production. A useful way to start is a "historical review" of a sample of problems traced to product development.

2. Identify specific projects and set up the machinery to pursue the projects as described in the scenarios.

3. Apply reliability engineering and other techniques to provide early warnings *before* specifications are released.

One manufacturer uses costs to measure the quality of specifications, drawings, and operating and maintenance manuals. Figure 2 shows the categories of quality costs in this engineering organization.

Software

Software is becoming an increasingly large percentage of the total cost of some products. One large electronics organization has traced the costs associated with correcting inadequate software. When an errror in software is detected in the original office preparing software, the cost of correction is $1.00. The relative cost of correcting that software, in the customer's office, is $110.00.

Purchased Items

For some products about 30% of the quality problems are due to purchased items. As a result, the industrial purchaser incurs extra costs. In addition, the supplier has quality costs that are included in the selling price and thus become a hidden cost to the purchaser. The potential is the extension of the purchaser's quality improvement program to the supplier.

Spare Parts

Providing spare parts to customers can be a source of errors and thus presents another potential area for improvement. One manufacturer measures quality costs for a spare parts operation. This system quantifies the cost of sending the wrong part of quantity to a customer, deterioration of parts during storage, and other types of errors or degradation in the quality of spare parts.

Marketing and Sales

An electronics manufacturer reports that 20% of the time of the sales force is spent on contracts which produce no revenue because the contracts are cancelled. Only 5% of sales proposals made by a process industry company result in an actual order. Both of these organizations understand the potential for improvement and are now reviewing the process of selling.

The failure to maintain a premium price on a product is a hidden quality cost. Gross (1) reports a case in which a chemical manufacturer achieves a premium price but the amount of the price premium decreases. A marketing research analysis reveals that the decrease is quality related. Subsequent efforts to improve the quality are successful and the amount of the premium is restored.

Nickell (2) describes how a marketing division of an electronics company has embarked on a project-by-project approach to quality improvement. Projects include increasing multiple sales of typewriters, improving customer satisfaction with marketing personnel, and reducing delinquencies in accounts receivable.

All activities in a company are subject to errors requiring corrective activities which become hidden costs of quality. These hidden costs can accumulate and become equal in size to the failure costs reported on the product. The potential return is akin to investing in electric energy at the dawn of the appliance age.

Other Areas for Application of Quality Costs

Quality cost concepts can also help to improve the effectiveness of information. Juran and Gryna (3) provide two examples from inspection activities. Another example involves the evauclation of proposed new equipment. Proposals for the replacement of four different equipments showed an average payback period (investment divided by labor saving) of 9.8 years. When the saving in quality costs with the new equipment was added to the labor saving, the payback period plunged to 0.9 years. Similar applications have been made to analyze the costs of lost gages, total cost of purchased parts, and manufacturing waste.

Finally, the service industries present more opportunities for quality cost applications. Rosander (4) discusses quality costs in service industries in general; Aubrey and Zembler (5) describe some examples in banking.

But the best is yet to come.

User Costs

Quality affects the economic picture of a company in two ways:
1. Costs. High quality costs represent an opportunity for internal cost reduction.
2. Sales income. An improvement in quality from the viewpoint of the user of the product can result in a higher market share or creation of a price premium because of superiority to competition. The result is a higher sales income.

Important as it is to embark on a quality cost and improvement program for internal operations, a larger potential exists by studying costs from the viewpoint of the user of the product. User costs over the product life can be high — often substantially higher than the purchase price of the product. The author has discussed this concept in a previous paper (6).

Studying user costs can lead to an improvement in fitness for use, rather than reduction of internal errors and defects. Improving the fitness for use, in the opinion of the author, can have a larger effect than improving the quality of conformance. A manufacturer must search for and identify opportunities where fitness for use can be improved and thereby user costs reduced.

To identify the opportunities, the manufacturer must be "stapled to the product" while it is being used by the customer. During this journey, the manufacturer observes exactly how the customer uses the product and keeps asking "how could we change the design of the product or service to make it better for the customer." An example of opportunities for an industrial product is shown in Figure 3.

One form of strategic planning aims to create a competitive advantage. Quality can be the parameter for the strategy. By studying user costs and the way in which the user applies the product, a manufacturer can create fitness for use superior to that of competitors. This superiority can differentiate the product from competition and thereby provide a unique competitive advantage.

A Final Thought

Improvement comes from deeds — not from scoreboards.

References

1. Gross, Irwin, 1978, "Insights for Pricing Research," a paper in "Pricing Practices and Strategies," edited by E. L. Bailey, Conference Board, New York, N. Y., 34-39

2. Nickell, Warren L., 1985, "Quality Improvement in Marketing," *The Juran Report Number Four,* Winter, 1985, 29-35

3. Juran, J. M. and Gryna, Frank M., 1980, "Quality Planning and Analysis," Second Edition, McGraw-Hill Book Company, New York, 17-18, 349-350

4. Rosander, A. C., 1985, "The Application of Quality Control to Service Industries," to be co-published by American Society for Quality Control and Marcel Dekker, Inc., Chapter 29

5. Aubrey II, Charles A. and Zimbler, Debra A., 1983, "The Banking Industry: Quality Costs and Improvement," *Quality Progress,* December 1983, 16-20

6. Gryna, Frank M., 1976, "User Quality Costs vs. Manufacturing Quality Costs," *20th EOQC Conference Papers,* Volume B, 11-21. Reprinted in *Quality Progress,* June 1977, 10-13

Sales: $11,900,000,000
Profit: 300,000,000 (or 15.8% of sales)
Cost of poor quality: $200,000,000 (or 10.5% of sales)
Total assets: $1,100,000,000
Asset turnover = $1,900,000,000 / $1,100,000,000 = 1.73.
The current ROI is 15.8% × 1.73 = 27.3%.
If quality costs were reduced from 10.5% to 5% of sales, the profit margin becomes 15.8 + 5.5 or 21.3%.
The ROI would then be 21.3 × 1.73 = 36.8%.

Figure 1 Calculation of Return on Investment

Appraisal — costs of checking, inspecting, and testing prototype units, drawings, specifications, and manuals.

Internal failures — costs of scrapping or reworking hardware and revising drawings, specifications, and manuals during product development

External failures — costs of correcting engineering errors after the release of the design and manuals to production

Figure 2 Categories of Quality Costs in Engineering

Stage	Opportunities
Receiving inspection	Provide data to eliminate inspection
Material storage	Design product and packaging for ease of identification and handling
Processing	Do pre-processing of material (e.g., ready mixed concrete); design product to maximize productivity in customer's process
Finished goods storage warehouse and field	Design product and packaging for ease of identification and handling
Installation, alignment, and checkout	Use modular concepts and other means to facilitate setups by customer
Maintenance — preventive	Incorporate preventive maintenance (e.g., self-lubricated bearings)
Maintenance — corrective	Design product to permit self diagnosis

Figure 3 Opportunities for Improving Fitness for Use

FOOD TECHNOLOGY

THE EFFECTS OF REGULATION ON QUALITY COSTS

(September Issue)

William A. Golomski
President
W. A. Golomski & Associates
Chicago, Illinois

Each nation in the world has its own approach to protect consumers where safety, economic considerations, and product claims are concerned. Occasionally, disagreement arises when the protection offered through regulations is viewed as having too high of a social cost.

Yale Brozen of the University of Chicago has provided a series of brilliant and incisive analyses concerning the cost of regulation. Brozen (1979) indicated that "either costs are already all internalized and no regulation is needed or they go far beyond the point of diminishing returns." Brozen (1978) said that behavior on the part of government officials and corporate officers often has its genesis in government intervention in markets. Rather than gaining protection from regulations, the citizen is swindled by being bewildered by the multifarious activities of government. "Citizens do not find it economical to invest in obtaining information on each of the many activities of their governments," he said.

Even on a journalistic level, we find that Joan Beck's June 10, 1977, column in the *Chicago Tribune*, titled "Our Family Can't Afford...All that Government," reflects the popular opinion that too much of our working day is used to pay for government.

We want protection against carelessness, technological naivete, violence, swindle, and fraud. We want services that can be provided by government. And we want the opportunities to achieve. But we do not want regulations to exceed their value. Fortunately, some regulators, such as the Food and Drug Administration and the U.S. Department of Agriculture are reducing the cost of government by relying on company records to assist in determining compliance. However, these activities do not go far enough to encourage "never-ending improvement" (Golomski, 1949). Furthermore, they do not consider the possibility of contracting for inspection or testing services to maintain quality. On the other hand, some regulatory costs reduce purchasing power and competitiveness across industries. They also reduce manufacturing margins and incomes.

In his early work on quality costs, Talacko (1948) was concerned about measuring the costs of poor-quality government regulations and operations. Golomski and Talacko (1950) later extended these concepts to industry and non-profit institutions such as universities, hospitals, and other service units.

In recent years, it has been more fashionable to eliminate some important costs such as the cost of lost sales due to poor quality, the cost of regulations on internal operating costs, and the cost of regulations on reducing purchasing power and competitiveness across geographical boundaries.

Basic Concepts

There are a variety of approaches to reduce the quality costs of any organization which provides products and/or services (Golomski, 1980; 1983; 1984a; b). One approach, optimizing total quality costs, does not lead to minimizing costs, nor does it lead to optimizing profits. Optimizing is in conflict with the concepts of never-ending improvement in consistency (quality of conformance). It also ignores the cost of lost business.

American executives have lost control of the management of quality in far too many companies. They have delegated the activity to quality specialists who are mainly interested in the technical aspects of products and processes. Quality improvement in every line and staff function in the company is ignored, and top management does not have sufficient sustained involvement and participation in the quality and quality-improvement process to be viewed as the quality champions of the firm; therefore leadership is lost.

The basic concepts in quality strategy are as follows:
- Prevention is preferred to detection of defects.
- Never-ending improvement is essential. Statistical methods will help. This involves more than just conforming to specifications.
- A customer (internal and external) focus is needed.
- Formulations and processes must be foolproof.
- There must be sustained involvement and participation in the quality and quality-improvement process from top management on down.

Powerful statistical methods and other problem-solving tools are not being used by the food and beverage industries to the extent that they should be (Golomski, 1984b).

When industry and government view the opportunities to serve the marketplace better through new quality control approaches, the political process will encourage a new philosophy of regulations. This will save the marketplace, government, and industry billions of dollars.

References

Brozen, Y. 1978. The ethical consequences of alternative incentive systems. In "The 1977 D. R. Sharpe Lectures of Social Ethics." University of Chicago, Chicago, IL.

Brozen, Y. 1979. Regulatory excess. Presented at Washington Spring Conference of the National Association on Manufacturers, March 29.

Golomski, W. A. 1949. Commonality matrices in production planning and control — seven proposals for profit improvement. Presented at Workshop on New Methods in Manufacturing and Distribution, Marquette University, Milwaukee, WI, Sept. 15.

Golomski, W. A. 1980. "Quality Control for Meat & Poultry Processing," Vols. I and II. MPI Training Center, U.S. Dept. of Agriculture, Fort Worth, TX.

Golomski, W. A. 1983. Experience with company officers — Five Fortune 500 companies. Presented at Inaugural Ellis R. Ott Conference on Quality Management and Applied Statistics in Industry, New Brunswick, NJ, Nov. 12.

Golomski, W. A. 1984a. Quality is not a spectator sport. Presented at Rocky Mountain Quality Control Conference, Denver, CO, June 7.

Golomski, W. A. 1948b. World class quality and productivity improvement. Presented at the American Statistical Association workshop on the Use of Statistics in Improving Product Quality and Industrial Productivity, Philadelphia, PA, August 12. (Available from author)

Golomski, W. A. and Talacko, J. V. 1950. Reducing the cost of poor quality in government, industry, and not-for-profit institutions. Mathematics Dept., Marquette University, Milwaukee, WI.

Hogg, R. V. 1984. Report of the University of Iowa Conference on Statistics for Scientists and Engineers, Statistics Dept., University of Iowa, Iowa City, IA, July.

Talacko, J. V. 1948. An approach to improving government and industry: Quality costs. Presented at Charles University, Prague, Czechoslovakia, January.

Based on a paper presented during the IFT Quality Assurance Round Table, "The Cost of Quality," at the 45th Annual Meeting of the Institute of Food Technologists, Atlanta, GA, June 9-12, 1985.

QUALITY PROGRESS

BUSINESS MANAGEMENT AND QUALITY COST: THE JAPANESE VIEW

(May Issue)

Hitoshi Kume
University of Tokyo
Tokyo, Japan

From a management perspective quality economics is more important than quality cost

Recently, the success of Japanese quality control has drawn worldwide attention to the difference between Japanese and Western quality control. One difference is in the attitude toward quality cost. The concept of quality cost was introduced in Japan by A. V. Feigenbaum's *Total Quality Control*[1] in 1961; since then, Japanese manufacturing companies have tried many ways to employ this concept for management. But while many Japanese companies record their failure cost correctly, few of them — except some foreign affiliate companies — can completely describe their quality cost. In Western countries, by contrast, many companies are so concerned about identifying quality cost that there seems to be an impression that quality control activities cannot exist without a formal quality cost system. An indication of the Western attitude is the special issue of *Quality Progress* devoted to the topic of quality cost.[2] Another is the current proposal that International Organization for Standardization Technical Committee 176 on Quality Assurance consider preparing a guide related to this subject as an international standard.

The author has tried in the past to apply the Western method of quality cost to several Japanese companies but failed to get a good result. Why? Because it is impossible to evaluate all of a company's quality control activities only through quality cost information, although it might be a useful measure in some limited situations. When failure cost is very high, quality cost information can be a useful piece of the picture — but it is not the whole picture.

Let us begin by looking at quality cost from a management perspective, the way a CEO or plant manager would. These top managers want to maximize profit, and they realize that they must consider a broad range of factors in order to determine the best way to reach that goal.

Perhaps the first point revealed by such a management analysis is that minimum quality cost does not mean maximum profit. A consideration of quality economics, as contrasted with quality cost, may reveal a different picture than quality cost alone would show.

Table 1 shows sales volume, failure cost, and profit in two divisions — the video tape recorder division and the television division — of a Japanese electronics company. These figures show that in the VTR division, the ratio of profit to sales is higher than in the

Some Principles of Quality Economics

- Minimum quality cost does not necessarily mean maximum profit.
- Minimum quality cost does not necessarily mean minimum product cost.
- Losses due to failure cannot be calculated only by failure cost.
- The cost of marketing research should be included in prevention cost.
- Quality of design cannot be evaluated by quality cost.
- The important thing about prevention and appraisal cost is not the total, but the way the money is used.

Table 1.

Profit and Failure Cost:
A Comparison of the VTR and TV Divisions
(in millions of dollars)

	VTR Division	TV Division
Sales volume	$540	$365
Failure cost (ratio to sales)	17.6 (3.3%)	2.4 (0.7%)
Profit (ratio to sales)	53.6 (10.0%)	20.8 (5.7%)

TV division, even though the ratio of failure cost to sales in the VTR division is very high.[3] Much of the failure cost is due to the fact that the VTR is a new product that has come to the market only recently; many elements in the process of design and production have not yet been refined. However, profit is increasing steadily because of strong demand for VTRs.

In the TV division, on the other hand, the process of design and production has been developed almost completely. As a result, the quality of the product is stable and there is very little loss due to quality problems. Nevertheless, it is getting harder to make a profit because TVs are in wide use throughout Japan, and the division has many competitors.

Generally, if a product is excellent, sales increase while the profit margin per piece decreases over the years. Why? When a product is very profitable, competitive items will appear in the market, unless the product has a patent or is under legal protection. In many cases, cost reductions resulting from mass production and from improvement in design and manufacturing processes cannot keep up with the reductions in price compelled by competition. That is why the profit margin decreases. In order to continue to grow and maintain profit rates, companies must develop, on a timely basis, new products that meet customers' requirements.

From the standpoint of qualty control, a new product is full of trouble. It contains many untried elements that cause unexpected problems. In most cases, the development of a new product increases quality cost, especially failure cost. However, the product is a success from a managerial, profit-oriented point of view as long as the increase in sales income exceeds the increase in cost. Reduction of unnecessary cost — including quality cost — is a problem to be solved in the future.

Many cases can be used to illustrate the point that minimum quality cost does not mean maximum profit. For example, Figure 1 plots the profit margin and failure cost (cost of reacting to customer complaints) of a TV camera that has been on the market since 1978. (The base year is 1978.) As we can see, both failure cost and profit margin are falling.

Let us take another and more obvious example. XYZ Company has a factory that produces 1,000 pieces of service repair parts per month. The current process percent nonconforming is 0.1%. An increase in demand makes it necessary to raise production to 1,500 pieces per month. Because the present production facility is at capacity, XYZ Company decides to use an old facility, too. That old facility produces five percent nonconforming. The variable costs — both direct and indirect — such as material cost, and others associated with operating the old facility, are $6 per piece, while at the present facility these costs are $5. The parts sell for $25. Table 2 shows gross profit (sales revenue minus variable production cost) before and after the production increase.

To calculate total product cost, fixed costs and general indirect costs should be added to variable costs. However, in whatever way we make this calculation, the increase of gross profit after the production increase is

$$28{,}850 - \$19{,}975 = \$8{,}875.$$

Quality cost, on the other hand (in this example, we are only concerned with internal failure cost) increases from $5 to $155 and the ratio of quality cost to production cost climbs from 0.1% to 1.94%.

The goal of business management is to increase profit, not to reduce cost. A cost increase will not be a managerial problem as long as the company obtains more profit to offset the additional cost. The most important thing about quality control is to determine whether the design, production, and marketing of a product that meets customers' needs are being carried out effectively. No matter how hard we try to cut the cost and reduce the percent nonconforming of a product, we will not be able to increase profit very much if our product itself is out of date.

It is a matter of fact that the lower the quality cost, the better — if other conditions are the same. Reducing quality cost is an ongoing managerial task. However, lower quality cost is not necessarily a sign of successful management. If a firm does not develop a profitable product and doesn't expand into new markets, but rather continues to do its business only with its old product in a limited area, the quality cost, generally speaking, is not large — but such a company can hardly be called a success.

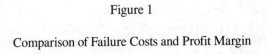

Figure 1

Comparison of Failure Costs and Profit Margin

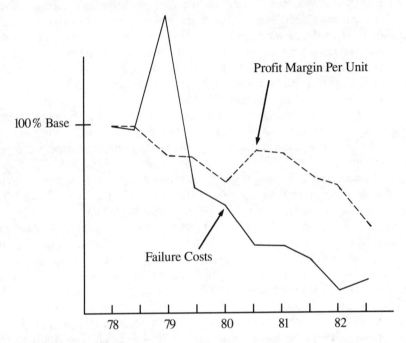

We have seen that minimum quality cost does not mean maximum profit. But we can go even further than that, because minimum quality cost does not necessarily mean even minimum product cost.

To gain a profit, the first step a company should take is to develop a product that customers really want, and in the process to set an acceptable price. There is, of course, another route to profit. We can define profit as follows:

Profit = Sales — Cost.

This equation makes it obvious that another means of gaining a profit is by reducing cost. Though cost reduction is not one of the main objectives of business management, it is among the most important activities to be carried out by a company.

Product cost includes many elements, and quality cost is one of them. The thing to remember is that those elements are interdependent and that reduction of one cost element does not necessarily result in the reduction of total cost. The same can be said of quality cost.

Take the example of a company that produces electronic computers. The company purchased semiconductors that had a fraction nonconforming of five percent — very high for effective use. The company asked the supplier to improve the quality of its semiconductor to an AQL of 0.1%. The supplier replied that it could not comply with the request unless it doubled the price. The customer could not accept such an increase; instead, it made an automatic classifier for the incoming semiconductors to obtain the required quality. Although quality cost (appraisal cost) increased considerably because of installation, operation, and maintenance of the classifier, total cost was reduced: the company would have spent more if it had agreed to pay twice as much for semiconductors with an AQL of 0.1%. Thus, an increase in quality cost helped to reduce total cost.

Minimum quality cost does not necessarily mean maximum profit, nor does minimum quality cost always lead to minimum total cost. In the management of cost, cost is broken down into various elements and analyzed. Because these elements are closely connected, an increase in one might reduce the whole, while a decrease in another might cause an overall increase. If you look at only one element of cost and ignore its relation to the whole, you run a great risk of producing unintended results.

A quality cost system will be useful when a department, which is limited in its scope, tries to reduce its internal costs. Such a system is not so useful, however, when an entire company tries to plan or evaluate its activities strategically from the standpoint of overall business management.

So far, we have been looking at quality cost from the outside, so to speak — at the limitations of quality cost in light of the need to increase profit or reduce total cost. When we look at quality cost from the inside, we find other limitations.

Let us start with failure cost. It is clear that losses due to quality failure cannot be calculated only by failure cost. Let us look back at XYZ Company. If the company operated its old facility, its fraction nonconforming would be five percent (25 pieces); failure cost would amount to $6 \times 25 = \$150$ per month. Can we say that the loss due to failure is $150 per month? Suppose the demand for this product is very strong. If XYZ Company cannot produce enough pieces to fill that demand, other companies will capture part of the market. Then the loss due to 25 nonconforming parts (remember that gross profit per piece is $19) becomes $19 \times 25 = \$475$. This loss is not an actual loss but a potential loss arising because the company did not have the opportunity to sell those 25 parts. In other words, this amount would be changed to profit if there were no failure.

The thing we have to remember is that loss due to quality failure includes loss of market, not only the cost of failures revealed by inspection or a customer's complaint. This invisible loss, which can't be calculated in many cases (as books on quality cost acknowledge), is sometimes larger than visible loss.

To see just how serious these invisible losses can be, let us look at one example. Thirty years ago, a powdered milk manufacturer with the largest market share in Japan shipped product containing an impurity of arsenic; the problem was due to a failure in the quality control of raw materials. The incident created a great social problem. The manufacturer incurred a large failure cost, of course, but the damage to the company went far beyond that. The company has never been able to recapture its old market share.

A manager of a copy machine company put it this way: "It's easy to say that you should develop a good product to raise market share, but rather hard to realize it. If a competitor gets involved in a quality trouble, however, we can get the market without any effort. Our competitors think in the same way."

This principle is also important in service industries. Suppose that employees at a hotel are not trained well enough and, as a result, standard services are not provided to customers. In this case, even if the customers were dissatisfied with the service, the

Table 2.

Gross Profit and Costs Before and After the Production Increase at XYZ Company

	Before	After	Increase
A. Production Costs (variable)	$5×1,000=$5,000	($5×1,000)+($6×500) =$8,000	$3,000
B. Sales Revenue	$25×1,000×0.999 =$24,975	$25 [(1,000×.999) +(500×.95)] =$36,850	$11,875
C. Gross Profit (B)−(A)	$24,975−$5,000 =$19,975	$36,850−$8,000 =$28,850	$8,875
D. Failure Costs	$5×1,000×0.001 =$5	($5×1,000×0.001) + ($6×500×0.05) =$155	$150
E. (D/A)×100%	(5/$5,000)×100% =0.1%	($155/$8,000) ×100%=1.94%	1.84%

hotel would suffer no failure cost. But these customers will never come to this hotel again. The hotel may not have incurred failure costs, but it has lost customers — in other words, part of its market.

It is obvious that failure cost is only one visible element of the loss caused by quality failures. Actually, the largest loss is sometimes the loss of market, not failure cost. It is clearly quite dangerous to forget this fact and discuss only visible failure cost.

Failure cost is not the only category that is often too narrowly defined. When we look at prevention cost, for example, we see that the cost of marketing should be included in this category. To see why, let's look at what happens when goods manufactured by a company are sold. In this situation, there are two ways to set up a product standard: the purchaser can set up a standard and require a supplier to design and produce a product that meets it; or the supplier can set up a standard and supply products to unspecified purchasers that will buy the product if they like it.

Let's call the former a contract-led product, the latter a market-oriented product. In the case of a contract-led product, the item should simply meet the standard that the purchaser set up; whether the standard is adequate for the customer's intended use is the purchaser's business, not the supplier's. In the case of a market-oriented product, on the other hand, the most important thing is whether the features of the product make it suitable for the customer's intended use; whether the product meets the standard set up by the supplier is less important. If the supplier sets a standard that is not adequate, complaints will arise, even if the product meets the standard. Let's look at three examples.

- When a Japanese TV maker exported its product to Canada for the first time, trouble occurred frequently in some areas. The reason was that the product was placed in warehouses where the temperature got down to $-40\,°C$; as a result, some parts of a circuit were broken. The company had designed the product to be stored at temperatures as low as $-25\,°C$; this standard was good enough in Japan but not in Canada.
- A camera maker encountered many market claims in a certain Asian country because the inside of the lens got musty. Why? A part in the camera gathered a mold peculiar to that area. To prevent mold, the material used in the part had to be changed.
- On some Japanese automobiles exported to North America, the lower frame of the car rusted from exposure to road salt. The problem never occurred Japan, where roads are not salted in winter; as a result, the quality standard of the car was not adequate for use in those new areas.

In all three of these examples, trouble arose because the product standards were not adequate to use or storage conditions. In Japanese manufacturing companies, in most cases, failure due to inadequate product standards amounts to more than half of the total failure cost. It is not easy to prevent such failure. Existing product or test standards are often inappropriate for a new market, but it is only after problems arise that a company recognizes the inappropriateness of the standards — only then can a company take measures to prevent a recurrence of the problem.

According to cerebral physiology, a man's ideas never go beyond his experience and knowledge. It is clearly difficult to foresee the unknown and take an action beforehand. Yet some failures could be prevented, to a certain extent, if the customers' needs and the conditions of use, transportation, and storage are investigated as carefully as possible. When a company is planning to extend its business to a new market, marketing research is definitely necessary to prevent external failures. Such research is often performed by the marketing department or the design and development department, not by the quality or reliability department. As a result, the cost of marketing research — an important means of preventing external failure — may often be assigned to marketing cost or design and development cost. More realistically, however, it should be classified as a prevention cost.

There is one more area that is extremely important, but cannot be evaluated by quality cost: quality of design. In a market-oriented product, an inappropriate product standard causes not only complaints about quality, but also a more fatal problem. The quality of a product — including price — is conformance to the requirements of customers. In a case of market-oriented product, when a product design doesn't meet the customers' needs, we find not a problem with complaints, but a problem with sales: the product won't sell. In the market, customers check the features of a product: if they like them, they will buy; if not, they won't. A product that is not sold doesn't generate complaints. Only a product that is sold, and then fails to meet the customers' expectations, brings a complaint.

Clearly, the quality of the design conception can't be evaluated by market claims or complaints. The loss caused by a design that doesn't meet the market's requirements is calculated in terms of sales volume, not cost. As a result, it cannot be calculated by quality cost, either.

Despite the limitations of quality cost systems with regard to prevention cost, such systems can be useful. There is one thing to remember: the important thing about prevention — and appraisal — cost is not the total, but the way they are used. Prevention and appraisal activities help reduce failure cost. This requires investment. Figure 2 shows the most economical level of investment in prevention and appraisal, and demonstrates the economic necessity of quality control activities. Conceptually, this figure is correct. However, failure cost and prevention and appraisal cost should be considered separately. Failure costs represent waste; they are genuine losses, because they would not be expended if quality were perfect. Therefore, if conditions of sales and production are the same, the lower the failure cost, the better.

Prevention and appraisal costs, on the other hand, are spent to reduce failure cost. Lower prevention and appraisal costs increase the cost of failure; however, an increase in prevention and appraisal costs does not always reduce failure cost. The important thing is the way prevention and appraisal activities are carried out, not the amount spent on those activities.

Usually, some are very effective, while others are wasteful. Quality — not quantity — is important. We can even say that, when analyzing prevention and appraisal activities, money spent ineffectively should be categorized as failure cost.

In Figure 3, monthly prevention and appraisal cost and failure cost of a Japanese factory are plotted over a period of time. Prevention cost is slightly increasing due to an increase in labor cost. The increase in labor cost is due to an increase in wages, not to additional personnel. Failure cost, on the other hand, is decreasing steadily. Thus, the figure shows that prevention activities are being carried out effectively and are resulting in fewer quality-related losses because problems are being corrected.

Quality, not quantity: that is an important principle to keep in mind when thinking of quality cost. There is much discussion about what the proper level of quality cost should be for a company. As we have seen, the "proper" level depends on what the product is; we can't say the ratio of quality cost to sales should be below a fixed rate for all companies. In the semiconductor industry, 10% may be good, but in the TV manufacturing industry, 3% is not good. It is a matter of fact that the lower the failure cost, the better — if the product and conditions of production are the same. The most important thing is to carry out activities effectively to decrease failure cost.

However, the reduction of failure cost is not the only thing to be done in quality activities. Quality control refers to those activities performed "to grasp the customers' requirements and to supply a product economically which meets them." Only some of those activities can be evaluated by quality cost; it is therefore inadequate to evaluate all of the activities only by quality cost. If we use only quality cost to measure quality control activities, we will narrow the range of quality activities.

If quality activities are not carried out effectively, many losses will arise. Some of them are visible and the others are not. A quality cost system always deals with visible cost. In business management, however, the invisible is also important. A company guided solely by the visible costs may very well be mismanaged.

Figure 2

Conceptual Relation Between Quality Cost and Quality

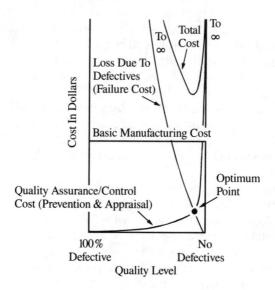

Figure 3

Quality Cost Trend by Month

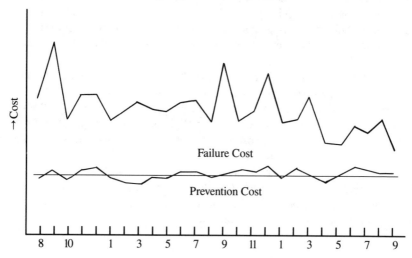

Acknowledgment

The author would like to thank Robert W. Peach, who suggested writing this article, and Jean G. Moreau, who offered many comments from the viewpoint of Western companies. The author also thanks Richard A. Freund, who offered many editorial comments on the English of the original version.

References

1. Feigenbaum, A. V., *Total Quality Control,* Third Edition, Chapter 7. New York: McGraw-Hill, 1983.

2. *Quality Progress,* April 1983.

3. It is worth noting that the failure cost for the VTR division — although higher than comparable cost in the TV division — would be considered low by traditional Western standards. A quality cost system may be more useful in cases where failure costs are significantly higher than in these examples.

QUALITY COST: BETTER PREVENT THAN CURE

(September Issue)

Alain M. Chauvel
Quality Director
Société de Contrôle Technique (SOCOTEC)
Yves A. Andre
Quality Department Manager
SOCOTEC Export

Quality must be measured in order to assess the performance of a company's staff and personnel. Quality cost is one of the ways to measure quality. A quality cost system does two things: it allows for an identification and valorization of the investments made to prevent nonconformities and errors; and it reveals loopholes and problems that prevent a company from attaining the necessary performance and competitiveness.

The quality cost concept is not an innovation. It is much argued about, and is misunderstood in most firms. There is contention between those concerned about quality cost and those who emphasize non-quality cost.

Yet, today, it is necessary to extend these two concepts. It does not suffice any more to assess only the costs attributable to such things as quality department operations, rework, scrap, repairs, and claims. Management errors must, in the same way, be incorporated in the global cost. All these are opportunities to be taken hold of and turned into profit.

Quality management cannot and must not be the job of one man, as good as he may be. It is the role of a management staff really concerned with the future of the company and of people who rely on them.

To get an idea of the range of things having an effect on the company's quality performance, we can examine and highlight the results obtained through a "quality diagnosis"[1] of small- and medium-size French firms (PME — Petites et Moyennes Entreprises). From the 113 diagnoses completed between September 1982 and July 1983, 54 were selected as being significant. Many of the other companies could not be included in this study because most PME do not operate an appropriate accounting system (as far as quality cost data acquisition is concerned). In other firms, the error margins related to the most important costs would have jeopardized the credibility of this analysis and were therefore not acceptable.

The data acquired and taken into account were selected by a team that included the company's staff and two SOCOTEC quality engineers. The first day of a diagnosis is spent introducing our engineers to the company's staff, defining objectives, and explaining the process to be implemented. A diagnosis lasts three to seven days, scattered over a period of three to eight months, depending on the size of the company and the budget allocated. Such an approach allows for most of the staff involved in various functions of the company to understand our procedures and gives them sufficient time to collect the information we are looking for.

Quality is a matter of a turn of mind based on communication, not on improvisation. It therefore takes time and patience.

The factors selected for this study are numerous but they conform to the principles of modern quality management, i.e., they satisfy a need, on time, and at the lowest possible cost. These factors have been classified as:
- Prevention cost
- Appraisal cost
- Failure cost.[2]

Prevention cost represents the deliberate investment that a company makes to prevent nonconformities and errors. Prevention cost is the very outset of quality. Sufficient time must be allocated to defining the product, the objectives, and the means necessary to master quality.

Prevention cost incorporates the following: administration of the quality function; design and planning of verification actions; preventive maintenance of equipment; review and updating of instructions, specifications, and procedures; surveys related to product warranty; and personnel training.

Appraisal cost represents an investment that may or may not have been made freely by a company. The company may decide on its own to control the quality of its production or a system of control may be imposed upon it by major customers. Depending on the company, such control may be centralized or decentralized.

Appraisal cost incorporates the following: appraisal of prototypes, new materials, methods, and processes; inspections performed by the company's quality department; equipment, supplies, samples, and premises necessary for the performance of inspections and tests; verifications by external laboratories or organizations; operators' self-inspection; and evaluation of competitors' products.

Failure cost represents an unintentional loss that is very often concealed. Compared to the two other types of costs, failure cost includes a set of factors that are sometimes difficult to assign value to. If a company does not appreciate the extent of such costs, it may limit or even preclude the possibility of taking the corrective actions necessary to improve its performance.

Failure cost incorporates the following: losses included in standard costs; manufacturing losses and extra consumption of materials; losses on stock and inventory variance; material substitutions; product recall; lost sales and penalties; disputes and visits to customers; sorting, rework, and repairs; disposition of scrap; inventory of defective and obsolete items; excess inventory; underused and standby equipment; losses of manpower efficiency; and insurance for civil liability.

The sample used as a basis for this paper — 54 firms — does not reflect the situation in industry as a whole. It covers some PME which have accepted — often thanks to the efforts of the French Ministry of Industry — the idea that a quality diagnosis could give them matters to think about to improve their performance and competitiveness.

Figure 1
Classification of Firms in Sample

In terms of activity
26% in the construction material sector
20% in buildings and public works activities
18% in mechanical activities
11% in carpentry
9% in the chemical sector
16% in others: medical devices, cosmetics, electricity, electronics, food products, and services

In terms of sales*
44% less than 25 million francs
48% between 25 million and 100 million francs
8% from 100 million to 400 million francs

In terms of workforce
81% having fewer than 200 employees
19% having from 200 to 800 employees

*Exchange rate as of June 1, 1985: French franc = $.1077.

These firms can be classified as shown in Figure 1. These figures are quite representative of small- and medium-size firms, but the results obtained cannot be extended to larger firms without excessive risk of error.

What conclusions can we draw from this study? Let us look first at the effect of prevention cost. As Figure 2 shows, there is a correlation between greater investment in prevention and lower total quality costs. With such data, it is safe to say that prevention efforts have a direct and positive influence on the company's profit margin.

The picture is not quite so clear when we add appraisal cost to our analysis. Figure 3 reveals a correlation between prevention and appraisal investment and the reduction of total quality cost. The correlation in this case is not as evident as it was for prevention cost alone. Some apparently contradictory situations show up. If we analyze these situations, the following points emerge:

- The situations involve firms that have very heavy inspection systems, often much final inspection and very little in-process inspection, i.e., a costly inspection scheme.
- These firms consider that sorting equals control because they have always operated that way, systematically. In reality, of course, inspection is only a remedy given the name of quality action.

Figure 2
Reducing Quality Cost: The Effect of Prevention

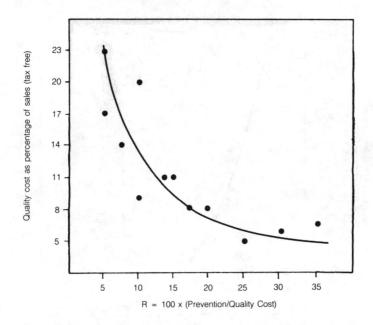

R = 100 x (Prevention/Quality Cost)

Figure 3
Reducing Quality Cost: Prevention Plus Appraisal

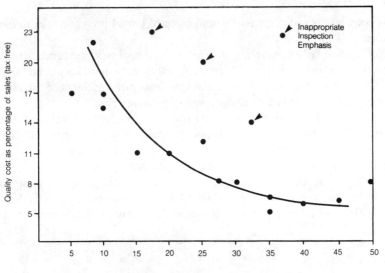

R = 100 x (Prevention + Appraisal/Quality Cost)

Figure 4
Failure Cost: Effect of Company Size
(Number of Employees)

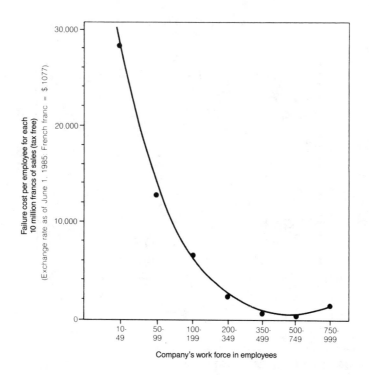

- Prevention efforts and corrective actions oriented toward prevention are rare or nonexistent.

Considering these factors, it is reasonable to think that both prevention and appraisal efforts are necessary to reduce quality costs. Nonetheless, we must keep in mind that appraisal or inspection efforts, without appropriate investment in prevention, may lead to a feeling of false security and may generate unnecessary costs.

In order to compare companies with one another and to assess the effect of problems and organizational loopholes in companies of different sizes, the firms have been classified in terms of number of employees. For each class, the cost of failures per year has been computed per person, for ten million francs sales. (See Figure 4.)

The correlation is outstanding: quality cost decreases dramatically with the size of the firms. One has to be very cautious about the three last values (350-499, 500-749, and 750-999) because the number of firms in these classes is not sufficient to make them significant.

Figure 5
Reducing Quality Costs: Effect of Prevention and of the Size of the Company's Workforce.

(Values shown in the boxes indicate failure cost (in thousand francs) per employee for each 10 million francs of sales (tax free).

Company's work force in employees

	10 to 49	50 to 99	100 to 199	200 to 349	350 to 499	500 to 749	750 to 999
< 3			11.2				1.9
3 - 7	29.7	20.0	8.2	4.3			
8 - 12	51.4	5.9	7.9	2.5			
13 - 17	30.0	11.9	7.7	2.3			
18 - 22	23.2	9.4	4.8		0.8		
23 - 27	11.3	5.6	4.0	1.5			
28 - 32		5.9	3.2			0.5	
33 - 37				1.4			

100 × Prevention/Quality Cost

(Exchange rate as of June 1, 1985: French franc = $.1077)

Two causes seem likely to account for this result:
- Larger companies are better organized. For instance, they probably have a product engineering department or a quality department.
- The cost of errors is spread over a larger sales amount.

We also analyzed the effect of prevention on companies of different sizes. Figure 5 shows the results of this analysis. It also highlights the fact that, whatever the size of the company, prevention efforts directly contribute to reducing quality cost.

It is quite obvious from this study that investments in both prevention and appraisal are necessary to reduce quality costs. But investing exclusively in appraisal is a mistake that may lead to unacceptable costs; and the protection it affords to the company's reputation is a fragile shield that will fall off anyway, one day or the other, after having taken a large bite out of the profit margin necessary to survive, develop, and stand in readiness for competition.

Figures related to quality cost are a useful performance indicator for a company's executive. This study leads to the conclusion that, whatever the size of the company, quality directly contributes to the profit margin.

It is a safe investment: better prevent than cure.

References

1. "Quality diagnosis" is the term used by SOCOTEC for a specific assessment method.
2. Details about these costs obtainable from the author.

1986

ASQC ANNUAL QUALITY CONGRESS TRANSACTIONS

COSTING QUALITY FOR SENSITIVITY ANALYSIS

Hugh E. Kroehling
Quality Control Engineer
Lawrence R. Petersek
Quality Control Engineer
AT&T Technologies
Richmond, Virginia

Abstract

The extraction of meaningful quality cost numbers in an accounting system that measures its cost on the basis of asset flow is difficult at best. The problem is compounded when a system is desired which will collect data on a regular basis and apportion costs according to standard designations of Prevention, Appraisal, and Nonconformance. The purpose of this investigation is to determine if a method of measuring costs in such a manufacturing environment can be developed that will consistently provide useful numbers without requiring an army of bean counters or a significant change in the present accounting system. In addition, the system must allow for sensitivity analysis so that results of changes in the manufacturing environment can be traced through changes in costs of quality.

At the outset, some costs will be easily extracted while others will be difficult to even estimate. The sensitivity of each "pocket" of quality dollars must be examined to see if it is worth the effort to reach for it. Some areas will require too much effort to get information and some will require such gross estimates as to remove sensitivity to change. It is hoped that results of this effort will lead to a method that will capture at least 85 percent of quality costs using current available accounting and manufacturing data.

The initial assumption in a study of the cost of quality is that every cost accounted for in operating has some portion that is present because of quality (or unquality). Every item of labor, material, and overhead is examined when reported to accounting and some means of extracting the cost of quality is considered.

The hierarchical order of investigation for this study is as follows:
1. Can a cost be totally assigned to quality? (Such as costs associated with an inspection organization or scrap costs).
2. Can the portion of a cost be logically broken out by the known function? (Like the number of process checkers in an operating department or the percent of the wage incentive rate included for checking work.)

3. Is an estimate of the proportion of expenses devoted to quality considerations the best that can be provided? (The percent of accounting expenses devoted to correction of errors, or reproduction expenses resulting from errors, etc.)
4. Once gross cost numbers are identified, two subsequent and parallel considerations are tackled:
   a. Costs must be apportioned or assigned to the categories of prevention, appraisal, and nonconformance. This is done by functional split in the organization reporting the costs.
   b. Quality costs not directly assigned to specific product lines should be apportioned by tons shipped, dollars shipped, or some other indicator.

Sensitivity to changes in quality is the prime consideration when deciding between alternative means of apportioning costs in the steps above.

A significant benefit of costing quality as described above is that one can easily, quickly, and accurately get answers to such questions as: what was the cost of the design error on model number 7253B? What was the cost of the water leakage in area 75? What was the cost of John Doe drilling 200 holes the wrong size? What is the ratio of the cost of conformance to nonconformance on product line A versus product line B?

Once the sources and formulas for extracting costs have been decided upon, the cost of quality is extracted according to the algorithms established.

Text

It is important early on to establish the philosophical approach to assigning costs to quality. Much arguing can result deciding if a certain cost is a cost of doing business or a cost of quality. Much of the disagreement can be attributed to the pejorative nature of the words "cost of quality." It is natural to not want costs over which one has little or no control assigned to quality, because of the automatic relation to numbers so designated. It *can* be legitimately argued that these are costs of doing business as they are always present and relatively constant.[1] However, it can also be argued that any cost that is there to prevent defects, measure defects, or results from defects is a legitimate cost of quality. The most important consideration here is that whatever interpretation is used, it should be consistently applied and that fear of overinterpretive "management" be controlled. The danger is that information will be watered down to the point that the cost of quality becomes another meaningless result to be manipulated through definitions rather than a tool for genuine cost and quality improvement.

The approach for this study is that all costs are initially considered as quality costs unless they can obviously and unquestionably be accepted as costs of doing business. Every item of labor, material, and overhead is examined for a means of extracting its quality content. This approach is tedious initially but forces a thorough examination of many items and assures that no important pocket of dollars is left out.

Prevention, Appraisal, and Nonconformance

When assigning costs to the categories of "prevention," "appraisal," and "nonconformance," it is important to keep the organizational picture in mind. Appraisal costs incurred in one department may be prevention costs when looked at from a downstream department's point of view. For example, checking the calibration of a measuring instrument is appraisal to the instrument shop but could be considered a prevention cost to the organization that depends on the instrument's accuracy to make good product.

The authors feel that the first priority for reporting cost of quality is to look at costs from the standpoint of the end product or service being produced. We have defined three distinct categories of quality costs; nonconformance, appraisal, and prevention, as follows:

1. Nonconformance costs are those costs directly attributable to failure whether it is the product itself, the breakdown of a machine, or the wrong part being ordered from the storeroom.
2. Appraisal costs are incurred when measurements or evaluations are done to detect defects or failure occurrences.
3. Prevention costs are the most difficult to identify *and* quantify and are potentially the most emotional. Francis X. Brown, Quality Manager of the Westinghouse Low Voltage Breaker Division, does not even accept the notion of appraisal and prevention costs as costs of quality. Mr. Brown believes that appraisal costs and prevention costs are just part of "doing the job" and are irrelevant to the problem of increasing profitability.[2] The authors find this view limits the usefulness of a cost of quality system as a tool for corrective action. It does not permit the development of algorithms for determining how much prevention is required to attain the minimum cost of quality.

Determine How to Measure Costs

Quality costs should be assigned to product lines. This allows concentrating engineering and shop efforts on high cost product lines. This can only be accomplished if accounting breaks up its operating cost centers by product line. This way, all repair, rework, touch up, inspection, test, maintenance, supplies, etc., which are normally accounted by cost center, will be broken down by product line.

What about engineering departments or managers that are not assigned to specific product lines? These groups must have their time apportioned to each product line. The basis for the apportion should be the one that most directly correlates with quality changes to the product. For instance, the Quality Control department may have engineers, managers, and inspectors that each work on several different product lines. These people's expenses should be apportioned to the product lines based on either production, sample size at final inspection, or should be assigned by the Quality Control department chief and reviewed periodically for any changes. If based on production, quality costs will increase with production increases which may not be valid. If based on product sample size at final inspection, quality cost will increase with increased sample sizes. Increased sample sizes are usually a direct relationship with increased quality problems, therefore, this would be a good means of apportioning quality costs.

Verification of Quality Costs

Once cost of quality is measured, experimentation should be done to see if cost algorithms reflect changes in quality. For instance, consider a chemical bath that is contaminated. Product is run through the bath without knowledge of the contamination. After subsequent processes, the product becomes noticeably defective. Once the cause of the problem (contaminated chemical bath) is identified, Maintenance can clean the tank, correcting the problem and production can begin again. What is the cost of the contaminated chemical bath? How do we know this is accurate?

The cost of the contaminated chemical bath is the cost of repairing or scrapping product run through the contaminated bath. It is also the cost of breakdown maintenance and all necessary supplies for the chemical bath, as well as the cost of Engineering work on the problem and any additional overtime spent by the shop trying to catch up on production. The cost of repair and scrap are calculated in a manner consistent with the Accounting Department. The cost of breakdown maintenance is that cost charged to the Chemical area by Maintenance. Included in this cost is the charge for any chemical supplies needed to get the bath back into production. The cost of overtime to the shop is charged directly to shop cost centers that had to work overtime. This allows tracking overtime by product line, since cost centers are divided by product line, as stated earlier.

What if overtime is not due to quality, but simply to increased demand? If production schedules are greater than what the shop is capable of producing on regular time with no quality problems, the amount of overtime necessary to meet schedules can be estimated and should be subtracted from actual overtime to arrive at overtime due to quality problems.

What if quality problems result in not meeting schedules? There are obviously costs of not meeting schedules, for instance, costs of not getting future business from schedule conscious customers, costs of bargaining with customers to accept product behind schedule, or lost profits if price must be reduced on product behind schedule. The costs are not included in our cost of quality program because they can be intangible and we were not able to develop an algorithm to capture these costs. As these costs are dependent on the customer, they are out of our control, hence there would be no consistent base for measuring these costs. Without a consistent measuring base, sensitivity analysis is impossible. Consequently, we chose not to capture these costs, so that we could benefit from more accurate analyses of the costs we do capture.

How do we know costs are accurate? The accuracy of costs is easy to verify. Direct costs are accurate as long as all costs are charged directly to the cost centers that were the cause of quality problems. For instance, the direct charge to the Chemical shop by Maintenance for cleaning baths is accurate if no unseen costs are evident. In other words, if two Maintenance men work eight hours each on cleaning tanks, and several different supplies are necessary, the costs of the two men's hours and supplies must be accurately charged to the shop. No more and no less should be charged. A quality control engineer coordinating quality costs should verify that the Maintenance charges are what they should be.

What if quality problems are not found until end of the line inspection? The cost of end of the line defects caused earlier must be charged to the shop cost centers where the defect occurred. This is done by analyzing defects so that cause can be determined. This is not always possible. We are able at AT&T in Richmond, Virginia, to pinpoint causes of 65 percent of end of the line defects.

What is the cost of quality when engineering replaces a machine in the shop due to its quality problems versus replacing a machine simply because it is old? Costing quality here is an Accounting function. The cost of quality is the book value minus the selling price of the equipment. Equipment with high book values which are replaced due to quality problems will show a high cost of quality unless they can be sold for a price close to the book value. This is an accurate assessment of the cost of quality, as there is no cost of quality, even for a brand new machine that performs unacceptably, if it can be sold for what it cost to buy. Similarly, the cost of quality is close to zero for equipment that is replaced due to age since its book value will be very small and its resale value will also be small, if not zero.

Measure Quality Costs by Department

Defining and extracting costs of quality for expense organizations where job functions are not clearly recognizable as being quality related is difficult without keeping complicated diaries. Diaries are subject to wide variations in accuracy and should be eschewed if possible. Constant percentages is one possible way of recording cost of quality but provides no sensitivity to the actual quality effort required during a selected period. The problem is to measure quality effort with some sensitivity without resorting to measurement techniques that are incursive on the employees' time (such as diaries). Accurate measurement of the cost of quality content in any expense department's charges will require a knowledge of the department's function and organization, coupled with an examination of the details of the charges that are available. This may lead to the apportioning of individual line items in the expense summary to quality at different rates depending on the extent at which they are made up of quality costs. This apparent overconcern with accuracy for comparatively small volumes of dollars stems from one of the desired goals of the quality cost system: to provide each organization with a regular measure of the previous periods' cost of quality assignable to itself. Ultimately, costs will be apportioned into three categories of prevention, appraisal, and nonconformance to provide a tool for allocating department resources.

An example of this type of analysis for a Richmond Product Engineering department is shown below:

Department function: To provide engineering support for the following functions:
1. Instructions for manufacture,
2. Process/product compatibility,
3. Design and quality considerations,
4. Customer technical concerns,
5. Product costs.

Of the people in the department, some number are exclusively dedicated to the investigation of product yield improvement. In addition, an amount of engineering travel is related to quality considerations. The remaining (non yield) engineering effort is involved to some extent in design reviews, customer interfaces, and decision making situations that are at least partly related to quality.

Combining the knowledge of the job organization, the descriptions of the individual line items, and the awareness of how quality costs are being extracted in other areas allow us to make the following first analysis of the expenses:

1. All repair costs must be extracted before any further allocation because they are accounted for in the maintenance summary and are considered as cost of nonconformance under that banner.
2. It is estimated that 50 percent of the general travel and living expense is quality related because of trips to the customer required when quality problems arise. The line item(s) for this expense should be removed from the remaining total and treated as a separate and variable cost of quality that is mostly a cost of nonconformance (because the trips are generally made in response to known defective conditions).
3. The remaining expense dollars may be apportioned as follows:
    a. Yield investigators = 12% of effort.
    b. It is estimated that 3% of the departmental effort is spent on some form of quality related customer trip at all times.
    c. Apart from traveling, approximately 6% of the department is working on quality related items at all times.
    d. Total apportionable cost = 21%.

The end result of a cost of quality program is for management to direct effort at reducing high costs of quality. No better method of reducing quality costs exists than displaying quality costs by department. Each department will have incentive to reduce quality costs. Incentive will come from management if not from the department's embarrassment of having high quality costs.

The cost of quality algorithms must be used to calculate costs and allocate them to shop, engineering, maintenance, and other professional and shop departments. It is essential that the calculation and allocation of quality costs be accurate (see verification of quality costs) if management is to use the system to direct efforts at reducing costs. Employees will not believe in the system, and hence lose incentive to reduce quality costs, if department costs are not accurate. If the quality cost program can be sold to all employees, prevention will increase, and quality costs due to scrap, rework, repair, errors,. . .will decrease.

Sensitivity Analysis

Sensitivity analysis allows us to measure cost effectiveness of corrections. For example, consider the cost of adding an additional process checker on a chemical bath versus the cost of not controlling the bath with the aid of a process checker. Assume the probability of having a contaminated chemical bath out of control is one hour out of every 100 of operation. Assume out of control conditions follow a binomial distribution. Further assume the out of control condition goes unnoticed until product is inspected two hours after entering a chemical bath. This results in circuit boards becoming defective before additional boards can be prevented from entering the bad chemical bath. The cost of quality for the out of control chemical bath is, let's say, $3,500. Adding a process checker who can check the bath every hour means that we will catch out of control conditions on average within a half hour of the occurrence. This will reduce the number of defective circuit boards resulting in a $1,300 savings. However, the cost of employing the process checking for 100 hours is roughly $1,000. Consequently, it would be to our benefit ($300 savings) to institute a process check every hour on the chemical bath.

Conclusion

Costing quality for sensitivity analysis is the tool management needs to direct activity for reducing high costs of quality. The program defined in this paper utilizes accounting cost figures, gives quick, easy, verifiable, traceable, and accurate accounts of 85 percent of all costs of quality by product line, by department, and by process. Estimation of costs are explained along with verification of estimates to ensure their accuracy. The paper explains how to measure costs of quality and gives several examples of quality costs in a manufacturing shop. Finally, sensitivity analysis is used to show how adding prevention costs can reduce overall costs of quality.

Appendix: Cost of Quality Elements AT&T Richmond Works Circuit Board Manufacturing

Appraisal: Cost figures are divided into three groups: appraisal, nonconformance, and prevention. Appraisal includes all inspection, test, process checking, and cluster auditing.

Behind Schedule Costs: Quality problems that result in product shipped behind schedule incur additional costs: costs of not getting future business from schedule conscious customers, costs of bargaining with customers to accept product behind schedule, lost profits if price must be reduced on product behind schedule, or overtime costs due to catching up to schedule. The only cost calculated in our cost of quality is the addition to overtime due to quality problems resulting in production falling behind schedule. The amount of overtime the shop requires to meet schedule is calculated. Any additional overtime is considered cost of quality.

Change and Repair: Maintenance orders classified as repairs are costs of quality. Changes only charged to quality organizations are costs of quality.

Chemical Stores and Supplies: Quality costs for chemical stores and supplies equal one minus the process yield for each product line multiplied by the amount of chemical stores and supplies designated for each product line. This designation is based on production dollar.

Costing by Cause of Unquality: Costs of defects to product are charged to the area where the defect is caused. This is done by identifying end-of-the-line defects by cause of defect and area where it is caused. Roughly 65 percent of end-of-the-line defects are identifiable to cause. The remaining defects are charged to the area where the defect is found. This is done through scrap audits as repair operators are not able to charge by cause of defect.

Design Review: Design review performed by the Tooling organization is a cost of quality. Parts programming performed by the Tooling organization is not a cost of quality unless it is due to a change that was required due to a quality problem. The amount of unquality will be charged as a percentage of work of the parts programmers. This percentage is assigned by the department chief. In other words, if the parts programming aspect of this department costs $100,000/month and 25 percent of their work is due to unquality, $25,000 is the cost of quality for parts programming. If the other aspect of this department, design review, costs $100,000/month, the total cost of quality/month for this department is $125,000. The cost of quality will change if this department changes in size or if the percent of work on quality done by the parts programmers changes.

Engineering: All quality control engineering costs are costs of quality. All raw materials and process control lab engineering costs are costs of quality. Engineering departments not directly related to quality but that, at times, work on quality, such as Product and Process Engineering, are included in cost of quality the same way the tooling group is costed (see design review).

Group Credit Hours: Time spent waiting for work; time spent waiting for machines. Industrial engineers categorized all group credit hours by cost of quality or cost of doing business. These quality costs are then reported by Accounting.

Machine Downtime: Machine downtime is included in cost of quality if operators must wait for machines that are down (see group credit hours) and as a maintenance charge for machine breakdown.

Maintenance: Breakdown maintenance, preventive maintenance, and repair maintenance are costs of quality. Maintenance due to change orders are not all costs of quality. The percent of maintenance due to change orders that is not cost of quality is the process yield on each product line. The department that Maintenance charges its time and supplies to is assigned that cost of quality.

Management: Except for the general manager, whose expenses are not covered by the Richmond Works, department chiefs, supervisors, and upper management are included in quality costs. One hundred percent of costs of management of quality organizations are included as a cost of quality. One minus the process yield multiplied by nonquality management is the cost of quality for other managers.

Nonbase Hours: All nonbase hours that fell out due to scrap are included in cost of quality by multiplying the percent of direct hours that are nonbase in a department (shop cost center) by the number of direct hours that fell out due to scrap. One hundred percent of purchased material inspection (PMI) nonbase are separated by product line based on the PMI department chief's evaulation of how much PMI is done on each product line. One hundred percent of training by the Training organization is included in cost of quality. Employee's time in training is not included in cost of quality unless the employee or section chief charges the time to Training. Consequently, shop employee time in training is charged as a cost of quality, but salary graded time in training is not. Not including salary graded time in training as a cost of quality underestimates cost of quality. However, this underestimate is compensated by including all Training organization costs as cost of quality even when some training costs are costs of doing business.

Nonconformance: Quality costs are divided into costs of nonconformance, appraisal, and prevention. Nonconformance includes all product conformance costs, scrap, hours that fall out due to scrap (see scrap hours), repair, and detail.

Prevention: Cost figures are divided into three groups: appraisal, nonconformance, and prevention. Prevention includes those categories not included in nonconformance or appraisal, such as preventive maintenance.

Process Checkers: All process checkers are included in cost of quality as an appraisal cost. Each process checker is charged to the department where he/she works.

Process Control Lab: All process control lab costs are considered costs of quality.

Product Conformance Costs: Scrap costs. Each product that is scrapped has a dollar value for material that is reported by cost center. All labor hours that fell out due to scrap are charged to labor product conformance costs by cost center.

Quality Control Department: All Quality Control Engineering personnel are costs of quality. Allocation of their costs by product line is assigned by the Quality Control department chief.

Raw Material Lab: All raw material lab costs are considered costs of quality.

Raw Material Stores: Raw material accepted by purchased material inspection or outside supplier inspection that is subsequently scrapped in the shop raw material storage areas.

Salaries and Wages: Expense salaries and wages, including shop management, are allocated by product line based on the department. The salaries and wages are multiplied by a percent of hourly labor that are process checkers, repair operators, or testers to arrive at those salaries and wages that are costs of quality. See management for cost of quality for nonshop management.

Salvage Value: The cost of quality for replacing equipment is the book value minus the selling price.

Securities and Accruals: Securities and accruals pay for social security responsibility of the company, health insurance, and other benefits. The percent of securities and accruals considered a cost of quality equals the percent of salaries and wages considered a cost of quality.

Supplies: One hundred percent of supplies charged to quality organizations are costs of quality. The percent of supplies charged to other organizations that is a cost of quality equals one minus the process yield for each product line.

Test: All test area employees are charged to cost of appraisal.

Training: All Training organization costs are charged to cost of prevention. Training done outside the Training organization is not charged to cost of quality.

Endnotes

1. *A Passion for Excellence.* Tom Peters. Random House, New York, 1985.
2. "What Costs Count?" *Quality Progress.* F. X. Brown. April 1983, page 35.

Bibliography

Brown, F. X., *Quality Progress.* "What Costs Count?" April 1983.

Crosby, Philip B., *Quality Is Free.* McGraw-Hill, New York, 1979.

Mendenhall, Scheaffer Wackerly, *Mathematical Statistics With Applications.* Wadsworth, Belmont, CA, 1981.

Peters, Tom, *A Passion For Excellence.* Random House, New York, 1985.

Small, Bonnie B., *Statistical Quality Control Handbook.* Western Electric Co., New York, 1958.

FOCUSING QUALITY COSTS USING THE BASICS

William O. Winchell
General Motors Corporation
Warren, Michigan

Abstract

We are currently witnessing an increasing interest in quality costs and for good reason. Quality costs may represent the last frontier of quantum productivity improvements. For many years, industry has concentrated on reducing productive labor which currently is under ten percent of factory cost for many companies. In contrast, quality costs may represent 25 percent or more of factory cost. This is largely in scrap, rework and warranty caused by products or services not conforming to customer needs. A smaller amount is used for inspection efforts and, typically, the smallest portion is devoted to preventing problems from occurring. Over the years many cost of quality reports were started with varying degrees of success. Many of the difficulties in the past were out of the control of those initiating the reports but centered on the "culture" within the company. For example, quality cost efforts in a highly structured company with many "fiefdoms" were of limited usefulness when improvements were suggested that changed power bases. Today, we find the "culture" of companies changing to compete in today's market place. Improvement teams and a cross functional approach allow improvements that are not possible without common goals among different functions. The environment for implementing quality cost concepts has never been better. But, we certainly must learn from what was done previously in order to improve our efforts. To achieve success with quality cost reporting, the most important thing you can do is to concentrate on building a strong foundation up-front, prior to starting the program. To do this, like in most anything else, you must address the basics. These basics are:
- WHY do you want quality cost reporting.
- WHERE are you going to do the reporting.
- WHEN do you do the report.
- HOW do you make it happen.
- WHO do you get to do the report.
- WHAT are you going to call quality cost.

A thorough analysis of the basics will lead you to the best strategy for your company.

Introduction

Over the years many cost of quality reports were started with varying degrees of success. Many of the difficulties in the past were out of the control of those initiating the reports but centered on the "culture" within the company. Now we are witnessing many companies changing their culture to compete in today's market place and the environment for implementing quality cost concepts has never been better. But, we certainly must learn from what was done previously in order to improve our efforts.

To achieve success with quality cost reporting, the most important thing you can do is to concentrate on building a strong foundation up-front, prior to starting the program. To do this, like in most anything else, you must address the basics. Before we discuss these basics, let me first caution you that there is no magic combination of ingredients that will guarantee success. But, a thorough analysis of the basics will lead you to the best strategy for your company.

Why

The first basic that should be addressed in laying the foundation is "Why" — just why do you want to do quality cost studies. A superficial analysis of this question may lead you to conclude that the objective is to "improve productivity and quality." Certainly, quality costs may represent the last frontier of quantum productivity improvements. Quality costs represent 25 percent or more of factory cost in contrast to productive labor which is probably under ten percent of factory cost for most companies. However, the facts are that many quality cost efforts in the past did not achieve this objective and were discontinued. They failed in meeting the objective because there was no associated improvement effort to determine the root causes of problems and implement irreversible corrective action. Also in many cases the quality cost report associated with this objective was a "control" type report intended to bring pressure on plant management to reduce quality costs. In essence, the "stick" of the "carrot and stick" approach was being used. History has shown us that if we feel threatened and don't like the message, the messenger may be expendable — the messenger in this case being the quality cost report.

A far more realistic objective for quality cost reports is to "support an improvement process that will improve productivity and quality." A properly designed quality cost report can point out the strengths and weaknesses of a quality system. Quality cost reporting uses dollars to define the relative amount of effort spent on each element of the quality system. Because of this, improvement teams having quality cost reports can describe benefits and ramifications of changes in terms everyone can understand. Improvement teams can also use "Return-on-Investment" models and other financial analyses constructed directly from quality cost data to justify proposals to management. Those involved on improvement teams can also use this information for prioritizing problems, seeking out root causes and implementing the most effective irreversible corrective action. Quality cost reporting can also track results to tell those involved in improvement if they are headed in the right direction. Such an objective recognizes that quality cost reports can be a big help to those making the improvements but that reporting will do nothing without someone "making it happen."

Where

The next basic to be discussed is "Where" — just where are you going to do the report. Some people call this defining the scope of the report. The important thing to remember is the scope must be large enough to be meaningful but not so large that you may not be able to complete the study. If you consider that quality cost can define what all the elements are in your quality system, it is very logical to initially do just that. Of course this would require coordinated studies of each function in your company — marketing, product engineering, quality assurance, manufacturing, materials management, and service. After completing the initial study, you will likely be able to evaluate which elements of the system are most sensitive to change and which elements should be restudied when certain improvements are completed. The driving force behind this should be the needs of those involved in the improvement process. Those in the improvement process can make better decisions if they are aware of all the relationships existing in your quality system. Not to be aware of all the major factors has serious implications for decision makers.

Let's now review an example of a quality cost reporting effort for which the scope was too small. It concerns an automotive component. The original scope consisted only of the activities on the factory floor. Major missing functions were the product engineering department and the service operation which performed repairs on the components when they failed in the field. Improvement efforts were not meeting the expectations of those on the improvement team because much of what they must improve was not being tracked. By enlarging the scope of the quality cost report to include the service operation, the improvement team could identify a significant cost for field failures that could justify needed improvements. Also by including product engineering, it could be verified if product designers were not actively pursuing problems during the production cycle and if their efforts should be redirected.

An example of a scope for a quality cost report that would be too large concerns a product design activity that only supports automotive components produced by outside vendors. Using the same logic as in the previous example, a strong case could be made for including the manufacturing efforts of the outside vendors in the quality cost reports. But the practicalities of sharing cost information among different companies, especially competitors, preclude this approach. Therefore, the scope of the study needs to be limited to the product design activity.

When

The next basic to be discussed is "When" — just when do you do quality cost reports. We have discussed using the initial quality cost report to allow the improvement team to understand the quality system. But, after the initial report, when do you repeat the study. If you accept that quality cost reporting is to support the improvement team, then it is logical to assume specific parts of the study should be repeated when the improvement team desires new information. This should be, for example, when an improvement is in place and a change in results is expected. To generate reports on a regular basis, not correlated to a change, may be wasted effort if no new useful information is provided. We should not minimize that regular reports in the past were perceived as "control" type reports with the inherent problems previously discussed. In addition, regular reports showing no real progress may tend to send the wrong message to those not familiar with the situation discouraging those that you may need for support. Repeating an appropriate portion of a study, when an improvement takes place, is a good way to communicate the good things an improvement team is doing to those concerned.

How

The next basic to be discussed is "How" — just how do we make quality cost reports happen. If you accept that a quality cost report is a tool for the improvement team and that the quality system encompasses all functions in an organization, then the first step is getting a cross-functional improvement team started. Organizing such a team requires a "buy-in" from all functions involved. There is really no better way to achieve a "buy-in" than to have all functions participate in the development of such a team and also the tools required for its support. Detractors traditionally do not feel part of the effort that they are not supporting. The most dramatic turn-around that I witnessed occurred on an effort that was floundering because of a lack of real support from the financial community. The fix was asking the controller if he would be leader of the project. He became our staunchest supporter almost overnight.

After you now have started cross-functional improvement teams, selling quality cost as a tool to those on a team is best done by showing and doing. It very well may take the active participation of a quality cost analyst to explain the quality system that has been defined by the initial report and to translate the needs of the team into specific types of future reports. Certainly as this knowledge and the techniques are learned by the team, and successes are obtained, quality cost will be recognized as a vital contribution to their efforts.

Who

The next factor to be considered is "Who" — just who should do the quality cost reports. Over the years, we have heard that the financial community must publish the quality cost report. After all they are the recognized specialists in sorting out where the dollars in a business belong. A minority has steadfastly maintained that quality cost should not be a financial report but needs to be translated by a quality specialist to put it in a format enhancing its use for analysis and evaluating improvements. Typically when the financial community produced a quality cost report, it was perceived as a "control" report with the potential problems we have discussed. Also only the easily identifiable costs usually have been reported leaving perhaps as much as 70 percent of the quality costs missing. Having the quality community prepare the report also had problems. It was difficult to find an individual that could analyze complex financial information like a "generalist" without the computer power available today and who then could inspire others in the organization to make needed improvements.

With the advent of cross-functional improvement teams, coupled with today's computer power, it seems logical that quality cost reports can be handled within such a team. As previously mentioned, a quality cost analyst could act as a facilitator for the team. Reports could be truly designed for the user — the improvement team. It would appear that this approach minimizes the problems found when preparing this report was assigned to a specific function.

What

The last, but not the least, fundamental is "What" — just what are the things you are going to call quality cost. For some, the driving force behind what they call quality cost is what is easily determined from existing financial records. Past surveys by the ASQC Quality Costs Technical Committee indicate that some companies have defined their only quality cost as scrap and their quality cost report as their scrap report. Most of us will agree that this is too narrow a definition. For others, on the opposite extreme, the driving force behind what they call quality cost is the elements in their quality system which encompasses every function in the organization. It could start in marketing and extends to product engineering, financial, materials management, production, service and is clearly not confined to quality assurance. As mentioned previously, knowing your quality system is vital to the decision process in cross functional improvement teams.

Many cost studies are somewhere in between these two illustrations. Typically, they include somewhere around 30 to 40 percent of the costs that would be present if they defined the entire quality system. Missing are costs that are not so obvious on normal financial reports. Major missing costs include the validation effort in product engineering. Also missing are the rework operations, inspection by the production operator, and process control by the set-up person which are all part of standard or productive labor. For warranty or field failures, often the costs are not projected to that expected at the end of the warranty period and are understated. Including these elements of the quality cost system in the initial study may be critical to the success of the improvement team. In one quality cost study, including validation effort by product engineering ultimately led to the conclusion that existing tests were really based on field problems that occurred as much as fifteen years ago and today's problems could be more thoroughly addressed through determining testing requirements by product failure mode and effects analysis. Other studies have shown that production operators are getting more money in total for inspecting products than the quality control department. This is a very significant factor in coordinating the transfer of inspection functions to the production operator.

After the initial study, which defines the quality system for the improvement team, the team can pick and choose what elements of cost it needs to support its future needs. Remember, that the definitions and terms used must make sense to you and especially to everyone in your organization. They must be in the language that is used by your company so that they are easily understood by everyone. For this reason, most quality cost reporting systems depart from the terminology suggested in ASQC literature.

Summary

In summary, the key to the success of any quality cost system is to concentrate on building a strong foundation up-front, prior to starting the program. To do that, like in most anything else, we must consider the basics. A thorough analysis of the basics will lead you to the best strategy for your company.

PREDICTING QUALITY COST CHANGES USING REGRESSION

K. S. Krishnamoorthi
Industrial Engineering
Bradley University
Peoria, Illinois

Abstract

Often times we want to know the interrelationship among the four components of the Total Quality Cost so that we will be able to say what changes will occur in one when another is changed. For example we may want to know how much of a reduction in failure costs will result from a certain increase in prevention or appraisal cost. Similarly we may want to know much the prevention budget must be increased in order to accomplish a certain reduction in the failure costs.

Even though the relationship among the component costs is as varied as there are companies keeping such cost systems, existence of some general pattern is hypothesized at least among similar companies doing similar business or producing similar products. Regression analysis is used to extract this relationship from data obtained from Q-Cost systems, and formulas are proposed as a rough approximation to describe this relationship. This model can be further refined, based on more data, to obtain more accurate models for specific industries such as: automobile, metals, pharmaceuticals, etc. Availability of such formulas will be helpful in estimating benefits from Q. C. investments while justifying budget proposals, and in recognizing when optimal condition is reached.

The relationship derived above is used in a simulation model that describes the cost system of a typical manufacturing company. The results from the simulator are used to show graphically how the unit cost of product changes with changes in investments in quality control.

Introduction

A quality cost study reveals considerable information about the health of a quality system. It tells whether the system produces good quality at the right price. If not, it indicates where improvement opportunities exist so that improvement projects could be pursued to achieve quality economically. A quality cost study is made by accumulating cost data in the four cost categories:
    1. Prevention Cost
    2. Appraisal Cost
    3. Internal Failure Cost
    4. External Failure Cost

Prevention and Appraisal are exogenous variables in that they are the components that can be controlled at different levels to achieve desired levels of the endogenous variables, Internal and External Failure costs. Details of definition of these categories can be found in ASQC's booklet: *Quality Costs — What and How* (1971),

When data has been collected in the above categories they are analyzed mainly in two ways:

1. Expressing total Q-Cost as percent of a base such as total sales billed.
2. Expressing the four component costs as percent of total Q-Cost.

The first analysis would indicate if the total Q-Cost is excessive and the second would indicate the direction where improvement projects could be undertaken. Guidelines for such analysis and other useful ways of analyzing Q-Cost data can be found in Juran and Gryna (1980) and Feigenbaum (1983).

While analyzing the components of the Q-Costs it becomes necessary to know how the four component costs are interrelated. For example a Q. C. manager may want to know how much of increase in appraisal is necessary to reduce the external failure by, say, 50%. Or, he may want to know how much a one percent increase in prevention will accomplish in reducing other three categories. Such estimates on effect of changing one component on the others are necessary when the Q. C. manager has to justify quality investments through cost benefit analysis. They may also be useful to predict what the quality cost would be in the coming year or when one has to know if optimum cost levels have been reached.

The interrelationship among the four components is quite complex. We may know the direction of changes. For example, we can say that external failure will decrease with increase in prevention and/or appraisal and that the internal failure will decrease with increase in prevention while it may increase with increase in appraisal. But, how much will the changes be in the failure costs due to changes in prevention and appraisal is very difficult to quantify. Not only is this relationship complex in any one given system, it will change from one system to another depending on nature of the business; and is dynamic in that it changes from time to time.

It is hypothesized, however, that such relationships can be considered to be uniform within systems in similar business or producing similar products so that if the relationships can be extracted from some systems within the group, they could be used for predictions in another belonging to the same group. At the present time when such predictions are needed some informal rules of thumb are used or ad hoc "guesstimates" are made. Thus there exists a need for investigating the possibility of a formal quantitative method to obtain these relationships among the Q-Cost components.

Quantitative Methods in Q-Cost Study

Several articles can be cited where authors have attempted to seek relationships among quality cost components and between quality costs and other business indices. Masao Kogure (1981) had obtained relationship between prevention costs and failure costs from past historical data of Japan Steel Works, Hiroshima plant. The article shows the impact

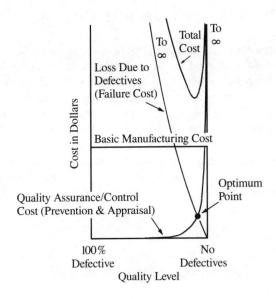

Figure 1
Graph Showing Q-Cost Changes

of prevention investment on the internal failure and external failure in the current year and future years. Similarly the effect of investments in appraisal on the failure costs are also shown. The method used in deriving these relationships is not available except the fact it was based on past data and experience. Yet the material in the paper points to the need and value of understanding the underlying relationship among cost components.

The popular quality cost curves are used by many authors to describe the relationship among components. The Figure 1 reproduced from Campanella and Corcoran (1982) shows the familiar graph showing relationship between quality level and quality costs. This graph indirectly gives the relationship between the input (prevention + appraisal) and output (total failure) components. In spite of the fact that the figure does not show any scale for the two axes of the graph it gives the broad picture of the direction of movement of the component costs.

In the same article we find the statement:"An ounce of prevention is worth a pound of failure." This is an effort to relate prevention cost and failure costs. In this, "ounce" and "pound" are used in a figurative sense rather than meaning that one dollar of prevention being worth 16 dollars of failure. Such statements however show that people are eager to find relationships between components of Q-Costs.

Brown (1978) was seeking to couple quality to profit. He used regression analysis to relate failure cost (as percent of Value Added Sale) to profit margin as percent of sales. He found that every dollar reduction in failure cost improved the profit by 5.45 dollars, indicating there were several 'intangible' gains when quality is improved and failures reduced. Such inferences have been possible through use of quantitative methods.

Predicting Q-Costs with Regression

Having recognized that the relationship among the components of Q-Cost is complex and it is not possible to develop a functional relationship logically based on the understanding of their behavior, we will consider the components Prevention, Appraisal, Internal Failure and External Failure as random variables jointly distributed, taking different values in different cost systems. These random variables represent the components expressed as percent of total Q-Cost. If values taken by these random variables in one system can be considered as one set of data and several such data points are available from several systems we could use regression analysis to fit a relationship among these variables. This is somewhat similar to economists using econometric models to find relationships among economic variables whose relationships are not too obvious.

Data and Regression Results

Data were available from 23 quality systems. They were taken mainly from published literature and some were made available from some confidential files. For this reason the data and their sources are not identified. The data comes from different types of industry and from systems in different stages of improvement. There are some that have failure costs as high as 96% of total and there are data points where the failure cost is only 32% of total. There are also a few data points that have failure costs accounting for about 50% of total. The set of data available for the present work was not the most ideal set in that it was not possible to categorize the data based on industry type. Quality costs as percent of sales would have been more desirable type of data. Yet because of scarcity of such information we had to be content with the data available and the following analyses were made using them. Figure 2 shows graphs of External Failure and Internal Failure against the input variables. These graphs provided guidance in determining which model would provide appropriate fit for the desired relationship.

One of the standard packages for regression analysis was used to regress the External Failure and Internal Failure percentages on Prevention and Appraisal. The fact that the four percentages added to 100%, introducing a dominant relationship among the components posed some difficulties in extracting the desired relationships. This prevented relating all four components together. The fact that the independent variables, Prevention and Appraisal are themselves related posed the problem of multicollinearity. In spite of these problems some interesting results were obtained that could still be useful. The two relationships that were fitted to the data are as follows:

$$E = \frac{5.9}{P} + \frac{298}{A} \qquad (1)$$

$$I = \frac{121}{P} + 0.213A \qquad (2)$$

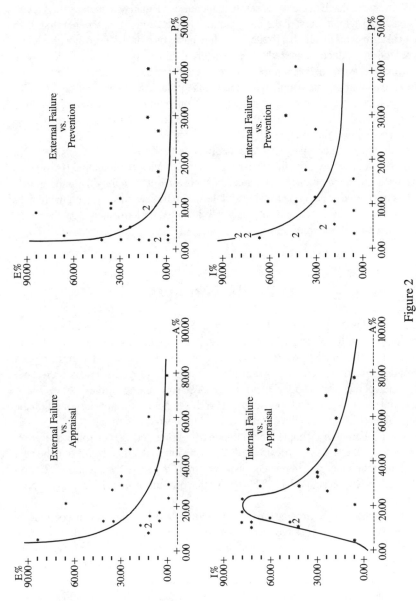

Figure 2
Graphs of Q-Cost Relationships

Equation (1) expresses External Failure (E) as a function of Prevention (P) and Appraisal (A). It shows that E is inversely related to P and A. The smaller coefficient for 1/P indicates that E decreases faster with increase in P than with increase in A. Also the equation says that most of the decrease in E has to come from A rather than P. In practical terms, this means that if external failure has to be reduced, even though a dollar of prevention produces more reduction in external failure, prevention can do only so much and we have to increase appraisal to reduce most of external failure. The equation says, at least that is what people have done in the past. If we look at the equation (2) we see that the internal failure cost gets reduced much more by increase in prevention than the external failure cost. Also we see that the increase in appraisal tends to increase the internal failure, though only slightly.

The above two equations give a reasonably good fit for the available set of data. In terms of explaining the variability in the dependent variables, the equation (1) explained about 63% and the equation (2) accounted for about 80% of variability. Thus the models can be considered reasonably reliable and these can be used to see quantum changes rather than predicting exact values of changes. With the gathering of more data, the model can be made more reliable.

Using the Formulas

The formulas can be employed to produce useful information in several situations. For example, say the percentages of the individual components are as follows: $P = 5$, $A = 6.5$, $I = 41.5$, and $E = 47$. It is desired that the E should be halved in the next year. Prevention investments take a long time to bear fruit and so appraisal is to be increased. How much of increase in appraisal is required? We solve the equation: $5.9/5 + 298/A = 24$, and get $A = 13.1$. This indicates that for this system in the present condition the appraisal has to be doubled in order to halve the external failure.

Let us take another example where the distribution of costs is as follows: $P = 2$, $A = 13$, $I = 53$, $E = 22$. It is desired to reduce the internal failure by 50% in the next year. How much increase in prevention is required? We solve the equation: $121/P + (0.213)(13) = 26$ and get $P = 5.2$. This means that prevention must be increased about 2½ times.

Caution must be used in using the above two formulas since they contain coefficients that are estimates with possible errors, which may vary in magnitude from situation to situation. We hope to narrow the error and come up with coefficients with more precision when the present effort in collecting and analyzing more appropriate data is completed. Our continuing effort is directed towards obtaining cost data from different systems and coming up with formulas (they may differ in values of coefficients) for different industries such as process industry, pharmaceutical, appliance, automobiles, etc. Also, if the data would be available as costs in percent of sales rather than costs as percent of total cost, some of the difficulties encountered in regression could be avoided and we will have more confidence in predictions using the formulas. Further work needs to be done in this regard.

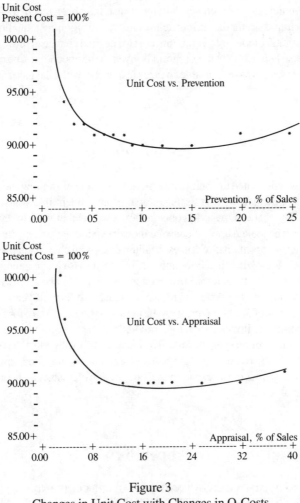

Figure 3
Changes in Unit Cost with Changes in Q-Costs

Simulating a Production System

A computer program was written to simulate a production system with a given production volume, cost of production, and a given break down of the four quality costs. The formulas obtained earlier were used in the program to record changes in the failure costs when prevention and appraisal are varied. For any given set of data the program will compute the unit cost of production.

Using this program the unit cost was computed for varying levels of prevention and appraisal and the graphs were plotted as shown in Figure 3. It can be seen that the unit cost decreases with increases in prevention or appraisal up to a point and starts increasing beyond that, indicating optimum levels of input costs have been reached. The program written in Fortran, asks for data on current volume, production cost, quality costs, etc., on a conversational mode and outputs unit cost for desired change in prevention and/or appraisal costs. The listing of the program is given in the appendix. The program can be used for any system to see possible changes in unit cost with changes in input quality costs.

Conclusion

A method was indicated to obtain a quantitative relationship among the four quality cost components which will be useful in quality investment decisions and project justification. The regression models obtained from data gathered from several quality cost systems can be looked upon as consolidation of experience of these companies and be used for seeing approximate changes in failure costs with changes in prevention and appraisal costs. Following the above approach, if formulas can be generated for any given industry from data collected from systems in that industry, the formulas could be quite helpful for another system in the same industry. If formulas can be generated from historical data of one company it can be used as a prediction tool for future trends in the same company. It will also help in predicting when optimum quality costs are reached. The computer program given will be found useful in this effort. It is emphasized that the method suggested is for purposes of estimation only and investments in prevention and appraisal must be made based on selection of improvement projects using cost benefit analysis.

Appendix

```
00100C THIS PROGRAM WILL GIVE CHANGES IN FAILURE COSTS AND
00110C UNIT PRODUCTION COST, GIVEN THE CHANGES IN PREVENTION AND
00120C APPRAISAL COSTS IN A QUALITY COST SYSTEM.
00130C THE PROGRAM REQUIRES ENTRIES OF CURRENT PRODUCTION COST,
00140C PRODUCTION VOLUME, AND CURRENT VALUES OF Q-COST ELEMENTS.
00150 PROGRAM COST (INPUT, OUTPUT)
00160 REAL IFC, IFCP, IFCPN, IFCN
00170 99 CONTINUE
00180 PRINT*, "ENTER CURRENT UNIT PRODUCTN COST IN DOLLARS"
00190 READ*, CPU
00200 PRINT*, "ENTER THE PRODUCTION VOLUME IN NUMBER OF UNITS"
00210 READ*, PARTS
00220 PRINT *, "ENTER SELLING PRICE PER UNIT IN DOLLARS"
00230 READ *, PRICE
00240 11 FORMAT (/, "THE CURRENT TOTAL PRODCN COST=", 1X, F10, 2, 1X, "$",//)
00250 PRINT*, "ENTER THE CURRENT PREVENTION, APPRAISAL, INTERNAL"
```

```
00260 PRINT*, "FAILURE, AND EXTERNAL FAILURE COSTS IN DOLLARS"
00270 PRINT*, "IN THAT ORDER, PUT A COMMA AFTER EACH FIGURE"
00280 READ*, PC, AC, IFC, EFC
00290 PRINT 60
00300 60 FORMAT (//)
00310 9 CONTINUE
00320 TMC=CPU*PARTS
00330 PRINT 11, TMC
00340 TQC=PC+AC+IFC+EFC
00350 PCP=PC/TOC*100
00360 ACP=AC/TQC*100
00370 IFCP=IFC/TQC*100
00380 EFCP=EFC/TQC*100
00390 QCPER=TQC/(PARTS*PRICE)*100
00400 PRINT*, "YOUR TOTAL Q-COST AS PERCENT OF SALES IS:"
00410 PRINT 112, QCPER
00420 112 FORMAT (F10.2)
00430 PRINT*, "YOUR CURRENT DISTRIBUTION OF Q-COST IS:"
00440 PRINT*, "PREV%      APPR%       INTF%      EXTF%"
00450 PRINT 111, PCP, ACP, IFCP, EFCP
00460 111 FORMAT (4F10.2,//)
00470 PRINT*, "ENTER THE DESIRED CHANGES IN PREVENTION AND"
00480 PRINT*, "APPRAISAL COSTS IN DOLLARS IN THAT ORDER."
00490 PRINT*, "TRY CHANGES NOT LESS THAN 50 DOLLARS."
00500 PRINT*, "PREFIX DECREASE WITH (-), PUT COMMA BETW FIGURES."
00510 10 CONTINUE
00520 READ*, DP, DA
00530 IF (DP. EQ.0) GOTO 221
00540 IF (DA. EQ.0) 221,222
00550 221 PRINT*, "DO NOT TRY WITH ZERO OR NEAR ZERO CHANGES"
00560 PRINT*, "ERROR IN MODEL GIVES AWKWARD RESULTS"
00570 PRINT*, "TRY AGAIN DIFFERENT CHANGES IN PREV, APPR"
00580 GOTO 10
00590 222 CONTINUE
00600 TQCN=TQC+DA+DP
00610 PCPN=(PC+DP)/TQCN*100
00620 ACPN=(AC+DA)/TQC*100
00630 IFCPN=(121.0/PCPN)+(0.213*ACPN)
00640 EFCPN=(5.9/PCPN)+(298.0/ACPN)
00650 IFCN=IFC*(IFCPN/IFCP)
00660 EFCN=EFC*(EFCPN/EFCP)
00670 PCN=PC+DP
00680 ACN=AC+DA
00690 TQCNEXT=PCN+ACN+IFCN+EFCN
00700 DQC=TQCNEXT-TQC
00710 TMCN=TMC-TQC+TQCNEXT
00720 CPUN=TMCN/PARTS
00730 DC=CPUN-CPU
00740 DT=TMCN-TMC
00750 DPC=PCN-PC
00760 DAC=ACN-AC
00770 DI=IFCN-IFC
00780 DE=EFCN-EFC
```

00790 PRINT 44
00800 PRINT 45. CPU, CPUN, DC
00810 PRINT 46, TMC, TMCN, DT
00820 PRINT 51, TQC, TQCNEXT, DQC
00830 PRINT 47, PC, PCN, DPC
00840 PRINT 48, AC, ACN, DAC
00850 PRINT 49, IFC, IFCN, DI
00860 PRINT 50, EC, EFCN, DE
00870 44 FORMAT (/,23X,"OLD",13X,"NEW",12X,"CHANGE")
00880 45 FORMAT ("COST/UNIT", 10X,F10.2.6X.F10.2,6X.F10.2.6X,F10.2)
00890 46 FORMAT ("TOTAL MANU. COST", 3X,F10.2,6X,F10.2,6X,F10.2)
00900 51 FORMAT ("TOTAL Q.-COST",3X,F10.2,6X,F10.2,6X,F10.2)
00910 47 FORMAT ("PREVENTION", 9X,F10.2,6X,F10.2,6X,F10.2)
00920 48 FORMAT ("APPRAISAL",10X,F10.2,6X,F10.2,6X,F10.2)
00930 49 FORMAT ("INT. FAILURE",7X,F10.2,6X,F10.2.6X,F10.2)
00940 50 FORMAT("EXT. FAILURE",7X,F10.2.6X,F10.2.6X,F10.2.//)
00950 PRINT*,"YOUR NEW DISTRIBUTION OF Q-COSTS IS:"
00960 PPCN=PCN/TQCNEXT*100
00970 PACN=ACN/TQCNEXT*100
00980 PIFCN=IFCN/TQCNEXT*100
00990 PEFCN=EFCN/TQCNEXT*100
01000 PRINT*, "PREV%   APPR%   INTF%   EXTFZ"
01010 PRINT 66,PPCN,PACN,PIFCN,PEFCN
01020 PRINT*, "YOUR NEW TOTAL Q-COST AS PERCENT OF SALES IS:"
01030 QCPERN=TQCNEXT/(PARTS*PRICE)*100
01040 PRINT 112,QCPERN
01050 66 FORMAT(4F10.2,//)
01060 PRINT 60
01070 PRINT*, "ENTER 0 TO QUIT, 1 TO TRY DIFFT CHANGES, 2 FOR DIFF PROB"
01080 READ*,M
01090 IF(M.EQ.1) GO TO 9
01100 IF (M.EQ.2) GOTO 99
01110 STOP
01120 END

Bibliography

1. ASQC Quality Cost-Cost Effectiveness Technical Committee, *Quality Costs — What & How* (1971)

2. Brown, Francis X., "Quality Cost and Profit Performance," Quality Congress Transactions (1978)

3. Campanella, Jack, and Frank J. Corcoran, "Principles of Quality Costs," Quality Congress Transactions (1982)

4. Feigenbaum, A. V., *Total Quality Control*, 3rd Ed., McGraw-Hill Book Co., N. Y. (1983)

5. Juran J. M., F. M. Gryna Jr., *Quality Planning and Analysis*, 2nd Ed., McGraw-Hill Book Co., NY (1980)

6. Masao, Kogure, "Factors Required for Japanese Quality Cost System," Quality Congress Transactions (1981)

SOLVING THE QUALITY COST EQUATION

David D. Kildahl
Quality Assurance Engineer
Honeywell Incorporated
Minneapolis, Minnesota

Abstract

The theme of this conference is "Quality: The Universal Equation for Excellence." Quality professionals are indeed working to solve an equation, and in fact there are a number of partial solutions available. A special case of the general equation of quality and excellence is the Quality Cost Equation, which can only be solved in particular circumstances. The author examines several versions of the Universal Equation, and then proceeds to a detailed consideration of the special case of the Quality Cost Equation and shows how it can be solved depending on the application and the assumptions in use. Finally there is a discussion of the obvious and subtle interrelationships among the variables considered in the Quality Cost Equation.

Introduction

"QUALITY: THE UNIVERSAL EQUATION FOR EXCELLENCE" is the theme for this Congress. In this presentation, I hope to show that using the model of an equation provides a number of useful insights. The "solution" of the equation is, of course, the key we all seek to realizing the many benefits inherent in a quality based strategy of business management. First we will look at a number of currently popular partial solutions to this universal equation. We will see that they all have features in common. One of these, we will call it the Quality Cost Equation, is a special case common to any meaningful attempt to solve the universal equation.

Text

Let's look at some of the current versions of the Universal Equation:
1. $Q = f(M + M + M + M)$

This model reflects the thinking of Dr. Ishikawa and is borrowed from his famous fishbone diagram. The four M's in this formula stand for Money, Manpower, Material, and Management. Any of you who have attempted to use a fishbone diagram in problem-solving know that these four labels are arbitrary and are only meant to provide a guide for labels more appropriate to the issue at hand. The model includes a money element, but note that cost may be a factor in any of the four (or more) elements of a fishbone diagram. Note too that this particular tool is most widely used in special case problem solving more than as a stand-alone management strategy for universal excellence.

2. $Q = 0$

This is the cost equation proposed by Phil Crosby in his book "Quality Is Free." As stated, this equation addresses the matter of the cost of quality, and says that quality is free, that is, that quality has no cost, that its cost is zero. This is strictly true when we recall that Crosby defines quality as conformance to requirements; once the requirements are set, it costs nothing to do just that — what costs is doing something else and having to make allowances where none should need to be made. We all know that there is more to it than that, and that I am taking liberties with Mr. Crosby's idea. I apologize for that, but there are people out there, not too many in quality, of course, who believe that this is exactly what Mr. Crosby had in mind. He is making a nice career out of explaining that there is more to it than that. "Quality Is Free" is a very well written book and I urge you all to read it if you haven't yet done so, and to reread it soon if you already have read it once.

3. $Q = 99.73\%$ CONFIDENCE

This version of the Universal Equation is intended to reflect the thinking of W. Edwards Deming and his belief in the importance of operating in a state of statistical control. A control chart is said to be in a state of control when it demonstrates that the process is operating within 3-sigma limits, and that all assignable causes have been dealt with. Since these limits represent 99.73% of expected behavior in a normal distribution, it seems this is a fair representation of Dr. Deming's model. Any given control chart is a special case, and there is no single all purpose control chart, but there are few tools more powerful and at the same time as simple as control charts for reducing costs and producing optimum performance. A thorough knowledge of control techniques belongs in the skills inventory of every quality professional.

4. $Q = P/I$

This is the classical, or "brute force," model of the Universal Equation. P stands for Product, any product; I stands for Inspect (or test, a kind of inspection). You might be trying to make sense out of "product over inspect," but if you read it as a fraction whose decimal value you wish to calculate, this equation becomes "inspect into product" and it represents a very popular solution to the Universal Equation. For a number of reasons, however, the solutions which derive from this model are less than satisfactory.

These models and others like them offer the working professional some valuable insights into the ways in which quality may be used (and abused) in the service of management. None of them are complete solutions, in fact one, the fourth, is almost foolish, like a fool's mate in a game of chess. Nevertheless we have all seen these and others like them used as models of good quality management. They are all solutions to the Universal Equation for Excellence.

The examples we have seen so far all deal with the whole of Quality. For the rest of this presentation, I'd like to focus on a somewhat narrower aspect of quality management, that of quality costs. In keeping with the theme of the Congress, there is an equation:

5. $\$ = P + A + F$

In this classic model, we see a description of how financial resources are allocated to the cause of quality. Here P refers to the money spent on Prevention, efforts and investments made to prevent the occurrence of nonconformance in any product or service; A stands for Appraisal in all its many forms; and F stands for Failure. The F element is often broken out as the sum of two components, I (internal), and E (external). Together these four categories serve to describe how resources allocated to quality are managed.

No general solution exists for the Quality Cost Equation, but particular solutions can always be generated for a given set of circumstances. We will look first at three equations involving only single elements of the general equation, and then conclude with comments which apply to the general case.

6. $\$ = P, \$ = A, \$ = F$

These three trivial solutions for the quality cost equation are descriptive, but not very useful. Partial solutions which focus exclusively on one cost component fail to convey the full extent and comprehensive applicability of the Quality Cost Equation. If only Prevention costs, $\$ = P$, are considered, the real total will be significantly understated. On the other hand, a close focus on prevention will result in a climate where constant improvement is fostered and achieved. The Japanese, for whom the cost of quality is a strange concept, have an intuitive appreciation for the benefits of a strong preventive climate. It forms the basis of their process improvement efforts.

Western manufacturing has favored the $\$ = A$ and $S = F$ models, probably because they are easier to quantify, and because for a long time quality has been an issue only on the manufacturing floor. The shift from manufacturing/product to administration/service in Western economics has helped reveal the inadequacies of "hardware-based" models of quality cost built around the ideas of appraisal or internal and external failure. When Hewlett-Packard took a close look at their own accounting activities, they found that 50% of allocated resources were devoted to finding and fixing errors. Experience like that can be a real "eye opener," as was related by H-P president John Young in his keynote address to this Congress one year ago.

The sad truth seems to be that American business manages quality costs by exception and perception. There is concern and awareness only for exceptional cases (usually disasters, rarely successes) and satisfaction is typically based on some vague notion that all is well, or that "there's never been a problem before." For too long we have been satisfied with "good enough" and with looking good at the expense of being good.

One of the large challenges facing our profession is to carry the quality message off the production floor into administrative and service areas where it has never before been a management issue. The Quality Cost Equation offers an excellent tool for dealing with this challenge, since it addresses a basic management issue, cost, in a context of improvement, which appeals to any responsible management.

The Equation is a flexible tool, applicable in virtually any context. By careful analysis all investments in quality can be accounted for. The categories of expense and their totals together provide a guide to decisions about where further cost improvements can be made economically. At the same time, the Quality Cost Equation offers a means to track and quantify the improvements which result, at least in dollar terms. Other measurements exist which illustrate gains in efficiency, productivity, and attitude.

Solving Your Quality Cost Equation

The idea of an equation for quality cost emphasizes the difference between MANAGING quality cost and simply COPING with quality cost. It is a self-fulfilling insight that, if you don't see and measure non-conformance as a cost, it becomes business as usual. This perception, that a given situation is "business as usual" (and that nothing can be done about it for that reason) is a major obstacle when it comes to actually doing something about managing quality costs.

It is not self-evident to most that quality costs are significant factors in the success of an organization. There are numerous reasons for this, but the big issues seem to be these:

1) Quality costs are typically not budget items as that term is commonly understood.
2) Quality cost items are not so much a matter of how much was spent, but rather what the spending did.
3) Quality is not typically seen as a contributor to the success of the organization, but rather as overhead/burden to be minimized.
4) The attitude that says, "If it isn't broken, don't fix it," is a recipe for disaster. If we don't make constant improvement a top priority, someone else will, and he will eat our lunch as he takes our markets away with better products at lower costs. Don't take my word for it, ask anyone in the steel industry; ask anyone who used to build VCRs or TVs.

Simply coping with quality costs can no longer be considered adequate management. For too long the quantification of quality cost information has been crude, cursory, simplistic, superficial, amateur, and approximate; these are lapses and luxuries no one can any longer afford.

Appraisal and failure costs tend to recur over time — we solve the same problems and fight the same battles over and over again; only the part number changes. The quality cost category of prevention offers an alternative which can break this vicious circle. The principle is simple, namely, fix the system of problem solving that permits the same problems to emerge time after time like a phoenix from the ashes of the last disaster. Piecemeal problem-solving (also known as firefighting) is not economical; rather the solution for any problem must preclude recurrence of the problem. This is the difference between working IN the system and working ON the system, and that is why quality cost is a management issue.

Conclusion

There are too many excellent guides to problem solving for me to say that any one is better than another. Each of you has experience with at least one, and they all work passably well. What I want you to do, though, is to take whatever techniques you are familiar and comfortable with and look for ways to apply them to the matter of managing quality cost more effectively, to the matter of solving your particular version of the quality cost equation. Here are some important points to keep in mind:

Take the time to establish a good approximation of existing costs in each of the categories. These are not good or bad, they just reflect existing circumstances. These will form the basis for measuring overall improvement.

Enlist the help of your accounting department. Any numbers you generate independently from them are meaningless. They are expert in what they do, and they can be a great help in keeping you from reinventing the wheel. They are usually quick to realize that their goals and yours are the same.

When the magnitude of the isssue is approximately clear, generate a strategy and goals for reduction of overall total quality cost. Your management must embrace this plan; if they don't, you and they will be working at cross purposes. This is not good quality career management.

Don't let a preoccupation with precision and detail distract you from the real goal of (quality) cost reduction and successful management of the business. There is no end to seeking more data. This is an excellent delaying tactic, rarely a good alternative to good management.

Monitor progress constantly, and use closely spaced milestones to keep interest and attention on the entire reduction process. When you're up to your ears in alligators it's difficult to keep in mind the goal of draining the swamp.

It takes time, patience, diligence, and a clear vision of the goal to show significant results. Quality cost reduction has to be a cooperative result produced by sustained effort.

Bibliography

Crosby, Philip B., "Quality Is Free," New York, New York, McGraw-Hill, 1979.

Crosby, Philip B., "Quality Without Tears," New York, New York, McGraw-Hill, 1984.

Deming, W. Edwards, "Quality, Productivity, and Competitive Position," Cambridge, Massachusetts, MIT Center for Advanced Engineering Study, 1982.

EUROPEAN ORGANIZATIONFOR QUALITY CONTROL
ANNUAL CONFERENCE PROCEEDINGS

QUALITY AND ECONOMY
MORE EMPHASIZE THE ROLE OF QUALITY ON SALES RATHER THAN ON COST

Noriaki Kano
Professor
Science University of Tokyo
Japan

In this paper, the following subjects are discussed:
1. Data-based study on the profitability of TQC companies, the external quality trouble cost and quality cost in Japan
2. Insufficiency of the existing accounting information system from the viewpoint of sales analysis
3. Proposal of a new tool for analyzing sales in relation with quality

1. Introduction

TQC, which has been enthusiastically promoted in many Japanese industries, is defined as business management which puts its priority on providing satisfaction to customers through products or services. Under the TQC philosophy, it is understood that the best way for an enterprise to increase sales and profit is to provide satisfaction to customers. We have more than 20 years experience in promoting TQC in Japan. As one of the evidences that this philosophy is effective, N. Kano et. al. (1983)(1) provided Fig. 1. This is the comparison between the profitability of TQC companies and the average one of the listed manufacturing enterprises where the TQC companies are represented by Deming prize winning companies after 1961. Deming prize is awarded to an enterprise which has enthusiastically promoted company-wide activities for establishing a system for quality assurance and has succeeded in performing remarkable effects in quality. Therefore, as shown in Fig. 1, quality is closely related to enterprise economy. However, the relation has not been clearly explained. In this paper, this point is paid an attention to and is discussed by data-based study.

The subject "Quality and Economy" could be replaced with "Quality and Profit." The relation between quality and profit would be more easily investigated by breaking it down into "quality and cost" and "quality and sales." So far, the subject "quality and cost" has been so frequently discussed under the name of "quality cost" in many papers. However, the subject "quality and sales" has been scarcely discussed. In this paper, the survey on external quality trouble cost and quality cost in Japanese industries is, first, reported, and then "quality and sales" will be discussed by proposing a new sales model and method which supplements the insufficiency of the existing accounting information system.

2. Quality and Cost

2.1. External Quality Trouble Cost

In 1983-84, the research committee for quality and economy of Japanese Society for Quality Control (JSQC) did the survey on external quality trouble cost borne by companies.

Fig. 2 shows the distribution of the ratio of external quality trouble cost to total annual sales, which was excerpted from the survey report by N. Kano et. al. (2) The questionnaires were sent to all the JSQC supporting companies (153 companies) and the companies (527 companies) of JSQC individual members and we received 217 companies and among them the 122 companies answered the ratio displayed in Fig. 1. It could be understood that the distribution in Fig. 1 shows the situation of the cost for external quality trouble in the companies which are enthusiastically promoting TQC, because JSQC is supported by such companies and members from such companies that are enthusiastically promoting TQC.

From Fig. 2, we can learn the following matters:
a) The ratio is distributed from "less than 0.01%" to "1 to 2%," but no more than 2%.
b) 80% companies answered "less than 0.5%."
c) The median which denotes the central position of the distribution is 0.12% and the 50% range around median which denotes the dispersion of the distribution is 0.05% to 0.3%.

2.2 Quality Cost

It has been said that there are very few companies in Japan which sum up quality cost. However, we had no data on this. In the above survey, we also put the question on quality cost. Then, we found 16 companies which possess quality cost calculating system, though it is a small number. Among them, 13 companies also answered the actual figures, which are displayed in Fig. 3 compared with the data of U.S.A. industries excerpted from Gilmore's report (1983).(3)

2.3. Conclusion

Compared with 3.9%, the average of current profit ratio to sales of the listed manufacturing enterprises in Japan in 1983, the values of Fig. 2 and Fig. 3 are not so small as we can neglect. Therefore, the activities for reducing the external quality trouble cost or quality cost is still one of the major pillars of TQC. However, this does not mean that we need no more additional quality activities. Particularly, in advanced countries, the sales aspect of quality has got important year by year in addition to its cost aspect. This will be discussed in the following chapters.

3. Basic Concepts for Quality and Sales

3.1. Insufficiency of Accounting Information System

For many executives, the most interesting material in their business is Profit & Loss Sheet (P/L Sheet) and, besides, for any companies, the most well arranged information system must be the accounting information system which provides financial statements including P/L Sheet.

The two major subjects related to P/L Sheet are sales and cost (including various expenses). It is very usual that an annual plan is set for each of the both sales and cost and every suitable period such as quarterly, semiannually or annually, the deviation of actual values from the plan is, if necessary, calculated and feedback action is taken. Apparently these managerial behavior seems to be the same for both sales and cost. However, there really exists a very large, essential difference.

For cost management, when actual cost is remarkably beyond planned value as displayed in Fig. 4, it is a problem. In such a case, we analyze the difference between the planned and actual costs by their elements compared with the details of a plan and then we found to which cost elements the major gap is due. Then, for investigating those cost elements, we check all the expense slips which are related to the elements. Finally, we could find how we made the gap, because all the cost data concerned with the actual gap from the plan are included in the accounting information system.

On the contrary, for sales management, the problematic situation is one as shown in Fig. 4, where the actual sales is far from the aimed value. Macroscopically speaking, we can easily calculate how many billions yens are lacking. However, it is common we cannot find the contents of shortage. Of course, although we can analyze sales data by market and product series or we can compare it with that of the previous year, we cannot reach the causes why such a shortage occurred. Because *sales data in the accounting information system are all successful data, but they do not include any lost sales data.* In order to find the causes of the shortage, we need the data of lost sales, which are not included in that system. *This is the vital insufficiency of the accounting information system.* However, many executives are ignorant of this and then they believe that the accounting information system is like something perfect. It is the biggest subject how to supplement this insufficiency. I believe this is a very important major task of TQC today.

3.2. One Hint for Solution

A department in a university decided to replace the copy machine because of its deterioration. After negotiating four makers A, B, C and D, the machine of A maker was decided in the department staff meeting. After the meeting, the phone call came from salesman of each company in the following way.

Mr. d from D Maker: How was your decision?
Dept Staff: Unfortunately, you lost it.
Mr. d: Next time, please.

Mr. c from C Maker: How was the result?
Staff: Unfortunately. . .
Mr. c: Which maker did you select?
Staff: We selected A maker.
Mr. c: Next time, please.

Mr. b from B Maker: How was your decision?
Staff: We didn't select yours, unfortunately.
Mr. b: Which maker did you select?
Staff: A maker.
Mr. B: Why? Do you have any clear reasons?
Staff: Yes, we have. If you like to know, why don't you come to my office?
Then, he visited the staff's office and was handed a list of comparison between A Maker's and his machine from the department viewpoint. He read it very seriously.

Mr. a from A Maker: Did you decide to buy the machine from us?
Staff: Yes.
Mr. a: Very good! Which company competed with us? And why did you select us?
Staff: B Maker. You can see the list of our comparison when you come to my office for contract.

Utilizing the above example, let's think the sales shortage problem in Fig. 4. Suppose all the salesmen in each of B, C, and D Makers behave themselves like each of Mr. b, c, and d respectively. Then, for D Maker, they have no data to be analyzed for shortage reasons. Therefore, it cannot take any actions except spiritual ones. For C Maker, it they have a system for summing up the competition data to the corporate level, they can recognize which makers they are beaten by. For B Maker, if it has a format for systematically recording "why" data and summing them up to the corporate level, it can recognize both which competitor it was beaten by and why it was beaten. It is very usual that most of "why" data are related with quality. Then, it can scientifically prepare which kind of actions should be taken. In this way for the company, people easily understand how much quality effects on sales and a scientific and systematic approach will be made even for sales management.

4. Relative Difference Analysis for Sales

4.1 What Is Relative Difference Analysis?

Let a variable, z, be equal to a product of two variables, x and y, such as:

$$z = xy. \tag{1}$$

If certain criteria on x, y and z exist as x_c, y_c, z_c, and if:

$$z_c = x_c y_c, \tag{2}$$

then it is well known that the following approximation holds:

$$\triangle z/z_c = \triangle x/x_c + \triangle y/y_c \tag{3}$$

where $\triangle x = x - x_c$, $\triangle y = y - y_c$ and $\triangle z = z - z_c$, and the approximation error for (3) is given by $(\triangle x/x_c)(\triangle y/y_c)$

Sales, S, is a function of market demand, D, and market share, m, such as:

$$S = Dm. \tag{4}$$

Applying (3) to (4),

$$\triangle S/S_c = \triangle D/D_c + \triangle m/m_c, \tag{5}$$

where, for example, each S_c, D_c and m_c denotes a planned value and each S, D and m denotes an actual values, the formula (5) shows that the relative difference of actual sales from a planned value is approximately broken down into the sum of relative differences of actual demand and market share from each of a planned value. The analytical method with use of the formula (3) is named *relative difference analysis (RDA) method*.

4.2. Case Study for RDA

1) Case

In this section, the relation of quality with sales is explained through a case study that RDA method is applied for the analysis of practical sales data. The data in Table 1(a) are ones for sales, demand, market share of a product series in a machine maker A.

2) Analysis of Sales with RDA Method

The analytical result of RDA method for the data Table 1 (a) is shown in Table 1 (b).
First, apply RDA method of "85 plan" with "84 actual" as criteria. Then, the first line (1) in Table 1 (b) is its result. This gives the idea that "85 plan" must have intended more than 20% increase compared with "84 actual" and this must have depended on 6% increase in market demand and 14% up in market share.

However, "85 actual" shows that "85 plan" was not achieved. Why? Then, apply RDA method for "85 actual" with "85 plan" as criteria. Its result is given in the second line (2) of Table 1 (b). This gives the reasons why "85 plan" was not achieved. The 10% shortage of sales was due to 14% down in market share although the demand was 5% increased compared with its prediction in "85 plan." These situations are more clearly expressed by radial diagrams in Fig. 5.

3) Analysis of Market Share with RDA Method
Definition
Business Approach is defined as activities of an enterprise to its customers for selling its products by such as visit, market channel arrangement, etc.

Business Approach Ratio, b, is defined as a ratio of the demand to which an enterprise makes business approach to the total demand of a certain product group. Some companies call this as "business talk participation ratio," "market coverage ratio," etc.

Successful ratio, s, is defined as a ratio of the demand which is met by products of a certain company to the demand to which the company makes business approach.

Market share, m, is a function of business approach ratio, b, and successful ratio, s, such as:

$$m = bs. \qquad (6)$$

Then, applying (3) to (6)

$$\triangle m/m_c = \triangle b/b_c + \triangle s/s_c, \qquad (7)$$

where each m_c, b_c, and s_c denotes a certain criterion of m, b, and s, respectively and each $\triangle m$, $\triangle b$ and $\triangle s$ denotes the difference of m, b, and s from m_c, b_c, and s_c, respectively. The formula (7) means that market share, m, can be analyzed by RDA method.

The data for market share (m), business approach ratio (b) and successful ratio (s) are given in Table 2(a) and the results of RDA and given in Table 2(b).

The results of RDA for "85 plan" with "84 actual" as criteria explain that "85 plan" must have intended to expand market share by sharp increase of success ratio with a slight increase of business approach ratio. It is estimated that the business approach ratio had already arrived at 80% level so that it would have needed a large amount of investment to further increase business approach. However, successful ratio was very low. Then, the company must have intended to sharply increase it by introducing a certain new product in 1985.

However, "85 plan" for market share was not actually met. Why? Then RDA method is applied for "85 actual" with "85 plan" as criteria. The results are given in the second line (2) of Table 2(b). These show that successful ratio is very much far from "85 plan" although business approach ratio is beyond the plan.

This situation is more clearly expressed by radial diagrams in Fig. 6.

4) Summary of the Analysis by RDA Method

By the previous analysis in 2) and 3), we can sum up in the following way.

"85 plan" must have intended to expand sales by sharp increase of successful ratio with almost constant business approach, being supported by the expansion of market. However, the actual sales did not achieve the plan in 1985 because of the steep fall in successful ratio although the demand expanded beyond the prediction and business approach was made over the plan. Then this analysis displays the necessity of investigation why successful ratio dropped down.

5) Investigation for the Decrease of Successful Ratio.

According to the analysis by RDA method previously described, it is clear that the shortage of the actual sales from the plan was due to the decrease of successful ratio. Then, the marketing department made the investigating for this and picked up the following points as its reasons:

(1) The sales force has not sufficient understanding of the new product released to the market in 1985 so that they could not provide clear explanation to these customers.

(2) The machine released to the market in 1983 had made quality troubles. Some customers who had bought the 1983 models replaced the machine in 1985, but most of them bought machines from the competitors because of the after effect of the quality troubles.

(3) The 1985 new product was remarkably enhanced in quality compared with the previous type as shown in Fig. 7, but its quality level was almost the same with the company C except a few quality elements and then the new product could not attract the users who had got accustomed with the rival machines.

(4) The price was set a little bit cheaper than the rival machine, however, it was not enough for filling the quality gap described in (2) and (3).

These results of (1) to (4) were reported to the top executive of the division. He reflected them on making the 1986 annual policy for the division in which he clearly directed to all the related departments the remarkable quality advancement for the new product to be released in 1987. He directed that it must surpass the rival machine at least in three quality elements. Without this analysis by RDA method, he would not have got dissatisfied with the 85 actual sales shown in Table 1(a) because it was beyond the previous year and then would not have a special concern about quality.

5. Quality in Sales

5.1. Sales Structure Model

The application of RDA method implies the necessity of introducing a sales model which is prescribed by the formulas (5) and (7). According to these formulas, the sales gap from a certain criterion is expressed in terms of
- demand gap,
- market share gap
   business approach ratio gap
   successful ratio gap

Then, if demand gap is dominant, the gap should be investigated from the viewpoints of:
(1) unexpected change of market structure like the oil crisis.
(2) inappropriateness of demand prediction method.

If business approach ratio gap is dominant, it should be investigated from the viewpoints of:
(1) sufficiency of customer list
(2) efficiency of picking up prospective customers
(3) sufficiency of sales force including market channels and number of salesmen
(4) efficiency of visiting prospective customers

If successful ratio gap is dominant, it should be investigated from the viewpoints of QPDSS, i.e., quality, prices, delivery, sales force and service as follows:
(1) quality
   - past satisfaction degree of a customer for the product which was bought from a company before and has been used.
   - quality comparison with rival products in function, size, newness, fashion, appearance, public reputation, handling, maintenance, reliability, safety, running cost etc.
(2) price
   - price comparison with rival products including payment conditions.
(3) delivery
   - easiness to see and touch and quick delivery compared with rival products.
(4) sales force
   - advertisement publicity
   - visits frequency
   - product explanation including quality of sale materials
   - human relations
(5) service
   - service turn-around period
   - service network
   - service skillness
   - service part arrangement

These situations can be illustrated by a radial diagram like Fig. 7.

5.2. Position of Quality in Sales Structure

The model based on the above-mentioned factors could be shown as Fig. 8. This model is named as Hierarchical Sales Structure Model (HSS model). By the introduction of HSS model, the role of quality on sales can be clearly positioned in sales structure.

5.3. Merit and Cost of the Application of RDA Method under HSS Model

Merit

The application of RDA under HSS model supplements the insufficiency of the accounting information system in the following points which the system cannot make approach.
(1) The role of quality in sales becomes clear.
(2) Systematic sales analysis is made and its results may often give a strong impact to executives for quality.
(3) The responsibility for demand prediction, business approach and successful/lost sales becomes clear.

Cost

The application of RDA under HSS model needs the new information system which is based on the data such as sales, demand, business approach, QPDSS comparison with rival products, lost sales report, etc. The collection of these data must need additional cost. When the executive makes decision for this, he is recommended to compare with how much money is paid for the existing accounting system.

Acknowledgment

To accomplish this paper, I was supported by several companies which I did consultation. Especially, I was obliged to Mr. Kenichiro Nakajima, TQC manager, the Overseas Business Division, Komatsu Ltd. Co.

References

1) Kano, N., Tanaka, H., Yamaga, S. (1983): "Economical Effects of TQC in Deming Winning Enterprises," Hinshitsu (Quality), Vol. 13, No. 2, pp. 67-75 (in Japanese)

2) Kano, N., Ando, Y., Mitsufuji, Y., Tanaka, H. (1985): "Data-based study on External Quality Trouble Cost and Quality Cost," Hinshitsu (Quality) Vol. 15, No. 4, pp. 88-95 (in Japanese)

3) Gilmore, Harold L. (1983): "Consumer Product Quality Control Cost Revisited," *Quality Progress,* Vol. 16, No. 4, pp. 28-33

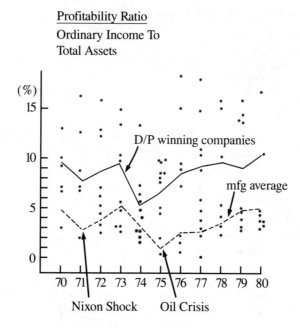

Fig. 1 Comparison between the profitability of TQC companies and the average of the listed mfg. companies [1]

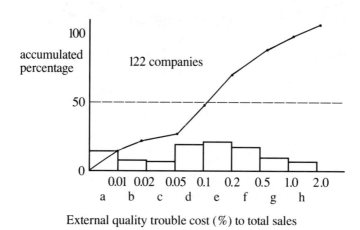

Fig. 2 Survey to JSQC supporting companies and JSQC members' companies in 1983-4 [2]

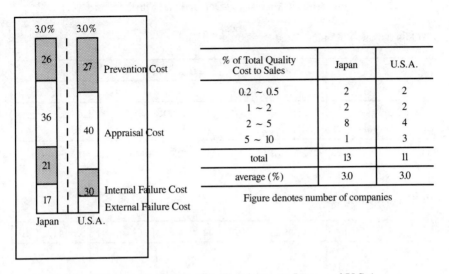

Fig. 3 Composition of Quality Cost between Japan and U.S.A.

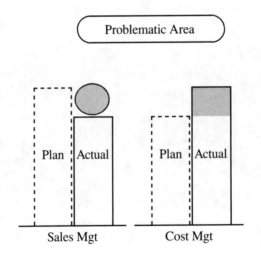

Fig. 4 Difference of investigating causes of unachievement between Sales Mgt and Cost Mgt.

Table 1 Sales Analysis by Demand and Market Share

(a) "84 actual," "85 plan," and "85 actual" data

	Demand (D)	Market Share (m)	Sales (S)
84 actual	20.3 bil. yen	14.0%	2.85 bil. yen
85 plan	21.5	16.0%	3.44
85 actual	22.5	13.8%	3.10

The numerical sales were modified because of confidential reason.

(b) Relative Difference Analysis

	Demand ($\triangle D/D$)	Market Share ($\triangle m/m$)	Sales ($\triangle S/S$)
(1) RDA for "85 plan" with "84 actual" as criteria	5.9%	14.3%	20.7%
(2) RDA for "85 actual" with "85 plan" as criteria	4.7%	−13.7%	−9.0%

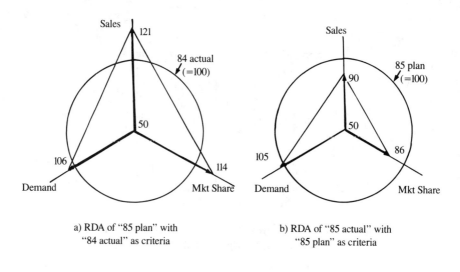

a) RDA of "85 plan" with "84 actual" as criteria

b) RDA of "85 actual" with "85 plan" as criteria

Fig. 5 RDA of Sales into Demand and Mkt Share

Table 2 Market Share Analysis by Business Approach Ratio and Successful Ratio

(a) "84 actual," "85 plan," and "85 actual"

	business approach ratio (b)	Successful ratio (s)	market share (m)
"84 actual"	79%	17.7%	14.0%
"85 plan"	80%	20.0%	16.0%
"85 actual"	81%	17.0%	13.8%

See the footnotes of Table 1 (a).

(b) Relative Difference Analysis

	business approach ratio ($\triangle b/b$)	successful ratio ($\triangle s/s$)	market share ($\triangle m/m$)
(1) RDA for "85 plan" with "84 actual" as criteria	1.2%	13.0%	14.3%
(2) RDA for "85 actual" with "85 plan" as criteria	1.2%	−15.0%	−13.7%

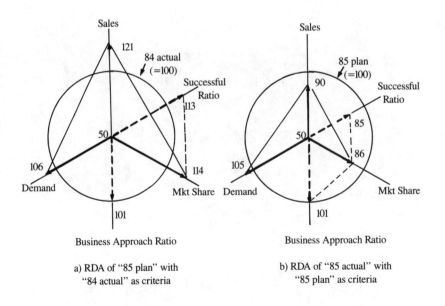

a) RDA of "85 plan" with "84 actual" as criteria

b) RDA of "85 actual" with "85 plan" as criteria

Fig. 6 RDA of Sales into Demand and Mkt Share (solid line) and RDA of Mkt Share into Business Approach Ratio and Successful Ratio

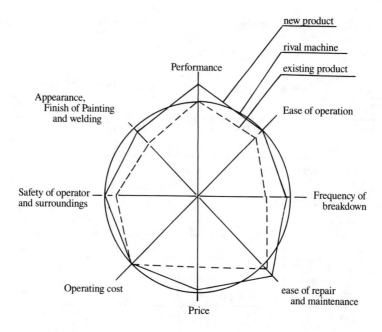

Fig. 7 Radial Diagram for Quality and Price Comparison with Rival Product

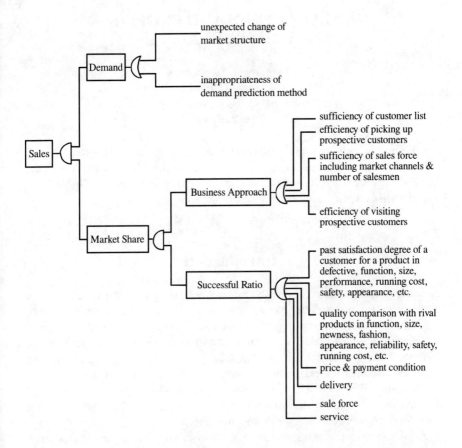

Fig. 8 Hierarchical Structure for Sales

QUALITY COSTS AND PROFITS — MYTH OR REALITY

Anthony R. Stephenson
Managing Director
International Quality Consultants Ltd.
New Zealand

Few Companies can — or wish — to substantiate their real costs and profits of Quality. Using a survey and case studies from a number of industries it is seen that the real costs are generally below the 20% of turnover often quoted. Another model of Value Added is proposed as a better parameter for cost analysis.

Introduction

The often quoted words of Lord Kelvin which essentially suggest that we know very little about a product/service unless we can measure it and express it scientifically is also echoed by the apparent words of Galileo Galilei who said "Count what is countable, measure what is measurable, and what is not measurable, make measurable." So we have the basic justification for counting Quality Costs and Profits. From our own experience it is often evident that the Accountancy profession (see also Claret 1985) does not have the ability — or willingness — to monitor quality performance in the most sensible ways, i.e., through the costs of Quality performance and the planning of future investment through, for example, sound investment analysis.

Admittedly as Quality Professionals we do ourselves a frequent disservice. How many so called Quality Professionals feel comfortable with financial matters? Too few I am afraid; we shroud ourselves in the mystique of statistics, specifications and measurement and totally ignore the financial burdens of the company. We need to help our accountants and costing staff to help us. To do this we must have a knowledge of basic accounting, data gathering, break even analysis, even discounted cash flow! Our experience of high inflation in New Zealand, for some years running at over 16% p.a., makes the use of discounted cash flow techniques worth examining — if only to confirm that some new technology is *not* affordable!

True costs — and profits — of quality are scattered throughout the organisation, from sales and service through to design, manufacturing, storage, and so on. Modern computing systems allow us to do much in the way of data collection and disbursement *but* we have to sit down and decide what we want to highlight from such information.

For many companies at the beginning of the road to enlightenment, I suggest a variation the 'Activity Sampling' theme, so popular amongst industrial engineers. It amounts to short-term gathering — *in depth* — of quality cost data in sections and departments. It may be only a few days, through perhaps coding of problem areas, but it is a start.

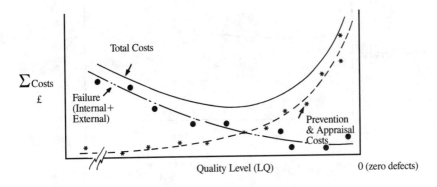

Fig. 1 Costs vs Limiting Quality

This, of course, leads us to another important area of concern found in many text books and standards on Quality Costs, namely the, so called, economic cost of quality. Through a series of reviews of data and models, the cost issues were explored in some depth by Stephenson (1981), Marriott (1982), and Marriott and Stephenson (1984). It had been expected that models of the form shown by Fig. 1 could be justified in some detail.

It will be observed that Limiting Quality (LQ — or LTPD) has been used since this is what effectively the customer sees in his own assembly/processing results. The first major lesson found was that as *consequential* costs of failure become charged to the producer, the so called optimum very rapidly moves towards zero defects! As with most modelling and reality, hard data becomes difficult to obtain. Another lesson appeared to be that for many situations the total cost curve is, in fact, quite shallow. In other words, there was no great advantage in trying to save a few cents at (say) 5% LQ when, in fact, one would deliver better than 1% LQ for virtually no extra cost!

In many instances there is concern over the allocation of quality costs between Prevention, Appraisal and Failure; Standards such as A.S. 2561-1982 have been evolved to assist. In many cases a simple comparison between expenditure to achieve a reduction in say warranty claims can be offset against the actual warranty claims. It is a simple exercise and allows us to show simply that a new measuring machine and facilities of $X can be paid for by reduced scrap/returns within say two years. That is what an accountant understands. Find then your cost of failure, and the costs incurred in reducing those failure, and then get on with it! This is how the Japanese have succeeded so well by looking at continually striving to reduce those final failures. Just ask yourself how reliable modern motor vehicles are compared with twenty years ago, then look at TV, audio systems, cameras, and so on. There is no reason why Fig. 2 cannot apply to all our endeavours, after all many electronic products are illustrated by Fig. 2.

Much has been written about Quality Costs and Profits; the material by Campanella and Corcoran (1983), Sullivan (1983), Dale and Plunkett (1983, 1984, 1985) and Jones (1978) are just a small selection. Before moving on to some survey results, a few of the concluding remarks by Plunkett and Dale (1985) are worth emphasising.

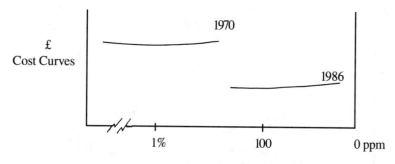

Fig. 2 Costs vs Limiting Quality

1. Much of the published cost data is not readily quantified by industry sectors.
2. Prevention, Appraisal and Failure headings are not — in themselves — always appropriate.
3. Not all available company data is seen as suitable or relevant for Quality Cost analysis.
4. Standard accounting practice needs to be modified.
5. Indirect personnel may spend a great deal of time on 'Quality' matters, e.g., supervisors.
6. Credibility of data must be guaranteed for acceptance.
7. Don't guess, obtain data.
8. Time delays may mask some important factors in improvement.

Some Survey Results and Industry Studies

Many of the following comments are based on a survey by Stephenson (1981), Plunkett and Dale (1983, 1984 and 1985) and Gilmore (1983). A number of observations are also based on private communications by the author with senior executives in a number of industries. The comments therefore are certainly not purely Antipodean or biassed compared with Europe and the USA.

In general there are three main areas of applying Quality costs for comparison (Stephenson 1981, Gilmore 1983) i.e., total costs of Quality as a percentage of:
- Total (Gross) Sales or Turnover
- Manufacturing Costs
- Value Added

Some of the key ratios developed in the survey by Stephenson (1981) were:

1. The value (book) of QC and QA test equipment compared with the book value of production equipment was generally less than 5%. There is a common failure to invest in high technology production equipment and then not to provide 'measuring sticks' capable of measuring to the appropriate accuracy.

Prevention			Appraisal			Failure		
% Cost	% Respondent		% Cost	% Respondent		% Cost	% Respondent	
<5	(38)	[36]	<20	(12)	[23]	<5	(38)	[32]
5-9.9	(25)	[18]	21-50	(50)	[57]	5-50	(38)	[61]
10-50	(12)	[46]						
>50	(25)		>50	(38)	[20]	>50	(24)	[7]

(Gilmore Data)   [Stephenson Data]

Table 1 Some Comparisons USA and NZ Data

2. The value of QC/QA staff to the value of total staff averaged about 5% with one company having a ratio as high as 15% and a number with a ratio less than 1%.

3. By combining the results of Gilmore (1983) and Stephenson (1981), Table 1 can be constructed.

4. Another significant factor on which it was pleasant to have independent corroboration, was the ratio of the total cost of quality to turnover. From the Gilmore (1982/3) survey the TQ cost ranged from 0.45% to as high as 8.2% for similar industry classifications as the NZ study. From the 40 respondents with NZ data — the TQ cost ranged from seven companies reporting less than one per cent to one company reporting greater than twelve per cent. The industry weighted average was just under 6% and this compared with a wide range of European data collected during the study by Stephenson (1981). In other words the number of companies with real costs of quality in the 15-20% bracket are in fact very few and are likely to be involved in 'Leading edge' design and development.

This does not mean that we should not be interested in potential savings. After all, a saving of 1% on a turnover of $10 million by reducing quality costs by 1% is still $100,000 and it can be traced rapidly to the bottom line or profitability.

5. We may, therefore, have done our profession a disservice on a number of occasions by quoting the very high percentages of costs such as greater than 20% of turnover. Most accountants would be relatively alert to such high figures distorting profitability. It is more in the 2-8% range of TQ cost turnover that the cost divisions may be hidden. Accountants need to see credibility in the figures, not emotion; then they will believe our message. In recent QA assessments we have been aware for example, of warranty cost accounts equal to 2% of turnover being set up and by no means being exhausted at the end of each year!

6. The Gilmore study (1983) went on to examine quality costs as a basis of % of manufacturing costs. One might have to examine the basic definitions of the parameters used in Manufacturing Cost but the most obvious would be direct labour, direct material (i.e., prime costs) plus the overheads associated with manufacturing such as supervision, planning, scheduling, stores, stockholding costs and the like. The similarity between Manufacturing Cost and Value Added is, in fact, quite close. From the Gilmore (1983) study we see that the TQ costs as a percentage of manufacturing range from one per cent to twelve per cent. It may well be that a manufacturing organisation could and should use more than one ratio, i.e., for manufacturing either a percentage of the manufacturing or Value Added costs should be taken in relation to — say — costs of prevention, appraisal and internal failure, while sales might well be largely interested in the costs of warranty or life cycle costs. Another factor common in some capital works is to take quality costs in relation to an item (e.g., motor coach, turbine, building).

It is now time to turn to the concept of Value Added as a quality measurement parameter.

Value Added — A Valid Quality Cost Parameter

Profit is not a particularly relevant measure of company output except, perhaps, in terms of return on investment. Turnover looks attractive but, for many industries with different lines of product, different mark-ups, sale prices, etc., the turnover figure is easily distorted. Then too there are inflation effects and over a full 12 months 16-20% inflation can play havoc with basic ratios. What then is Added Value and how can we use it as a quality cost parameter?

Added Value — The Basic Concept

Basically it is the difference between bought in materials/services and sales revenue, provided the time scale differences between these two are not excessive (i.e., 4-6 months) then inflation factors are present in bother terms. It is then literally the value the company has added to the materials/services by processing them, and this represents the company's net output in real economic terms. It is spent on wages, salaries, capital charges, dividends and profit retention.

We can define Added Value as:

$$S = M + W + P \qquad \text{Eq. 1.}$$

where S = Net Revenue after deducting discounts, external freight.

M = Total cost of all materials and services bought in.

W = Total personnel costs — wage, salaries, levies, training costs, pension schemes, etc.

P = What is left and is the gross profit to cover capital charges, depreciation, interest, taxes, rent, insurance, dividends and, hopefully, retained profit.

Rewriting Eq. 1. we have $S - M = W + P =$ Added Value Eq. 2.

As an example to illustrate where it may be used as a tool of Productivity, and Quality is part of that productivity package!*

Example 1

Let the net sales (S) be $1,000,000

Subtract the total cost of raw materials and services (M) such as fuel, components, phone, travel, etc., items which are other people's Added Value $ 650,000

The Added Value is $ 350,000

Let our wages bill (W) be $ 210,000

Let the capital costs and profits (P) be:
Rent	$ 5,000		
Depreciation	$ 27,000		
Interest	$ 14,000		
Insurance	$ 8,000	$54,000	
Rates	$ 8,000		
Taxes	$48,000	$56,000	
Net Profit		$30,000	
			$ 140,000
			$ 350,000

One useful ratio might be labour productivity, i.e., Added Value per Dollar of employee cost.

In the example this is 350,000 divided by 210,000 or $1.67/$ of wages. For every dollar on wages we are adding $1.67 to the value of our materials. The real advantage to the technique now becomes valid when we compare, from published statistics, these same ratios for different industries. Some industry figures in the New Zealand context are shown in Table 2.

	1973/4
Food Manufacturing	1.19
Beverages	2.25
Furniture	1.35
Power	1.43
Rubber	1.40
Machinery	1.36
Electrical	1.36
Transport	1.45

Table 2 Added Value for Selected Industries (NZ)
(Recent trends will be discussed at the Conference based on 1985/6 figures)

---

*I am indebted to Brian Melville (1977) for some inspiration at this point.

It will be apparent to many that the ratio is also sensitive to capital invested in — say — new high technology plant, i.e., we may have a low wage bill but be paying high interest on our borrowed capital to achieve the required output. In the earlier example, for $350,000 of Added Value, another company in that industry might have a wages bill of only $170,000 so the added value ratio per labour Dollar becomes 2.05:1. We must not forget these are essentially labour ratios. What about Quality costs.

Quality Costs and Value Added

How would you feel as a manager being told that the costs of prevention, appraisal and failures were up to 20% of your added value! Yet that is just what the survey of Gilmore (1983) and Stephenson (1981) tended to show. Gilmore's data showed a maximum percentage of about 26% while the Stephenson survey (of 66 companies) showed seven companies with data greater than 14% and four reporting over 25%.

Going back to Example 1, it is evident from the Left Hand Side of the equation that say excessive use of material/components through scrap, etc. reduces the available Added Value.

Example 2 (Using Example 1 as the base)

S	$1,000,000
M	$ 665,000
Added Value	$ 335,000

The wages bill is increased by overtime and site work to sort out problems

W	$ 220,000
Other costs)	$ 54,000
remain the same)	$ 56,000

Then, to balance, the net profit is reduced to $ 5,000
Added Value $ 335,000

and — naturally — the ratio of Added Value to labour has also dropped ($335,000 divided by 220,000) = 1.52:1. If the above facts don't make your accountant sit up and ask searching questions, nothing will!

For further work in this area the articles by de Guerin (1977) and Jones (1978) are helpful.

Conclusions

1. There appears to be an increasing amount of data which indicates lower TQ cost ratios than have previously been published.

2. Such data seems to indicate that total costs of quality are, on average, between 6-8% of turnover, 8-12% of manufacturing cost, and up to 25% of Value Added.

3. Value Added has some merits as a statistic since we can determine what are appropriate figures for different industry sectors by looking at published data.

4. Within our own specific industry sector we can then monitor our own ratios of TQ cost against Value Added on the basis that we need to know *the impact* of increased costs to rework, scrap, fix warranty claims, inspect inwards goods and so on.

5. The paper has presented a scenario of costs and profits which should help QC/QA personnel become familiar and confident with accounting terminology.

Acknowledgments

To Walt Hurd and Merv Burt for their continued support and guidance, to Roger Atkinson for reminding me to keep the cost analyses simple, to Guenter Arndt for driving me hard on the analysis and to Peter Brown for some timely advice on the real costs of services and life cycle analyses.

References

AS 2561-1982 "Guide to the determination and use of Quality Costs"

BS 6143-1981 "Guide to the determination and use of quality Related Costs"

Campanella J. & Corcoran F. J. (1983) "Principles of Quality Costs" *Quality Progress*, April

Claret J. (1985) "Mind the cost — but feel the Quality" *Quality Assurance News* Vol. 11, No. 7 July

Dale B. G. & Plunkett J. J. (1983) "Quality Costing: a study in the pressure vessel fabrication sector of the process plant industry" *Quality Assurance News* Vol. 9, No. 4 December

Dale B. G. & Plunkett J. J. (1984) "A study of audits, inspection and quality costs in the pressure vessel fabrication section of the process plant industry" *Proc. I. Mech.* Vol. 1983 No. 2

de Guerin D. (1977) "Our manufacturing industry — are we getting better or worse?" *C. Mech. Eng.* February

Jones J. S. (1978) "Quality Costs and Quality Improvement" *C. Mech. Eng.* February

Marriott R. (1982) "The practice and cost of Quality Assurance in New Zealand" *NZOQA Journal*, May

Melville B. J. (1977) "Added Value measurement of Productivity" Private communication with the author

Plunkett J. J. & Dale B. G. (1985) "Some practicalities and pitfalls of quality related cost collection" *Proceedings, I. Mech. Eng*, Vol. 199 No. B1

Stephenson A. R. (1981) *"The Economics of Quality Assurance in New Zealand"* M.E. Thesis, Univ. of Aukland

Stephenson A. R. & Marriott R. (1984) "Some economic quality cost models" *EOQC/ASQC/IQA World Quality Congress*, Brighton

Sullivan E. (1983) "Quality Costs: current ideas" *Quality Progress*, April

Sullivan E. (1983) "Quality Costs: current applications" *Quality Progress*, April

INSTITUTE OF ELECTRICAL AND ELECTRONICS ENGINEERS

TRANSACTIONS ON ENGINEERING MANAGEMENT

FINDING THE COST OF SOFTWARE QUALITY

Charles P. Hollocker
Senior Member, IEEE
AT&T Network Systems
Lisle, Illinois

Abstract

The growth of software use in recent years has focused attention on the problems long associated with its development. Uncontrollable costs, missed schedules, and unpredictable quality have all been regrettable trademarks of the software industry. Moreover, these problems appear to vary in some proportion to the size and complexity of a development project. The maintenance of quality leadership and the reduction of development costs are key managerial issues.

Cost has always been a major engineering management concern. As we enter the information age, the cost of information is tightly coupled to the cost of the software that processes and directs it. More and more system solutions include embedded software.

Quality affects a company's economics through its effect on income and its effect on costs. Although there is a growing need for a quantitative study of these effects in the software industry, there is an apparent inability to apply quality cost reduction programs proven successful in other high-tech industries. To assist software engineering management in the identification and control of quality costs, a guide to the application of Juran's quality improvement program is now presented.

Note: Manuscript received November 7, 1985; revised February 7, 1986. Some of the ideas for this paper were developed during the author's participation in research on "Measuring the Effectiveness of the Software Development Process for Manufacturing," being conducted by the program of Research on the Management of Research, Development, and Innovation (POMRAD) at Northwestern University. That study is being sponsored by the National Science Foundation (NSF). The content of this paper is not meant to reflect the views of the NSF.

I. Introduction

Unfortunately, there is increased confusion accompanying the "increased concern" for quality in the software industry. Specifically, the difference between software quality assurance programs, quality improvement programs, and total quality control programs appears less distinct than we might like. To provide adequate background for putting the costing issues into perspective, the family of quality programs merits a few words.

As an overview, quality improvement programs [1] generally refer to improving the efficiency and effectiveness of product development (or if you prefer, design and manufacture). From an economic perspective, however, we know that factors external to product development can effect our company's financial situation; hence total quality control is an issue [2], [3]. Though these two program approaches vary more in scope than in general philosophy, one assumption is integral to both. Each program can be applied only where reasonable and appropriate technologies are applied and controlled.

Software quality assurance programs [4], [5], as defined by the IEEE [6], "provide a planned and systematic pattern of all actions necessary to provide adequate confidence that software conforms to established technical requirements." This is of increased importance where critical software is concerned. The IEEE [7] defines critical software as: "...where failure could impact safety or cause large financial or social losses." Developing software that conforms to its technical requirements and is delivered on time and within budget requires a model [8] that includes:
- An understanding of the development process [9]-[11] that includes tasks and responsibilities, deliverables and completion criteria, tools and techniques, and the application of verification and validated on a per-phase basis.
- Three control components: project management [12], [13], configuration management [14], and quality assurance [15], [16].

With a software quality assurance program in place, program tuning can be made through the application of a quality improvement program. Since there is a need, however, to identify and control quality costs, the following sections are intended to help identify quality costs for software.

II. Quality's Effect on Income

Although the principal issue here is quality's effect on cost, a general understanding of quality's effect on income is required. Specifically, quality differences map to price differences if the user recognizes a quality difference and interprets it as superior fitness for use. Thus, the user's interpretation of value, not the development costs, is the basis for any price differential. Pricing takes into account the alternative of converting quality superiority into higher prices or increased market share. Profits increase in proportion to price increments or sales volume. With adequate marketing, quality leadership can be maintained through continuing product and process improvement.

Software maintenance can be broken down into three categories: corrective, perfective, and adaptive maintenance. Corrective and perfective maintenance meet the

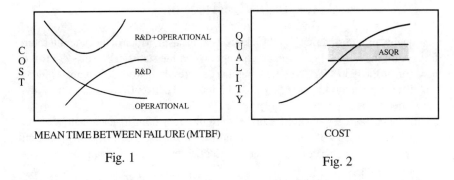

MEAN TIME BETWEEN FAILURE (MTBF)　　　　　　COST

Fig. 1　　　　　　　　　　　　Fig. 2

product's need to be fit-for-use at delivery and during its normal life span. Adaptive maintenance can require massive effort for large or complex software, and is intended to add features or functionally to ward off product obsolescence. For software products of great longevity, since they may require extensive adaptive maintenance, variable costs will drift upwards with time. This is owing to the Law of Increasing Unstructuredness, and decreases profits late in the product life cycle. The Law of Increasing Unstructuredness, paraphrased, states that a program undergoes structural deterioration with prolonged maintenance activity.

Although improving reliability and other quality factors [17]-[19] reduces the operating and maintenance costs, incremental design improvement costs must be weighted against the resulting "cost of usage." One way of looking at these costs (Fig. 1) is to graph Research and Development (R&D) and operational costs against reliability as represented by Mean Time Between Failures (MTBF). This shows that an optimal quality level exists for minimized costs.

In order to control life cycle costs, knowledge of failure rates, downtime, repair or service costs, and other elements of cost of usage must be maintained. Failure analysis must differentiate hardware and software failures, and map each defect to its development phase of insertion. The value of such a database to the company is related to risk avoidance, possible lost opportunities, and strategic planning. Since product failure does occur, regardless of the defect avoidance and removal methodologies employed, evidence of reasonable and prudent development activities can be used to prevent consumer or court overreaction. Possible quality-related grievances include: misrepresentation of product quality, product failure, and inability to get satisfaction when products do fail. Application of accepted software engineering practices [20], and the maintenance of supporting data can provide this evidence.

Two popular myths exist that merit immediate extinction:
- To prosper in the marketplace, we must produce quality products without regard to cost.
- Quality leadership means having products of the best possible quality.

Though intrinsically attractive, these two statements are fallacious; strategic quality leadership involves the identification and maintenance of an Acceptable Software Quality Range, or "ASQR" (Fig. 2). There will be more on this in the next section.

III. Quality's Effect on Cost

A quality program cannot be sold solely on enthusiasm. Much of the older literature focuses on the hidden "gold mine" concept behind the statistical quality control of hardware in the 1950's. The argument was that costs resulting from defects are avoidable and that the potential for savings is a gold mine. The concept was used to show that programs for defect production can provide a good Return on Investment (ROI) for the additional staffing costs involved. Current software engineering management, however, should be concerned with several issues central to having a successful quality program:
- The ability of the company's accounting system to present appropriate cost summaries.
- The capability of identifying all costs associated with the quality function.
- The quality function's ability to control quality costs.

These issues are amplified by advancements in technology, an abundance of standards-making organizations, and labor-intensive maintenance over extended product life spans. Although the magnitude of the challenge presented by these issues can be discouraging, one company's "road blocks" can be another company's milestones. Tomorrow's success stories will be based on successful quality cost control. The problem is getting started!

A model (Fig. 3), consisting of three phases (e.g., Selling, Project, and Control), has been suggested (21, pp. 2-4] to integrate snapshot and prolonged control views for quality cost studies. Although this paper addresses predominately the first phase (and is, therefore, more concerned with providing initial estimates) certain trends are predictable. As Selling Phase progress reports are prepared and issued, emphasis will shift from program justification to measuring managerial effectiveness. This alone is probably enough to kill most programs for the software industry; it appears that software development managers, like many other people, have an aversion to being held accountable.

The Selling Phase ends and the Project Phase begins when managers accept being held accountable for their efficiency and demand that more precise information (not rough estimates) be used as a basis for evaluation. Although the reporting function is still the responsibility of the quality organization, significant progress has been made in integrating the accounting department into the overall program; they can now verify the data. Improvement projects are tracked and progress is stimulated by the quality organization.

At all levels, managerial accountability and commitment to quality must precede entry into the Control Phase. This phase begins with an agreement of what "normal" quality costs are, and sees the accounting department publishing reports based on their own data. The quality organization adds comments and any further interpretation to the reports, while directing the control of quality costs.

Early entry into this phase will result in continued "avoidable cost" losses since control standards will be too loose, or commitment too weak. Overly ambitious programs can fail for this reason. Late entry into the Control Phase wastes time and resources that could have been spent more wisely.

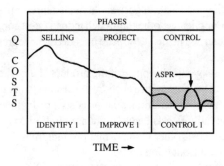

Fig. 3

In both the Project and Control Phases, we need to design scorekeeping to minimize arguments over estimate validity. Cost of quality expenditures believed typical for this model, when presented (Fig. 3), can be aggravating: *we desire the lower costs and greater predictability provided by the third phase, but have not suffered the necessary learning curve associated with traversing the first two phases.*

Recall that an optimal range, the Acceptable Software Quality Range (ASQR) exists. Thus, the objective of the Control Phase in a product assurance program is to provide quality software as defined by the ASQR, but within optimal cost constraints (Fig. 3) named the Acceptable Software Productivity Range (ASPR). This further accents the relationship between quality and productivity.

IV. Quality Cost Categories

Quality cost elements, discussed here by category, are identified by answering the following question: *If we produced a product with no defects, what costs would be eliminated?*

The first step in any control effort is to decide just what is to be controlled. Since an outsider cannot establish an extensive list appropriate for your own product and process characteristics, the suggestion is made that a few key supervisory personnel review this paper and attend a brainstorming session to define applicable cost elements before your specific data collection, reporting, and process control techniques are defined. After element identification, a one-time study should be made to formulate initial cost figures from:

- Established cost accounts such as training, guarantee/warranty, etc.
- Existing accounting records and organizational charts.
- Temporary records kept to help in estimation.
- Best-guess estimates (when all else fails).

For each of the following cost categories, elements representing software costs are identified and, where applicable, other high-tech industry cost concerns are discussed. The cost elements identified in this section are intended to be representative rather than all-inclusive.

External failure costs would disappear if the product had no defects. This category is distinguishable from internal failure costs by the discovery of defects *after* shipment to the customer. Transfer/installation quality costs are included in this category. Specifically, external failure costs include, but are not limited to:
- Customer support costs other than normal product transfer or installation costs.
- Excessive product introduction costs.
- Cost of classifying defect as hardware or software related.
- Software failure or anomaly identification (isolation) cost.
- Problem reporting and change initiation costs.
- Field update costs.

The software industry shares the following concerns with other high-tech industries: complaint adjustment, returned products, warranty charges, and allowances.

The following internal failure costs would disappear if no defects existed in the product before shipment to the customer:
- All in-process Problem Reporting and Corrective Action (PRCA). This includes "debugging" and test environment (execution time) costs.
- Rework of specifications.
- Code rewrite.
- Rework of any prior software development phase.
- Reapplication of appraisal costs (see next section).
- Lost opportunity owing to delayed product introduction or deployment.

Other high-tech industry internal failure cost concerns include scrap, production system downtime, yield losses, and the usability of nonconforming products.

Appraisal costs are incurred during product development (and adaptive maintenance) for verification and validation. Verification comprises all techniques used to assure that each phase in a software development life cycle encompasses the intentions of the previous step. Validation, on the other hand, comprises those techniques used to ensure that a software product functions correctly and contains the features prescribed by its requirements. Appraisal costs include:
- All testing efforts and support (lab) costs except integration.
- Simulation.
- Management reviews.
- In-process development inspections.
- Technical audits.
- Compliance audits.

Other high-tech industry appraisal cost concerns also include incoming material inspections, maintenance of test equipment accuracy, materials and services consumed, and the evaluation of stocks for storage degradation. This should imply that the cost of tape archives might also be included.

Prevention costs, expended to minimize appraisal and failure costs, include:
- Specific training, application, or support costs for analysis and design methodologies such as Yourdon, Jackson, Finite State, etc.
- Any assurance program related training.
- Computer-aided design and manufacturing support.

Although Juran's scheme [21, pp. 5-6] for prevention costs also includes components of control (i.e., quality planning, process control, quality data acquisition and analysis, and quality reporting), these elements are factored out in this scheme into a separate "CONTROL" category. This is suggested since control issues are apparently the hardest to manage in the software industry.

Control costs are incurred to assure product quality. Normality and written charters separate this category from preventions costs.
- *Quality Assurance:* including manual maintenance and overall program support; definition of data acquisition, analysis and reporting standards; monitoring the operation and maintenance activity levels of products in use; etc.
- *Quality Control:* including program support at the Product Center level; the collection, analysis, and reporting of quality data; etc.
- *Project Management:* including the definition of standards and the maintenance of liaisons for project selection and general scheduling; System, or Product Center, coordination; setting general organizational efficiency (productivity) and effectiveness goals; Product Center project definition and initialization; etc.
- *Process Control:* including project element tracking and coordination; status and progress reporting; acquisition, analysis and reporting of productivity data; etc.
- *Configuration Management:* including definition and maintenance of standards for configuration item identification, change control, status accounting and auditing; auditing of configuration change control procedural compliance and of configuration status; etc.
- *Change Control:* including coordination of configuration changes; general support to the technical staff; the reporting on change control activity levels and configuration status for each project at the Product Center level; etc.

As stated earlier, general industry considers this category a subset of prevention costs. Should the stated function not be provided by department charter, then the cost elements should return to the "PREVENTION" category account.

A. Categorical Relationships

Juran [21, p. 12] has shown that quality costs can be minimized based on the ratios for the various cost categories (Fig.4). On closer examination of the total quality cost curve (Fig. 5), three zones [21, pp. 13-15] relative to the curve optimum (minimum) come into view. Each zone represents a software development process, or methodology mix, with various elements and attributes.

In the *Quality Improvement Zone,* failure costs are typically excessive when compared to total quality costs, with prevention costs being nominal. Management's goal should be to identify and support projects targeted at quality improvements. Emphasis shifts to control and maintenance of the current position in the *Zone of Indifference.* Failure and prevention costs are balanced and profitable improvement projects are more difficult

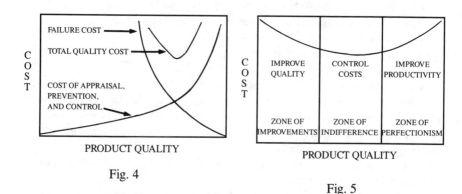

Fig. 4

Fig. 5

to identify. When appraisal costs are greater than failure costs, existence in the *Zone of Perfectionism* is established. Here emphasis should shift to improving productivity. Quality standards should be reviewed for necessity and their contribution to fitness for use. Removing or relaxing some standards may improve productivity. Review the methodology mix for possible relaxation of appraisal cost elements. Fewer reviews or less rigorous testing might be in order. Another example is the high cost of in-process documentation. If such standards are inflexible or too restrictive, excessive costs will certainly be experienced. Further perfectionism issues will be raised later.

B. *Cost Minimization by Category*

Although there is a critical relationship between cost categories, and it is insufficient to minimize each category alone, Juran has presented some heuristics [21. p. 14] relative to the individual categories:

- *Failure Costs:* are minimized when profitable failure cost reduction products cannot be identified.
- *Appraisal Costs:* are minimized when they have been brought down to a minimum on their current curve, no profitable projects can be identified to put the appraisal costs on the next technological curve, and good process standards are documented and followed.
- *Prevention Costs:* are minimized when the prevention work itself has been subjected to rigorous improvement analysis and is controlled by sound budgeting.

V. Cost Data Collection

Certain cautions are warranted to avoid ambiguity and to preserve data integrity. Perhaps the most common cause of quality cost overstatement in general industry, *Nonquality Cost Inclusion* should be avoided. Possible examples from the software industry include integration testing or reviews held to examine "off-the-shelf code" for fitness for use in a given new development. Both relate to "building" the software product, and would be done regardless of product defect levels. Moreover, *Hidden*

Quality Costs, although avoidable, are sometimes perpetuated by unrealistic standards. Over-emphasis on variances can include avoidable costs in standards merely by accepting some level of quality or cost as a common practice. For example, if we target the defect identification efficiency for in-process development software inspections at 50 percent, and use this as a basis for costing the quality element on inspections, then quality control will only signal alarm when efficiency falls below 50 percent. As the technology improves and personnel become more experienced, the targeted efficiency should be revised so as not to "hide" defect identification costs.

When developing a cost accounting scheme, care should taken to avoid "perfectionism" in the scheme itself. Attempts at identifying and controlling all quality cost elements can make the scheme unmanageable and encourage disputes. Remember that data reporting consistency and major cost element control are the main issues. Data collection should be as unobtrusive to the technical staff as possible; if it's difficult, it will not get reported. Data summaries should be made available by:
- Product, process step, defect type, etc.
- By project.
- By organizational responsibility.
- By quality cost category.
- By calendar period.

VI. Cost Control

"Zero Defects" is a wonderful concept, but since unattainable, a poor goal. Although an optimum level for quality costs exists, that level is not zero. Care should be taken not to set any *False Limits* on quality costs.

Overdesign, or *Perfectionism in Quality of Design,* is an avoidable cost. In software development, overdesign pertains to an excessive investment in the many "Abilities" [17]-[19] that are known as quality factors. Examples of overdesign include:
- Designing a product with emphasis on maintainability when technical obsolescence is pending.
- Aesthetic improvements that remain unobserved. This includes functions above and beyond approved specifications ("creeping functionality"), or any costly finishes on internal surfaces.
- Unnecessarily detailed specifications (tolerance or attributes that exceed needs for fitness of use).
- Excessive attention paid to development of software that will never be used (unlikely paths in fault recovery software).

Before designs are published, elements of overdesign should be removed. This might be done by including overdesign issues on checklists for the managerial reviews and peer group inspections called for by the product assurance program. Not just major products, but also scaffolding falls prey to overdesign. Many software tools are developed with capabilities that are infrequently used, or that far exceed needs.

A second type of perfectionism, *Perfectionism in Quality of Conformance*, also can be controlled. Examples of over-conformance include:
- Process standards that cause the product to possess quality attributes not called for by its specifications (or in excess of stated levels).
- Setting standards for stylistic issues unobservable by the users and irrelevant to later product support.
- Emphasizing conformance to process and performance standards despite the established, long-term success of nonconforming efforts.

Though the identification and reduction of quality costs is the immediate goal, care must be taken not to increase total costs. To improve overall company economics, data must not be reported out of context with known quality and productivity relationships.

VII. Quality Cost Reporting

When reporting quality and productivity data, a *basis for comparison* must be provided. Emphasis in the software industry should shift from "getting the numbers" to reporting and interpreting the relationships. Without a reasonable set of indicators, and management personnel who know how to use them for decision support, companies will remain data-rich and analysis-poor.

One relationship, however, that is *not* suggested is "Per dollar of sales." Although it can be used when close tracking of sales trends is desired, the cost dollar does not include marketing expenses, general overhead, and profit like the sales dollar. Though attractive, this scheme should wait until the program is mature enough (i.e., reliable and metric-rich) to provide further interpretations.

To start a program, some short-range level of activity, such as "per hour of direct labor" or measures that relate to product size, such as "per thousand normalized lines of noncommentary source code" is recommended. If longer range measures are needed that are indifferent to inflation, then "per dollar of direct labor" or its extension, "per dollar of standard manufacturing costs," might be used. The latter measure includes overheads and is more stable because it is insensitive to the degree of automation.

Published quality cost reports take three major forms (i.e., tabular, graphic, and narrative) and are provided by calendar element (e.g., yearly, monthly, etc.), as driven by development phase completion, or as required by special managerial request. Different levels of management require different information, and thus, different reports. Management should receive reports at a frequency related to the inverse of their specific management level. The responsibility for reporting is best divided between the quality function and accounting, with QC providing graphic reports, and accounting concentrating on the tabular.

VIII. Making Program Adjustments

Periodic quality program evaluations (i.e., audits, reviews, surveys, etc.) will provide the local Quality Assurance department with information needed to evaluate the program's effectiveness and to identify areas to be added to, or released from, scrutiny. These evaluations can be done by the department itself or by a corporate quality assurance entity.

Positive program adjustments are of two types: incremental and magnitude order. Incremental improvements take place when the figure for a cost element is moved closer to its curve minimum by applying available technologies. Cost element graphs, however, really represent a family of curves. Thus, a magnitude order improvement moves a cost figure to a lower cost element curve but, close enough to the new minimum for control. This is accomplished through the application of new and emerging technologies.

Expanding on the concept of magnitude order improvement, consider a family of total cost curves. Identifying an initial curve, we know that any "fine-tuning" of a given quality program will bring us closer to the specific curve minimum. Should we make substantial improvements in controlling failure costs, a transition to a different curve would be made possible. Similarly, significant improvement in controlling only appraisal and prevention costs would also result in a curve change. The most significant improvement would be reached by improving both "failure" and "appraisal and prevention" cost controls, either jointly or in any sequence.

Though exhorbitant claims have been made, Quality/Productivity Circles are no substitute for competent, well-informed software development process management. Disappointment will result if circles are relied on to provide magnitude order changes, or even to provide fine tuning for an ill-defined methodology mix. However, circles have exhibited great success in providing incremental improvements along a cost curve and can assist in maintaining operation in the Acceptable Software Productivity Range.

IX. Summary

On examination of quality's effect on corporate income and cost, a three-phase model (originally proposed by Juran et al.), was augmented for software quality programs. Since operating ranges for acceptable software quality and software development productivity were suggested, the model accounted for both quality and productivity issues. Moreover, it was proposed that through model application and the analysis of quality cost categories, the operating ranges are attainable.

General application of the model, with serious attention paid its cost definition steps, would significantly advance the cost effectiveness of quality programs throughout the software industry. Program results have been proven for sibling high-tech industries.

References

[1] J. M. Juran, "Quality improvement," in *Quality Control Handbook*, 3rd ed., J. M. Juran, F. M Gryna, R. S. Bingham, Eds. New York: McGraw-Hill, 1974, Section 16.

[2] A. V. Feigenbaum, "Total quality control," *Harvard Bus. Rev.*, vol. 34, no. 6, pp. 93-101, June 1956.

[3] A. V. Feigenbaum, *Total Quality Control*, 3rd ed. New York: McGraw-Hill, 1983.

[4] J. J. Greene, *et al.*, "Developing a software quality assurance program based on the IEEE Standard 730-1981," in *Proc. IEEE Computer Software Applications Conf.*, Nov. 1982, pp. 257-262.

[5] G. G. Gustafson and R. J. Kerr, "Some practical experiences with a software quality assurance program," *Commun. ACM*, vol. 25, no. 1, pp. 4-12, Jan. 1982.

[6] IEEE Standard 730-1981: A Standard for Software Quality Assurance Plans.

[7] Forward to the IEEE Standard 730-1981: A Standard for Software Quality Assurance Plans.

[8] C. P. Hollocker, "The standardizations of reviews and audits," in A Handbook for Software Quality Assurance, Schulmeyer and McManus, Eds. New York: Van Nostrand Reinhold, (to be published).

[9] G. D. Bergland, "A guided tour of program design methodologies," *Computer*, vol. 14, no. 10, pp. 13-37, Oct. 1981.

[10] B. W. Boehm, "Verifying and validating software requirements and design specifications," *Software*, vol. 1, no. 1, pp. 75-88, Jan. 1984.

[11] D. J. Reifer and S. Trattner, "A glossary of software tools and techniques." *Computer*, pp. 52-60, July 1977.

[12] R. H. Thayer, A. Pyser, and R. C. Wood, "The challenge of software engineering project management," *Computer*, pp. 51-59, Aug. 1980.

[13] R. H. Thayer, A. Pyser, and R. C. Wood, "Major issues in software engineering project management," *IEEE Trans. Software Eng.*, vol. 7, no. 4, pp. 333-342, July 1981.

[14] E. H. Bersoff, "Elements of software configuration management," *IEEE Trans. Software Eng.*, vol. 10, no. 1, pp. 79-87, Jan. 1984.

[15] F. J. Buckley and R. Posten, "Software quality assurance," *IEEE Trans. Software Eng.*, vol. 10, no. 1, pp. 36-41, Jan. 1984.

[16] R. M. Poston, "IEEE 730: A guide to writing successful SQA plans," *Software*, vol. 2, no. 2, pp. 86-88, Feb. 1985.

[17] J. P. Cavano and J. A. McCall, "A framework for the measurement of software quality," in *Proc. 1975 SQA Conf.*

[18] J. A. McCall, "An assessment of current software metric research," *EASCON '80 Rec.* IEEE Electron. Aerospace Syst. Conventions, 1980, pp. 323-333.

[19] J. A. McCall *et al.* "The automated measurement of software quality," in *Proc. Computer Software Applications Conf.*, Nov. 1981.

[20] F. W. Neitzke, *A Software Law Primer.* New York: Van Nostrand Reinhold Co., 1984, ch. 9.

[21] D. M. Lundvall and J. M. Juran, "Quality costs," in *Quality Control Handbook*, 3rd ed., J. M. Juran, F. M. Gryna, R. S. Bingham, Eds. New York: McGraw-Hill, 1974, Section 5.

NORTHEAST QUALITY COSTS CONFERENCE

GUIDE FOR MANAGING VENDOR QUALITY COSTS

William O. Winchell
Quality Consultant
General Motors Corporation
Warren, Michigan

Introduction

The subject of our talk today is managing vendor quality costs. This is an important subject to all of us because in most cases we not only buy from vendors but we are a vendor to someone else. Key to first understanding and then managing vendor quality costs is recognizing that much of these costs are hidden. By hidden, we mean that these costs are not really easily identified or perhaps not even seen as vendor related quality costs.

For the purpose of this talk, we will look at these hidden costs from just one perspective — that of the buyer. From the buyer's viewpoint, hidden vendor quality costs occur in three basic ways.

Vendor Quality Cost

The first type of cost is quality cost inside the vendor's facility. For the most part it is much the same as the buyer's quality cost. However, it is included as part of the piece price the buyer pays for the part. The vendor, perhaps rightly so, protects the identification of this investment for competitive reasons. The buyer is furnished the piece price alone, and if this price is competitive, he's usually satisfied with this knowledge.

The second type of hidden vendor quality cost is that which buyer spends, but does not usually segregate, to insure the quality of a product at the vendor's facility. This may be in the form of sending quality engineers to the vendor's plant to help him solve a crisis.

The third type of hidden vendor quality costs is that which the buyer spends in-house, again not usually segregated, specifically for vendors. An example of this may be process engineering required to repair or correct a purchased part not quite up-to-specification. Many of us call this fire-fighting to keep the line going.

Of course, there are also visible costs that the buyer can, and sometimes does, segregate to specific vendors. An example of this type of cost is receiving inspection effort.

For a typical manufacturing firm, these vendor related costs are a huge chunk of money — perhaps in the range of 10-20% of the selling price to your customer. Through proper control of these hidden and visible costs that are vendor related, we should achieve several results:
- The first is to optimize our profits
- The second is to insure a product that is "fit-for-use" by our customer. "Fitness-for-use" of our products will insure our success in the marketplace. If we don't meet our customer's requirements, our competitors will, and we will not survive in the long run.

The basic purpose of my talk is to suggest to you methods by which you can control both the hidden and visible quality costs incurred in your vendor relationships.

Very simply, two ways will be emphasized. The first way is to talk your vendors into adopting a quality cost program. However, discretion must be used since all vendors cannot support a quality cost program. They may be too small or special circumstances may prevent successfully accomplishing it.

The second way is to use your quality cost program to identify high magnitude vendor problems for resolution. Before we discuss these methods in detail, let us first review the basic principles of quality costs and how they apply to our vendors.

The American Society for Quality Control publishes a document titled "Quality Costs — What and How," authored by the Quality Costs Committee. Included in this publication are the essential elements that make up a quality cost system, recommended definitions and methods of installing and gaining acceptance.

In this publication the ASQC recognizes four basic categories of cost. They are identified as:
- Prevention — which primarily are the costs resulting from planning to prevent defects in products.
- Appraisal — costs relative to the conformance to quality standards.
- Failure — costs are the results of poor planning and include the failure of products to meet the established or required quality standards.
    - failure costs are either *internal,* which are mainly manufacturing losses, or *external,* which are due to defective products shipped to the customer.

Each category of prevention, appraisal, internal and external failure contains many elements. Let's look at a few examples of elements in each category.

Prevention includes:
- Quality control administration
- Quality systems planning and measurement
- Vendor quality surveys
- Quality data analysis and feedback

Appraisal includes:
- All inspection and test costs
- Laboratory acceptance testing costs
- Vendor quality audits and surveillance

Internal Failure includes:
- Rework
- Scrap
- Retest and reinspection

External Failure includes:
- Returned material processing and handling
- Warranty replacement

The first part of the quality cost program, *that of measuring*, is the most difficult. Most businesses do not enjoy the sophisticated accounting system that permits interrogating for the cost of quality elements.

A system for collection of this information has to be designed to fit each specific company.

Once we have collected the information and categorized it, we can now *analyze*. The most obvious is to examine those quality elements for which the greatest costs have occurred. It is here that the most effective and dramatic cost savings can be realized. But we must also note the quality elements having little or no costs, since insufficient effort may be occurring. For example, little action in vendor evaluations may result in high receiving inspection costs.

Action, of course, is the result of analysis. All the good cost reporting in the world is wasted unless someone takes corrective action. The logical next step is to take action in areas needing improvements against a time scale to assure effective results.

Finally, we *control* the improvements in a continuing effort to do even better over the long term.

Now that we have reviewed quality costs in general, let us now discuss how they apply in the buyer-vendor relationship. Previously we discussed how the vendor related costs can be classified as either hidden or visible.

Now I would like to discuss each of the categories in detail. Hidden quality costs, as you may recall, are in three parts:
- The first are those which are incurred by the vendor at his plant.
- The second are incurred by the buyer in fixing problems at the vendor plant.
- The third are those costs — which are not usually allocated to vendors — incurred by the buyer as a result of potential or actual vendor problems.

Those quality costs incurred by the vendor at his plant are, for the most part, unknown to the buyer and, therefore, hidden. However, even though the magnitude is hidden, the type of cost is not. They are the same type quality costs as the buyer incurs. For example, the vendor certainly has prevention effort. If he manufactures the product, he has expenses related to the quality engineering of the product. Even if the vendor is a small shop, this task must be done by someone and may very well be handled by the Production Supervisor if the plant lacks a Quality Engineering Staff. Certainly effort is expended in the appraisal area even in the smallest vendors. Someone must inspect the product prior to shipping to the buyer. In a one man shop this is done by the man who made the part. Unfortunately, we all have costs whether we are a large shop or a small one. When the vendor makes a mistake in manufacturing they must either rework the part or scrap it causing an internal failure expense. If the vendor sends it to the buyer, it may be rejected with the resulting cost being considered as external failure.

The second type of hidden quality cost — that which is incurred by the buyer in fixing problems at the vendor's facility — is usually not specifically allocated to vendors. Except for an awareness of our troublesome vendors, there is usually no tabulation of the cost of the effort expended. Therefore, the actual expense is hidden. One example of this cost is the buyer sending a Quality Engineer to a vendor to resolve a crisis. Another example is the auditing a buyer may do at the vendor's plant to insure that he has the proper quality controls to build a product to the buyer's specifications.

The last type of hidden quality cost is that at the buyer's plant incurred on behalf of a vendor. Like the second type of hidden cost, we are certainly aware of it, but we probably do not keep track of it. The magnitude of this expense is not known for specific vendors. Again, the actual expense is hidden. This type of cost may include, in the prevention area, the following:

- Preparing the specifications for packaging the product that is shipped by the vendor to the buyer.
- Specification and design of gauges that must be used in the buyer's receiving inspection and perhaps as well by the vendor prior to shipping.
- Designation of appropriate specifications that the vendor must follow in the manufacture of the product.

For the hidden appraisal costs, we may incur expense for:

- Certain inspection operations and quality control effort in the buyer's production line related specifically to a vendor product.
- Review of test and inspection data on vendor parts to determine acceptability for processing in the buyer's plant.
- Calibration and maintenance of equipment necessary in the control of the quality of vendor parts.

In the internal failure area we may do troubleshooting or fire-fighting in order to use a vendor part that is not quite up to specification.

Finally, regarding external failure, there may be field engineering required to analyze and correct a problem caused by a vendor.

It must be remembered that this discussion of the types of vendor related hidden costs is by no means exhaustive. There are many more. Some of the types of costs not pointed out may be very significant depending on your individual situation.

There are, as discussed initially in this talk, vendor related quality costs that are apparent and perhaps much easier to identify in magnitude and assign to various vendors by the buyer.

These are called visible costs. They are primarily in the appraisal and failure areas.

Included in the visible quality cost category are the following:

- Receiving inspection
- Laboratory acceptance testing
- Scrap and rework that is vendor responsibility
- Warranty replacement on parts supplied by vendor

The visible costs, if tracked, are perhaps most significant because they can be good indicators of problem area.

Now that we have looked at the various types of hidden and visible quality costs, let us discuss the possible methods through which we may optimize these expenditures.

The first step for the buyer in optimizing his vendor related quality costs is to determine what costs are important to him. It is suggested that comparing the relative magnitude of quality costs by category could be a good start. To do this, your own quality cost program could be invaluable. But, if you don't have a quality cost program, a special study could be initiated to determine this information. The ASQC publication "Quality Costs — What and How" would be a valuable guide in accomplishing this.

This slide illustrates a hypothetical situation in which vendor scrap and rework are the biggest problem areas for the buyer. If the buyer makes the assumption that through improvements in vendor caused scrap and rework warranty will be lowered, then for this company vendor responsible scrap and rework are the important items in the vendor relationship.

The next step is determining which vendors are causing the problems. Very likely, you will find that 20% of your vendors are causing 80% of your problems in scrap and rework. Now the buyer can narrow his effort to the vital few of the vendors and take appropriate action.

What is appropriate action? First and foremost, I would suggest that you promote that the "vital few" vendors start a quality cost program using the ASQC publication "Quality Costs — What and How," if appropriate. Discretion must be exercised before insisting on this. Some companies may be too small to support a quality cost program or to accomplish special studies. Special circumstances may exist in other companies that would prohibit this action. However, if a vendor finds that launching such a program or study is feasible, the costs most visible to the buying company will most likely be reduced by doing so. If these are reduced, the hidden costs expended by both the buying company and the "vital few" vendors should also be lowered. The result will be that the vendor's product and the buying company's product are more "fit for use." This should increase profits for both which is very desirable. Also, improved profitability for the vendor may eventually result in lower prices for the buyer in a competitive market.

What other action can be taken? Keeping track of vendor related costs — both hidden and visible — may be a gigantic undertaking for the buying company. However, some way must be found to insure that progress is being made. It is recommended that the buying company track the important visible costs for each of the "vital few" vendors. It is submitted to you that, in most cases, you will find that if progress in reducing the visible costs is demonstrated, hidden costs are also decreasing.

What other action can be taken if we know the magnitude of the visible costs? It is possible that this information can be incorporated into a buyer's vendor rating system. The Guide for Managing Vendor Quality Cost describes a system utilizing the cost of the purchased part and the identifiable quality cost. Although not theoretically perfect, this system is simple and has achieved significant improvements in quality. It is well worth your consideration.

The ASQC Procurement Quality Control Handbook can also supply valuable help to you in accomplishing this. A vendor relationship depends upon more than the traditional measures of meeting deliveries and receiving inspection rejections. It also must recognize the "fitness for use" of the vendor's products for which visible quality cost can be a valuable indicator.

Visible quality costs can also be used in other ways. Some companies debit vendors for the scrap and rework occurring in the buyer's plant to put the responsiblility for failures where it hurts most — in the pocketbook. However, in the long run this may be counterproductive in that some vendors may ask for a price increase to cover this situation. On the other hand, many buyers find it very effective to reduce the amount of business given to the offender and reward the good performer with a greater share of the "order pie."

Perhaps a far more positive method is to use the visible costs to identify vendor quality improvements that are needed, the buying company would initiate projects jointly with vendors to resolve the problems that are the source of high quality costs. The problem may be solved through an action by the buying company — perhaps the specifications are not correct or the vendor really doesn't know of the application of his product in the total package.

On the other hand, it may be that the vendor's manufacturing process needs upgrading through perhaps better tooling and the vendor must take some action. Through joint projects, using visible quality costs as facts, we can solve those problems and get better products. If quality costs are collected in a regular fashion, the results will document lower cost to both the vendor and buyer.

Let's now summarize what we have discussed. Basically, use quality costs in your vendor relationships. Through this tool find out what costs and vendors are most important to focus upon. Suggest that these vendors, if appropriate, adopt a quality cost program in order to obtain improvements in "fitness for use" of their products. Use your visible vendor quality costs as a basis for starting joint quality/reliability improvement projects with your suppliers.

Most important of all is that any quality costs program is not complete without an effective correction action program. The mere act of collecting quality costs alone will do nothing but add cost for you or your company. Only through pinpointing the important problems and solving these can we make progress.

QUALITY PROGRESS

ACCOUNTING FOR THE REAL COST OF QUALITY

(January Issue)

Dean-Michael Lenane
Senior Reliability Engineer
United Technologies Automotive Products Division
Dearborn, Michigan

Despite the fact that U.S. managers seem to recognize the need for better quality systems and are implementing quality improvement programs at a greater rate than ever, their companies' corresponding manufacturing and accounting structures are not being changed to reflect new quality requirements. Managers still are rated largely on how much — not how efficiently — they produce. Quality improvement programs generally are handed to quality departments as additional responsibilities, and process control systems assume a secondary role.

Because these systems have not been altered, the manufacturing culture stays the same. Current quality improvement and process control programs thereby are cast as what Juran calls, "the bee flying into the window."

Quality personnel are faced with new demands and constraints, yet must work within a manufacturing environment virtually identical to that of the '50s and '60s, and quality departments end up treating problems symptomatically through programs, rather than by reaching the systemic origin of problems. Stated another way, quality often is used tactically rather than strategically. Without an overall strategy, tactics are of limited value.

Current systems must be modified to direct the strategic emphasis away from product control toward process control; away from "how much" toward "how efficiently;" away from a unitized, programmatic approach toward the broad, strategic implications that effective quality systems represent.

Failure to integrate an overall defect prevention philosophy throughout an organization results in a corporate clash of philosophies. Manufacturing, accounting, and engineering managers must understand what quality departments are trying to accomplish. The cooperation or resistance of these groups will determine the success or failure of any quality effort.

It is clear that the success of the Japanese in penetrating our markets, coupled with the admonitions to top management by figures such as Deming and Juran, has resulted in a superficial, albeit genuine, consideration of quality.

Management's lack of understanding, too, is a product of the system. Immediate financial considerations, as well as problems concerning markets, diversification, securities, and business legalities constantly occupy the greater portion of management's time. Management has difficulty finding time to consider the deeper ramifications and arcane subtleties of total quality control.

These problems require quality departments to *sell* quality philosophy to semiautonomous and frequently hostile groups. Quality departments, therefore, must demonstrate the effectiveness and the value of quality systems in explicitly *quantifiable* terms.

If any lesson has been learned by embracing statistical methods, it is that decisions must be based on measurable units, not opinions or conjectures. "Units," to the quality professional, mean almost anything quantifiable, but to management, units mean only one thing: dollars.

It therefore is imperative to convert costs associated with the quality system into dollars to document statistics' successes. This is initially a simple task. Classic quality cost systems have been written about extensively. The four categories — prevention, appraisal, internal failure, and external failure costs — have been defined thoroughly and the elements within these categories enumerated. However, these classical cost of quality categories show only a portion of quality's total cost.

A typical problem arises when a statistical process control system is applied to the classic system of quality cost documentation. Prevention costs can jump significantly, with no readily apparent decrease in appraisal or failure costs. The result can be that management sees an expense, not a cost benefit from the SPC system. This can help perpetuate the myth that achieving high quality is costly.

The classic COQ system does not reveal the cost of unquality or document the hidden production capacity of a line or plant; it does not show the magnitude of uncontrolled indirect labor and materials cost. When standards and scrap allowances are applied, they create the illusion of efficiencies. Numbers generated may appear acceptable on paper, even while waste and rework are being generated at high levels. This situation results in an increased piece price to the customer that, in turn, makes the industry noncompetitive with its rivals.

Part of the solution to the problems of management misunderstanding and accounting system errors can be achieved through implementation of a system to document the actual cost of operation-by-operation quality (or unquality). Such a document can quantify a production line's hidden capacity and quickly allows a Pareto analysis of nonconformities occurring on individuals to determine what factors contribute to the cost of quality on an individual assembly. This system, combined with closed-loop statistical process control, complements the total quality program and helps to justify the increased prevention costs associated with total quality control implementation.

Understanding of this system begins with a simple definition of quality costs. Quality costs may be defined as "the difference between current actual operating costs and operating costs if there were no product and system failures and no staff mistakes." Standards and allowances defeat this definition, make the acceptance of product nonconformities and inefficiency inevitable, and lull management into a false sense that everything is operating properly and at maximum profitability.

Often, when standards are applied in the measurement of quality costs, cost figures can appear minimal and their active pursuit will become a secondary goal.

By describing the operation-by-operation costs of production lines, allowing for no nonconformities, the COQ figures become significantly greater and the system more accurately reflects the reality of the situation. Management interest is stimulated and quality becomes more important as a long-term corporate strategic goal.

The measurement of production rework, operation-by-operation nonconformity levels, and production inspection must be simple and practical to be manageable and stimulate interest. Because typical quality improvement programs are front-end loaded (i.e., prevention) and the magnitude of internal quality costs (i.e., appraisal and internal failures) is less apparent in traditional systems, it is difficult to convince management of the need for change. A good method for accomplishing this is to define quality costs to management, then offer the current disparity between typical corporate cost of quality in the U.S. (10% to 20% of the sales dollar) and the cost of quality figures offered by the Japanese automotive industry (2.5% to 4.0% of the sales dollar). It should be emphasized to management that this program can reveal hidden production capacities and help allocate resources in the most efficient manner. Once management is convinced of the need for such a program, a starting point must be established.

To begin a quality cost analysis program, the quality manager and the controller should study the standard categories of the classic quality cost system and the elements that fall under each of the four categories. Once these categories and elements are identified, accounting data that are currently available can be utilized to develop a base cost-of-quality figure. This number is readily obtainable and is somewhat useful in apprising management of the significance of part of the operational cost of quality.

However, the costs associated with nonconforming production, rework, work-in-process inventories, excessive material handling, and production in-process inspection are generally overlooked or obscured. These costs, which can increase the quality costs associated with particular product lines or plants dramatically, must be revealed and highlighted.

A system that addresses these costs should begin with the selection of a target line — one that has a record of extensive rework or a high percentage of in-process nonconformities. Once a target is established, three steps must be taken to provide a foundation for an easily manageable, accurate, and controllable operation-by-operation cost system.

The first step is to audit the line itself to the assembly breakdown or fully burdened bill of materials. The object is to determine if any off-standard operations exist and to evaluate the accuracy of the assembly breakdown itself. On-line rework areas may or may not be included in a particular cost center, and any existing rework areas must be compared to the assembly breakdown to see if they are accounted for. Any off-standard operation then must be time studied to determine handling and repair costs of nonconforming products. Standard operations also must be studied to determine the labor variance incurred in handling a nonconforming product and any appraisal inspection being performed on good products.

One common practice that reduces the apparent magnitude of rework costs involves putting the rework figures into a separate account or cost center. This must be stopped. Every line's rework costs must be applied to its own cost center or ledger account.

The second step involves implementation of a method to tabulate production and nonconformities on an operation-by-operation basis. The amount of product dropping out of the line then can be determined. This information can help identify problem operations and develop a true first-run-capability percentage. The use of an efficiency ticket that tabulates the amount of product passing into and out of an operation helps accomplish this second step.

The third step is analysis and classification of nonconformities at each operation to determine major categories of nonconformity.

Once these three steps are taken, a figure approximating the average cost of a nonconformity at a particular operation is generated.

Using a quality cost worksheet, (See Table 1.) a major nonconformity category is evaluated by adding the labor variance generated in handling a nonconforming unit to the material cost of lost parts or subassemblies and the direct labor lost up to that point in the assembly line. A Pareto analysis of the major nonconformity categories then is made, and the volume and costs of the major nonconformity categories are calculated using a weighted average to provide an average cost-per-nonconformity by operation. After this is completed, daily production and rejection figures provide a fast, accurate estimate of a production line or cost center's daily internal quality costs on an operation-by-operation basis. This reveals losses in production capacity due to nonconformity production.

Once this reporting system is established, it must be utilized in an efficient and effective manner to apprise management of the total magnitude of in-process unquality. This will make the short-term increase of prevention costs more palatable and demonstrate both the lost production capacity of a line and the profit improvement potential (when compared against a 2.5% to 4.0% of sales dollar minimum quality cost).

In developing an appropriate management report, the average cost of a nonconforming unit at a particular operation is coupled with the yield percentages documented through the use of the efficiency tickets. (See Table 2.) By utilizing a data-base management software system, an effective tool for use by management can be generated quickly.

The management report contains three sections:

(1) A fixed input section containing relatively stable information on the operations, parts used, and standard minutes per piece. (See Table 3.)

(2) A weekly input section listing the number of pieces run weekly, number of rejects, and the actual time used to build a product at a particular operation. (See Table 4.)

(3) The report output section, which manipulates the data and presents it in several different ways to express the impact of costs related to unquality. (See Table 5.)

The fixed-input section is simple and straightforward. It contains appropriate report headings and seven information categories: cost center, operation number, final assembly, and operation name, standard minutes per piece allowed by industrial engineering to build a part, and average cost per nonconformity. (Please note that this assumes that, unless a permanent corrective action is implemented, the nonconformity cost on an operation remains more or less static. If, however, the Pareto analysis of common failure modes at an operation reveals an assignable cause that is then eliminated, the average cost of a nonconformity must be restudied to determine a new average cost of a nonconforming unit.)

The seventh input category is appraisal cost — documented costs incurred in any retest or operator inspection of assembled components.

The second report section, the weekly input, consists of information garnered through the use of efficiency tickets. The efficiency ticket is a daily operation-by-operation recap

Table 1. Quality Cost Worksheet

Plant _____
Cost center/operation ___65/60___

Cost center ___65___ Operation no. ___60___ Date ___Jan. 12, 1984___
Product name ___Ignition switch___ APD P/N ___Operation 50 assembly___
Operation name ___Terminal block assembly___ 200-508 "B" term
Standard minimum product cost ___$.0542___ 200-322 "S" term

Calculation

Defects: defective incoming material

Cost
(1) Cost to handle a nonconforming part
 "B" term — .1050 min $.0148
 "S" term — .1050 min $.0148
 OP 50 assembly — .1600 min $.0296

(2) Material cost
 "B" term — $.0099
 "S" term — $.0527
 OP 50 assembly — $.1082

(3) Direct labor
 OP 50 assembly — $.0124

Cost to fix _____ "B" term. $.0247
 "S" term $.0675
 Cost to total scrap ___ Operation 50 assembly $.1502

Concurred by: _____ Plant manager _____ Division QC

ASQC Quality Costs Committee

Table 2. Daily Efficiency Ticket

Name　　　　　　　　　　　Clock Number　Shift　Cost Center　Date

Operation Number	Station Number	Remarks	Total	Total Pieces	−	Total Failures	=	Net Production

If operation number is unknown contact your supervisor for number and standard.
All down time over five minutes each time must be signed by your supervisor.
Use reverse side of ticket if more space is needed.

Table 3. Fixed Input

Cost Center/ Operation	Part Name(s)	Operation Name	Standard Minimum Per Piece	Throw Away Cost	Appraisal Cost Per Piece
65- 10	Ignition Sw.	Carrier Assembly	0.0631	0.0634	0.0045
65- 20	Ignition Sw.	Carrier Assembly	0.0631	0.0133	0.0045
65- 30	Ignition Sw.	Carrier Assembly	0.0631	0.0697	0.0045
65- 40	Ignition Sw.	Carrier Assembly	0.0631	0.0985	0.0045
65- 40 +	Ignition Sw.	Carrier Test	0.0582	0.0365	0
65- 50	Ignition Sw.	Term. Block Assy	0.0542	0.0471	0.0012
65- 60	Ignition Sw.	Term. Block Assy	0.0542	0.061	0
65- 70	Ignition Sw.	Term. Block Assy	0.0542	0.0376	0
65- 80	Ignition Sw.	Term. Block Assy	0.0542	0.0309	0
65- 90	Ignition Sw.	Term. Block Assy	0.0542	0.0329	0
65-100	Ignition Sw.	Stake Terminals	0.1084	0.3796	0
65-110	Ignition Sw.	Housing Assembly	0.0626	0.1587	0
65-120	Ignition Sw.	Housing Assembly	0.0602	0.1391	0.0024
65-130	Ignition Sw.	Final Assembly	0.0602	0.1797	0
65-140	Ignition Sw.	Final Test	0.1247	0.5896	XXXXX

Table 4. Weekly Input

Cost Center/ Operation	Total Parts Run	Actual Minutes Used	Aways
65- 10	44234	4470	312
65- 20	54234	4470	210
65- 30	51127	4350	166
65- 40	48968	4020	129
65- 40 +	82200	4100	245
65- 50	50147	2880	159
65- 60	50147	2880	115
65- 70	50148	2880	182
65- 80	50037	2850	196
65- 90	50148	2880	120
65-100	50136	3000	200
65-110	54697	3255	131
65-120	51992	3225	491
65-130	51592	3240	409
65-140	57523	6690	1555
Totals		51170	4491

of number of parts run, number of rejects, and number of hours spent by a particular operator of a particular operation. These daily sheets are compiled to provide weekly numbers used in the input section.

The third and final section, the output section, is the most complicated and powerful section of the report. Its power is derived from the fact it quantifies problem operations, shows actual first-run capability, and demonstrates the magnitude of in-plant quality costs. This report quickly gives management an accurate and effective scorecard as to operation and line performance and problem areas. When this report is combined with a classic quality cost system, the true magnitude of total quality costs becomes apparent. Important targets can be seen by management, and the promise of reduced total quality costs through increased prevention costs becomes logical, explainable, and, most important, quantifiable.

The final output section begins with a column entitled "cost for nonconformities." This number is the product of the total number of rejects multiplied by the average nonconformity cost identified in the report's fixed input section. This number shows, by operation, the costs incurred over one week's time in the production of nonconforming units. A summary of the line's costs appears at the bottom of the column.

The next column is appraisal cost. Appraisal costs on an operation-by-operation basis consist of any inspection or test performed on a subassembly or final product. It may be on or off standard, but each operation must be studied to determine if any appraisal is being done, and, if so, what time is spent to inspect. Any testing is, of course, entirely an appraisal cost, and any testing station on the line is relegated entirely to appraisal cost. The appraisal column is derived from operation-by-operation time studies to determine the appraisal cost per piece multiplied by the weekly total gross production figure to yield the dollar figure lost to operational appraisal.

Table 5. Output

Total Number Rejected	Cost For Nonconformities	Appraisal Cost	Cost Per Piece
312	$ 19.78	$199.05	$0.0004
210	$ 2.79	$244.05	$0.0001
166	$ 11.57	$230.07	$0.0002
129	$ 12.71	$220.36	$0.0003
245	$ 8.94	$ 0.00	$0.0001
159	$ 7.49	$ 60.18	$0.0001
115	$ 7.02	$ 0.00	$0.0001
182	$ 6.84	$ 0.00	$0.0001
196	$ 6.06	$ 0.00	$0.0001
120	$ 3.95	$ 0.00	$0.0001
200	$ 75.92	$ 0.00	$0.0015
131	$ 20.79	$ 0.00	$0.0004
491	$ 68.30	$124.78	$0.0013
409	$ 73.50	$ 0.00	$0.0014
1555	$ 916.83		$0.0159
4491	$1,230.00	$858.13	$0.0220

Quality Cost Percentage	Annual Cost Projection	Cost Center Operation Percentage Efficiency	Cost Center Operation 1st Run Percentage Capability	Cost Center Operation Throughput Percent
0.051	$ 890	62.4	99.3	99.3
0.006	$ 126	76.6	99.6	98.9
0.026	$ 521	74.2	99.7	98.6
0.030	$ 572	76.9	99.7	98.3
0.013	$ 402	116.7	99.7	98.3
0.017	$ 337	94.4	99.7	98.0
0.016	$ 316	94.4	99.8	97.8
0.016	$ 308	94.4	99.6	97.4
0.014	$ 273	95.2	99.6	97.0
0.009	$ 178	94.4	99.8	96.8
0.174	$ 3,416	181.2	99.6	96.4
0.044	$ 936	105.2	99.8	96.2
0.151	$ 3,073	97.1	99.1	95.3
0.164	$ 3,307	95.9	99.2	94.5
1.832	$ 41,257	107.2	97.3	92.0
2.533%	$55,340	96.2%		

The next column, cost per piece, is relatively simple. It represents the cost for nonconformities divided by the total gross production figure. This number gives us an operation-by-operation piece cost for the production of a nonconforming product.

Quality cost percentage is a figure representing the quality cost per piece divided by the manufacturing cost. In other words, this percentage equals the amount spent per piece building nonconforming units divided by the amount of money it costs to build the part. The resulting percentage shows the proportion of quality cost per manufacturing dollar. Note that the manufacturing cost could easily be replaced with the selling price of a unit achieving equally satisfactory results.

Annual cost projection is another number. It consists of the weekly operational nonconformity cost multiplied by the number of weeks production will be run over a year's time. In this case, we assume 45 weeks. The number helps to spot expensive operations quickly.

The next column is cost operation efficiency. This number is extremely useful in judging the accuracy of the standard. It is arrived at by dividing the standard minutes earned (gross production multiplied by minutes per piece) by the actual minutes used at the operation (information acquired from the efficiency tickets). This number helps to spot individual operations that have problems or unrealistic standards. This number also is useful in finding underestimated standards.

The next column, cost center first-run capability, is again derived from the information that appears on the efficiency tickets. The operation's net production is divided by gross production to determine a first-run capability. This operation-by-operation first-run capability is especially interesting because the final test first-run capability is generally reported as the entire line's first-run capability (in this case 97.3%). The fact is, final-test first-run capability is only applicable to the final-test operations. The true first-run capability is illustrated by the next column, entitled cost center operation throughput percent. This column charts, by operation, the rate at which parts drop out of the line. For example, if the first operation begins with 1,000 subassemblies and loses 80 pieces in the course of final assembly production, the net throughput percentage is 92%. The reported first-run capability is 97.3% because the final-test number is used. However, true first-run capability is only 92%. Use of the throughput percentage number will identify operations that lose large amounts of subcomponents and give management an accurate representation of true first-run capability on a given operation and line.

A final summary is provided at the bottom of the report. (See Table 6.) The quality cost percentage is expressed as a dollar figure. The appraisal cost is added to the final test costs (pure appraisal) to express total line magnitude of appraisal costs. The actual hours used are expressed, and the failure cost per actual hours is calculated by dividing the total cost for nonconformities by the actual hours used. The appraisal cost per actual hour is calculated and expressed by dividing the period appraisal cost by actual production hours. Finally, internal failure and appraisal costs are added together to reveal the total failure and appraisal cost on line per actual hour.

To summarize, when previously hidden appraisal and internal failure costs are combined with a classic analysis of other quality costs, the magnitude of quality's real cost is revealed. In addition, a line's problem areas, along with cost savings and improvement potentials, are highlighted.

Table 6. Report Summary
(in Management's Language)

$0.0253 is the cost lost per sales dollar for this period.
$1,791 is the appraisal cost for this period.

853 Standard hours used
$1.44 Failure cost/actual hour
$2.10 Appraisal cost/actual hour
$3.54 Failure + appraisal cost/actual hour

A common complaint about quality cost systems is that they demonstrate a high cost number with no real opportunity to reduce it and no incentive to realign cost elements. An operation-by-operation cost system shows where and how monies are being spent and gives management the opportunity to marshal corporate resources toward the investigation of specific high cost areas. Statistical methods and thinking can then be applied to high cost areas and progress documented and quantified. The slight increase in prevention costs associated with the implementation of a quality system then becomes insignificant compared to the possibilities for cost reductions in appraisal and internal failure costs.

Statistical methods are potentially the most important management tools of the last quarter century, but, to gain wide acceptance, they must be tied to dollars, the language of management. It is up to quality professionals to translate into dollars the gains made through statistical methods. We must speak both languages.

AN ENGINEERING ORGANIZATION'S COST OF QUALITY PROGRAM

(January Issue)

Lawrence J. Schrader
Product Assurance Manager
FMC Corporation, Ordnance Division Engineering

Establishing a cost of quality (COQ) program in an engineering organization is challenging but it can be done with effective planning and cooperation. At FMC Corporation, a highly diversified conglomerate, we view COQ as a means to measure improvement.

One of the major differences in COQ for an engineering versus a manufacturing environment is due to the products produced. Engineering departments may produce hardware products to evaluate or demonstrate concepts, but their real products are technical data packages that someone else will use: drawings, specifications, designs, analyses, ideas, and software. Manufacturing produces hardware: parts, subassemblies, and systems. Hardware scrap, rework, and repair is visible — it can be seen and touched. How visible is a scrapped idea or analysis? If you fail to make a calculation, it may not come to light unless it results in a failure.

Establishing engineering COQ requires new definitions. An error in engineering can cost thousands of dollars to correct once the product is in manufacturing or the hands of the user. To measure progress, one needs a base. (One must measure improvement or even the boss won't believe it.)

To review how costs escalate, it is best to use a hypothetical case. If an engineering error is found on the drafting board, it costs $1 to correct. If the error is found at checking, it will cost $10 to correct. If that same error is found when the piece parts are produced, it costs $100 to correct. If found in an assembly prior to release it costs $1,000 to correct. But, if not found until the product is in the hands of the user, the cost is $10,000 to correct. If the error requires field retrofits, it will cost $100,000 to correct. If the engineering error results in a lawsuit, the cost is in the area of $1,000,000. If the judgment is against the company, the cost can be even greater.

Over the past two-and-a-half years, FMC has utilized COQ as a cornerstone measurement for the following reasons:
(1) It provides common measurements for management and labor.
(2) It helps develop quality awareness as a focal point.
(3) It provides a common language for management and labor in terms of dollars.
(4) It's a signal of management support of quality.
(5) COQ moves to quality improvement process and to total quality assurance (TQA) (through waste reduction and employee involvement) and the understanding that each of us is responsible for the quality of our work.

During the past two years of the program, FMC, Ordnance Division Engineering (ODE) reduced its quality cost by 5%. How this was accomplished and the results obtained require some chronological explanation. The first step was that upper and middle management demonstrated support by transferring a COQ advocate from corporate staff to engineering. They then identified manpower needs and began recruiting qualified quality engineers to implement COQ and to revise the systems and procedures necessary to improve performance.

Management was convinced of the need, and COQ became a basic measurement cornerstone.

The groundwork had been completed, and support from upper-level management was present. FMC was now, as Ken Lawrence, manager of engineering, said, "100% started in creating an effective quality management team." Lawrence assembled a task force to review which costs should be gathered and monitored. The task force would also establish the various COQ functions.

During this same period the engineering management staff agreed to a quality policy stating, "It is the policy of Ordnance Division Engineering to provide products and services that meet our customers' requirements and conform to our internal standards."

The policy statement complements the belief within engineering that quality starts with product design. To quote Bradley T. Gale, managing director of research at the Strategic Planning Institute, "The internal view of quality focuses on statistical quality control and conformance to specification. Close conformance to an inferior design may not help customer-perceived quality. But high relative quality of conformance pays off when relative quality of design is also high."

Next the task force agreed to the following COQ categories:
- Prevention costs: steps taken to assure tasks are done right the first time, every time.
- Appraisal costs: cost associated with evaluating whether a task has, in fact, been done right the first time.
- Internal failures: costs associated with not doing tasks right the first time, every time, but are caught and identified within the Ordnance Division Engineering realm.
- External failures: costs associated with any failures experienced by our customers after release of the Technical Data Package.

Having a basic understanding of which areas quality costs impact, a program was defined for each department to determine what costs would go into each of these categories.

The engineering organization consists of functional groups, some of which may be assigned to specific projects on a full-time basis. Examples of FMC Ordnance Division Engineering's functional departments are: armament systems, design integration, engineering tests, product design, drafting/CAD, design assurance, product assurance, operations, integrated logistic support, reliability, human factors and safety, and system engineering. Additionally, there are material control, purchasing, inspection, tech publications, producibility, independent research and development, and others.

For tracking purposes, separate quality costs were defined for each of the major departments. Table 1 provides examples of some quality costs we decided to include for engineering, product assurance, and integrated logistic support.

The major impacts of generating quality costs in an engineering environment were the support of management, the recognition that COQ was measurable, and that the total cost of quality involved all departments. At the beginning of 1984, management

Table 1. Quality Costs

Product Assurance

	Quality Control	Quality Engineering	Supplier Quality	Inspection Engineering
Charter	Provide project quality control engineering and inspection service	Provide quality assurance engineering and audit service	Provide vendor surveillance and rating service	Develop inspection documentation
Prevention	Quality plan, QC cost estimate, QC instructions	Quality manual audit plan	Vendor survey plan	Establish QA requirements
Appraisals	Proofread inspection plans, inspection-audit	Proofread quality auditor evaluation	Source inspection, audit vendor rating audit	Checking QARs, FIRs (inspection instructions)
Internal Failure	Rewrite inspection plan errors found by QC supervision	Errors found by QED supervision	Errors found by supervision material review board	Rewrite or update QARs, FIRs
External Failure	Defects or squawks on subsequent operations and/or found on fielded products	Errors reported after publication of audit	Errors found during use or audit of vendor rating system	Required change proposals, update QARs and FIRs

Engineering

	Definition	Design	Technical Staffs	Engineering Test
Prevention	Investment to minimize problems	Design manual	Design manual	Plan tests, devise test methods
Appraisal	Evaluate, test, inspect, audit	Design reviews and checking	Design reviews and checking	Inspection
Internal Failure	Revise the tech data package, solve problems before release	Redesign problem solving prior to drawing release	Corrective problem solving and redesign of a project not released	Failure of test equipment during test
External Failure	Revise the tech data package, solve problems before release	Redesign problem solving after drawing release	Corrective problem solving and redesign of a released project	Incorrect test data that was reported

Integrated Logistics Support

	Maintenance Engineering	Technical Publications
Charter	Provide quality technical documentation products on schedule and within budget	Provide quality technical documentation products on schedule and within budget
Prevention	Quality plan — personnel training/cross-training desktop procedures	Quality plan — personnel training, desktop procedures
Appraisal	Quality audit/assessment processes	Quality audit/editing
Internal Failure Prior to Release	Reprograms, rerun programs, rewrites, redistributions, reillustrations, re-edits, reprints, retypes/retypesets, rechecks	Rewrites, redistributions, reillustrations, re-edits, reprints, retypes/retypesets, rechecks
External Failure After Release	Reprograms, rerun programs, rewrites, redistributions, reaudits/assessments, reillustrations, re-edits, reprints, retypes/retypesets, rechecks, review	Rewrites, redistributions, reillustrations, re-edits, reprints, retypes/retypesets, rechecks

decided to develop specific improvement programs that would result in reducing COQ while simultaneously identifying process deficiencies. These programs were:
- The engineering manager established a design improvement committee made up of each functional and project manager.
- Quality improvement goals were required of each direct report of the Ordnance Division Engineering manager.
- The operation manager established a material flow improvement committee made up of purchasing, quality engineering, receiving inspection, receiving, hardware management, and supplier quality control.
- Design engineering established methods of measuring quality performance tied to the number of engineering changes.
- The product development process improvement planning committee was established.

The net result was a reduction of 5% of the total direct labor dollars. Ordnance Division Engineering employs more than 1,700 people. The 5% reduction in total cost of quality resulted in a $4,000,000 savings.

Quality costs shifted from failure correction to prevention and appraisal. (See Table 2.) Additional effort has been applied to engineering design improvement in 1985. The goal for 1985 is to further reduce our total cost of quality by 10%. Initially, in 1983, the total cost of quality was 18.9%. In 1984, it was at 14.1%, and the projected COQ for the end of 1985 is 12.5%.

To reach our 1985 cost avoidance goals, engineering started to analyze current drawing errors found at checking. By categorizing these drawing errors, problem areas could be identified. After a short period of reviewing drawings, errors were categorized as either Type I (design) or Type II (drawing). (See Table 3.)

By conducting a Pareto analysis of the breakdown of drawing errors, errors accounting for 80% of the total were identified. Applying specific training efforts to these problem areas enabled management to correct the cause of the error and reduce recurrence.

For design errors, training concentrated on call-outs, general notes, and dimensions/tolerances (the three types of errors that accounted for 80% of the errors). Design errors almost always resulted in engineering changes. The top five drafting errors account for 82% of all drafting errors. Training drafters to review revision blocks, title blocks, word crafting, external detailing, and lines/arrows improved the quality of the drawings.

The average cost to change a released drawing (which, at times, includes retrofitting fielded products) is about $2,000. The cost of correction increases substantially as the correction is delayed. Managing these costs puts a dollar amount into a "pay me now or pay me later" decision. Cost of quality in a research and development engineering environment is possible with proper support and implementation. The cost of correcting engineering errors following final assembly and delivery is excessive. Implementing a cost of quality program in the engineering phase can reduce costs and improve quality for the whole of the organization.

Table 2. Quality Cost

	1983 % of effort	1984 % of effort
Prevention	25	35
Appraisal	34	44
Internal Failure	20	14
External Failure	21	7

Table 3. Drawing Errors Breakdown

Type I: Design Errors

	Drawings	Notices of Revisions	Total
1. Callouts	279	50	329
2. General notes	156	47	203
3. Dimensions/tolerances	97	40	137
4. Parts list	80	15	95
5. Symbols	41	6	47
6. GD&T practices	21	2	23
7. Diagrams	2	0	2

Type II: Drafting Errors

	Drawings	Notices of Revisions	Total
1. Revision block	312	53	365
2. Title block	143	156	299
3. Word crafting	236	55	291
4. External detailing	172	45	217
5. Lines/arrows	202	7	209
6. Subcontract/format	141	34	175
7. Sectional views	65	9	74
8. Clarity	34	6	40
9. Internal features	14	6	20

Significant Items:
- The top three design errors account for 80% of the Type I errors.
- The top five drafting errors account for 82% of the Type II errors.

Note:
- The Drawings column includes revised and new drawing data.
- This data is from the previous six months of 100% inspection of checkprints.

Table 4A. Engineering — Costs of Quality

	Definition	Design	Technical Staffs	Engineering Test	Producibility, Support to Production, UECP	RAM-D	CADD, CAE, Drafting	Human Factors, Product Safety	Configuration Management/ Checking/ Design Review	Reproduction
Prevention	Investment to minimize problems	Design manual	Design manual	Plan tests, devise test methods	Plans and procedures	Engineering manual	Documentation and validation	Engineering manual	Configuration management plan, drafting manual, engineering manual	Operating plan
Appraisal	Evaluate, test, inspect, audit	Design reviews and checking	Design reviews and checking	Inspection	Design review	Evaluate designs for RAM-D considerations	Checking	Design review	Review released drawings	Inspection
Internal Failure	Revise the tech data package, solve problems before release	Redesign problem solving prior to drawing release	Corrective problem solving and redesign of a project not released	Failure of test equipment during test	Method failure found during design review	Problem solving prior to release as a result of a design evaluation	Correct drawing errors, debug programs	Participate in problem solving and failure investigation as a result of a design review	Rechecking of rejected drawing	Scrap rework
External Failure	Revise the tech data package, solve problems after release	Redesign problem solving after drawing release	Corrective problem solving and redesign of a released project	Incorrect test data that was reported	Method failure found during production	Problem solving after release	Correct drawing errors after release	Participate in problem solving because of customer complaint	Rechecking of released drawing found in error	Returned work

Table 4B. Engineering Activities

Category		Activity
Prevention	Time spent by ODE-operations personnel in planning, developing systems, and performing internal checks to **prevent** quality problems and costs.	**Quality circles:** time spent by members, leaders, and facilitators in quality circles activities (assumes 50% related to quality matters).
		Training: time spent by employees, supervisors, and managers in training relating to quality or quality costs.
		Procedures preparation: time spent in developing and/or administering procedures (formal and informal) relating to quality or quality costs.
		Vendor quality surveys: time spent in surveying vendor quality systems.
		Internal activities: includes internal error prevention, preventive maintenance, and checking.
Appraisal	Time spent by ODE-operations personnel in evaluating, documenting, and, if necessary, resolving problems originating outside their immediate control.	**Scrap/rework tracking/reporting:** time spent estimating, documenting, and reporting scrap/rework costs.
		Vendor quality tracking: time spent evaluating and documenting vendor quality levels.
		F/O tracking system audits: time spent by M.C. personnel in auditing accuracy of information in F/O tracking system.
		Appraisal/resolution: time spent evaluating work of other departments and resolving problems, if necessary.
		Production test: time spent by ODE-operations personnel in hardware testing prior to inspection by product assurance personnel.
		Department/function quality measurement tracking: time associated with measuring the quality of output of a department or function.
Internal Failure	Costs incurred because of problems not found or corrected during prevention/appraisal process.	**Scrap/rework costs:** time spent in reworking discrepant material, or time spent working on material which is ultimately scrapped.
		Vendor DMR processing: time spent resolving problems associated with discrepant material received from vendors.
		Documentation revisions due to errors: time spent revising documents because of errors discovered after "official release" but before hardware impact.
		"After shop" revisions: time spent revising documents (including change incorporation) after hardware has begun.
External Failure	Costs associated with correcting quality problems not discovered until after the product reaches the customer.	**Warranty claims:** time and material costs and other expenses associated with correcting problems discovered by customers.

Table 5. Measurement Bases for Quality Costs

Base	Advantages	Disadvantages
Direct labor hours	Readily available and understood	Can be drastically influenced by automation.
Direct labor dollars	Available and understood; tends to balance any inflation effect	Can be drastically influenced by automation.
Standard manufacturing cost dollars	More stability than above	Includes overhead costs both fixed and variable.
Value added dollars	Useful when processing costs are important	Not useful for comparing different types of manufacturing departments.
Sales dollars	Appeals to higher management	Sales dollars can be influenced by changes in prices, marketing costs, demand, etc.
Product units	Simplicity	Not appropriate when different products are made unless "equivalent" item can be defined.

Table 6A. Drawing Quality Log CADD

	Frequency	Cumulative Frequency	Percent	Cumulative Percent
Appearance of drawing conforms to FMC accepted practices	68	68	32.69	32.69
Title block is complete & correct	65	133	31.25	63.94
Dimensions & tolerances allow for maximum economies	29	162	13.94	77.88
All necessary dimensions are shown	13	175	6.25	84.13
GD&T properly employed modifier	7	182	3.37	87.50
Spelling, punctuation, word order are correct	5	187	2.40	89.90
Part fits properly into assembly	4	191	1.92	91.83
Extension & leader lines are to correct surfaces	4	195	1.92	93.75
GD&T properly employed tolerance	3	198	1.44	95.19
GD&T properly employed datum	3	201	1.44	96.63
Dimensions & tolerances allow for ease of inspection	3	204	1.44	98.08
Draft angles, fillets, and corner radii are given	2	206	0.96	99.04
Agreement between drawing & general notes	2	208	0.96	100.00

$$\frac{208 \text{ errors}}{38 \text{ drawings}} = 5.47 \text{ errors per drawing}$$

Table 6B. Drawing Quality Log Manual

	Frequency	Cumulative Frequency	Percent	Cumulative Percent
Spelling, punctuation, & word order are correct	77	77	41.40	41.40
Title block is complete and correct	43	120	23.12	64.52
Apppearance of drawing conforms to FMC accepted practice	35	155	18.82	83.33
Abbreviations are correct	10	165	5.38	88.71
All necessary dimensions are shown	8	173	4.30	93.01
Agreement between drawing and general notes	6	179	3.23	96.24
GD&T properly employed datum	4	183	2.15	98.39
Dimensions agree with design layout	3	186	1.61	100.00

$$\frac{186 \text{ errors}}{13 \text{ drawings}} = 14.31 \text{ errors per drawing}$$

OPTIMUM QUALITY COSTS AND ZERO DEFECTS: ARE THEY CONTRADICTORY CONCEPTS?

(November Issue)

Arthur M. Schneiderman
Director, Quality and Productivity Improvement
Analog Devices
Wilmington, Massachusetts

As defect levels drop, failure costs decline while appraisal plus prevention costs increase. This apparent tradeoff suggests that an optimum quality level exists and that attempts to further improve quality above this level will increase total cost and decrease financial performance. Proponents of this view therefore argue that striving for zero defects (ZD) through a program of continuous improvement is not in a company's best economic interest.

J. M. Juran discusses the concept of optimum quality in his *Quality Control Handbook*.[1,2] Figure 1 depicts his model for optimum quality costs. Juran also defines three quality zones relative to the point of minimum total quality costs. The "zone of improvement projects" lies below the optimum quality level, while the "zone of perfectionism" lies above it. Between them, and in the area of the minimum, lies the "zone of indifference." It is the zone of perfectionism that most troubles proponents of zero defects, for here Juran suggests relaxing prevention efforts and allowing (even encouraging) increased defect rates. Furthermore, he identifies the boundary of the zone of perfectionism as lying, typically, at a quality level where failure costs amount to 40% of the total quality cost. Applying other rules of thumb, this translates into a defect level only half that which exists in the zone of improvement.

Coming from anyone other than Juran, this apparent heresy might go unnoticed. Are the ZD movement and the concept of continuous improvement wrong? Or can these two apparently disparate views be reconciled? To answer this question, consider the "physics" involved in quality cost optimization. First, the mathematics of optimization:

Let $f(q)$ = Total (internal+external) failure costs
$p(q)$ = Total (appraisal+prevention) prevention costs
$T(q)$ = Total quality cost = $f(q)+p(q)$
q = quality level (0 to 100% good product)

Then, $T(q)$ is minimized[3] when $dT/dq = 0$ or $dp/dq = -df/dq$.

Juran's Model of Optimum Quality Costs

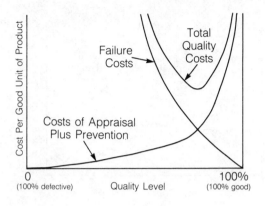

From J.M. Juran's *Quality Control Handbook*, Third Edition (New York: McGraw-Hill, 1979) p. 5-12.

Figure 1

In other words, at the point of minimum total quality costs, an *additional* dollar invested in prevention will produce exactly one dollar's worth of reduced failure costs. Below the optimum, this incremental dollar's worth of prevention provides more than the proverbial dollar's worth of cure. Above it, the opposite is true. There are two other crucial lessons to be learned:

1. Optimum quality costs depend on incremental, not total, elementary costs. At the optimum, nothing in general can be said about the relative levels of prevention and failure costs.

2. There is no mathematical requirement that the optimum occurs at q < 100%. There may be no optimum in the range of q = 0 to 100%. There might be a minimum rather than an optimum, and it could very well be at q = 100%. Figure 2 shows an example.

The optimum (or more correctly, the minimum) quality cost could lie at zero defects (q = 100%) if the incremental cost of approaching ZD is less than the incremental return from the resulting improvement. Juran asserts that "prevention costs rise asymptotically, becoming infinite at 100% conformance."[4] This implies that the incremental cost is also infinite. Since the incremental return is not, it follows from his assertion and the above mathematics that the optimum lies below 100%. The question now is, "Does it really take infinite investment to reach zero defects?"

Optimum Quality Level Equals Zero Defects

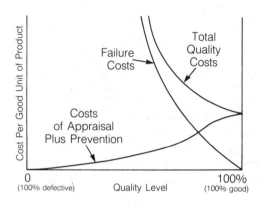

Figure 2

Two Contrasting Improvement Processes

	Kaizen vs.	**Innovation**
FOCUS	Design, Production and Marketing	Science and Technology
TARGETING	Broad: Quality, Cost Safety, Efficiency Product Development	Narrow: Feature Technique
EXPERTISE	Conventional Know-How	Leading Edge, Breakthrough
CAPITAL NEEDS	Very Modest	Major Investment
PROGRESS	Small Steps	Big Jumps
RESULTS	Continuous	Spontaneous
VISIBILITY	Not Dramatic	Very Dramatic
INVOLVEMENT	Everyone	Selected Few
COOPERATION	Group Activity	Individual Effort
RECOGNITION	Effort, Process	Results
	⬇ **Evolution**	⬇ **Revolution**

Figure 3

Zero defects advocates endorse continuous improvement. This is the never-ending effort to totally eliminate all forms of waste (the Japanese call it "muda"), including reworks, yield losses, unproductive time, over-design, inventory, idle facilities, safety accidents, and the less tangible factors of unrealized individual and societal potential. The methods used in this process are widely misunderstood and in fact may lie at the root of the issue. The Japanese word for continuous improvement is "kaizen." Figure 3 compares this method to an alternative improvement process — innovation.[5] While innovation is characterized by costly major events, kaizen represents inexpensive and almost imperceptible continuous improvement.

Kaizen is much like the tortoise in the fable of the tortoise and the hare. It often beats innovation in the race for competitive advantage. One striking example was the reduction of dip soldering failures at Yokogowa Hewlett-Packard (YHP), shown in Figure 4.[6] For a little more than two years, the continuous improvement process on average produced a 50% reduction in the failure rate every 3.6 months. Defects were reduced by a factor of over 250. The process eventually slowed, probably due to equipment limitations. Interestingly, that equipment had purportedly been discarded as obsolete by a sister plant in the U.S. It would not be at all surprising to find that the equipment had become obsolete because of an innovation that resulted in improvements of a factor of *only* two or three.

What was the incremental cost to YHP in going from a defect rate of 3 ppm to 2 ppm? What was the incremental return? A detailed cost analysis could probably capture all of the costs and benefits, but the results can be guessed. The incremental costs are essentially zero. Why? At a minimum, one could argue, there are the labor costs associated with the time spent working on the improvements. But these were not incremental or increased costs. They were fixed costs based on a process that encourages everyone to spend about 5 to 10% of their time working on improvements.

What if less time were spent? Evidence suggests that this would result in backsliding (Figure 5).[7] After 20 months of continuous improvement (at an improvement rate of 50% each 5.1 months) and a tenfold reduction in scrap, the problem was declared solved and all efforts toward further improvement abandoned. The result: the gains could not be held, and the scrap rate increased until the continuous improvement program was reinstated. Quality is not a stable property. Without constant effort from everyone, the organization naturally drifts toward poor performance: higher cost and lower quality.

If the incremental labor costs are indeed zero, what about capital costs required for these improvements? Again, they are probably negligible. Kaizen-type improvement is usually the result of better methods or small equipment changes or additions.

The direct incremental benefits of continued improvement are clearly small in going from 3 to 2 ppm. However, there are some major cultural advantages: organizational pride, reputation, spillover into other areas, and experience in problem solving, to mention a few.

The correct way to view quality cost optimization is on the basis of incremental economics. However, as ZD is approached, it becomes harder to quantify any increased costs or benefits as less tangible issues enter the equation. A program of continuous improvement does not necessarily introduce increased costs as the quality level approaches 100%. *Any benefit at all* could produce a minimum quality cost at zero defects. The apparent contradiction therefore disappears once the underlying economics

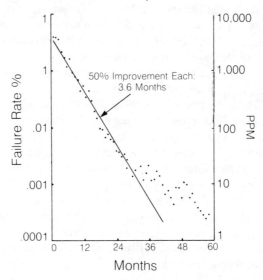

Figure 4

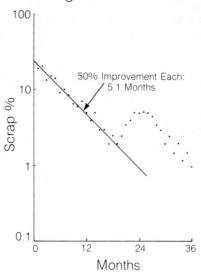

Figure 5

of both concepts are clear. Perhaps the best test of this view is the competitive performance of firms that believe in continuous improvement and zero defects. That group includes not only Toyota and Sony, but also IBM, Hewlett-Packard and an ever-increasing number of successful U.S. firms.

References

1. J. M. Juran, *Quality Control Handbook,* Third Edition (New York: McGraw-Hill, 1979), p. 5-11.

2. J. M. Juran, "Optimum Revisited," *Quality Progress,* April 1986, p. 10.

3. The minimum value of a function is determined by the point at which its slope is zero, or using calculus, when its first derivative vanishes.

4. Juran, *Quality Control Handbook,* p. 5-12.

5. This comparison was first made by Masaaki Imai, president, The Cambridge Corporation, Tokyo, Japan.

6. Kenzo Sasaoka, "Our TQC Experience as a Partner of America," presented at the 1984 Seminar and Plant Tour to Study Productivity of Japanese Industry, sponsored and organized by The Cambridge Corporation, Tokyo, Japan.

7. J. M. Juran, *Management of Quality,* Fourth Edition (New York: J. M. Juran, 1982), p. Jb9.

CIM REVIEW

ACCOUNTING FOR QUALITY COSTS — A CRITICAL COMPONENT OF CIM

(Fall Issue)

Wayne J. Morse
Chairman
Accounting and Law Department
Clarkson University
Kay Poston
Assistant Professor of Accountancy
University of Missouri
Columbia, Missouri

Improved quality is a major objective behind any implementation of CIM technology. However, without effective methods for tracking and reporting quality-related benefits, it is impossible to know the actual effect of these new manufacturing technologies and techniques on quality. Accordingly, the development and implementation of better quality cost measures and reporting practices should be an integral part of a company's CIM plan. To help manufacturers establish such practices, the authors of this article discuss the quality cost factors that must be accounted for and offer guidelines for accurately measuring and reporting those costs.

Today, more and more companies are searching for ways to improve the quality of their products and services. This search has led many manufacturers into investments in automated manufacturing technologies that incorporate advanced quality assurance and control techniques and computer-based quality systems (e.g., coordinate measuring machines (CMMs), automated testing equipment (ATE), machine vision systems). Companies are also investing heavily in other areas. For example, employees are being trained in statistical process control methods, variance analysis, experiment design, and other quality monitoring techniques. In addition, many companies are establishing quality circles to promote the communication of quality-related issues among their employees. Such efforts clearly represent a substantial investment. Still, companies are willing to spend these large sums of money because they realize that poor quality has a much greater cost.

Given this high level of investment, feedback and evaluation of results are an important part of any quality improvement program. Management must know whether the program is producing the desired results. Commonly used measures such as defect levels and rejection rates can be useful indicators of progress. However, the company's quality costs, though often overlooked, are probably the most important indicator of the effectiveness of the quality program. Quality costs serve as the financial register of the company's quality achievements; management must carefully assess whether investments in quality improvement efforts are yielding benefits in excess of the costs.

Statistical process control or the lowering of rejection rates does not adequately show whether the financial commitment has generated a sufficient return. Unfortunately, few companies really know what their quality costs are because, until recently, few have cared.

The Definition of Quality

The word *quality* can have many meanings. Accordingly, when a company determines that someone wants a quality product, there is usually some uncertainty as to how that desire should be fulfilled because quality depends on any number of product characteristics and personal perceptions and values. For example, one individual might view black-and-white television sets as being inferior to color TVs. Someone else might claim that small-screen televisions, whether black-and-white or color, have poorer picture quality than large-screen sets. A third person might feel that sound is the key factor and insist that any nonstereo TV is of lesser quality than a stereo set. Although these three individuals perceive differently what constitutes a high-quality television set, they would all agree that any type of new TV that doesn't work when it is first plugged in is of unacceptable quality.

Quality Categories

Recognizing that individual values and perceptions differ, quality professionals have defined quality as fitness for customer use. In this context, quality is determined by three factors: customer expectations, product specifications, and product performance in the field. The terms *grade, quality of design,* and *quality of conformance* are applied to these three factors, respectively.

Grade refers to the difference between products that have the same functional use. For example, black-and-white and color TVs are two products designed to meet the same basic need. However, color TVs are of higher grade than black-and-white sets. A similar comparison could be made between TVs with 13-inch screens and those with 26-inch screens or between stereo and nonstereo TVs.

Such grade variations are the producers' response to the different levels of purchasing power in society. Presumably, as individuals acquire affluence, their needs and expectations change. Thus, people who are willing to spend more on a television set are likely to purchase a higher-quality television. The TV is of higher quality because it is of a higher grade, not because it works better than any other television set on the market. It fulfills more needs because it is a color set or it has a large screen or it offers such special features as cable readiness, remote control or stereo sound.

After a company identifies its target market, it can assess customers' grade expectations and develop product specifications accordingly. This is when quality of design enters the picture. Quality of design is the degree to which the producer's design specifications for the product meet the customers' expectations. For example, if a producer designed a television set with a remote control unit that worked only when it was within six inches of the TV, the set would most likely be rejected by the market. The remote control capabilities simply would not meet customer expectations for the feature, and consumers would be unwilling to pay more for this variation in grade. Thus, the quality of design is low. In contrast, consumers are willing to pay a premium for the remote control capabilities that exist in today's marketplace because television manufacturers have produced a remote control television with high quality of design.

A product of high grade and quality of design is not necessarily a high-quality product, however, if its performance in the field, or quality of conformance, is unsatisfactory. Quality of conformance is the degree to which a given product meets its design specifications. Any product, regardless of grade and quality of design, must have high quality of conformance to be fit for customer use. Most individuals would rather have a black-and-white TV that has a clear picture than a color set that has a snowy picture.

Breakdown of Quality Costs

Quality costs are the costs associated with quality of conformance. They are incurred either before or after the product's manufacture is complete. Quality costs can be attributed to preventing or identifying quality problems before the product reaches the marketplace or correcting or compensating for quality problems once the product is in the marketplace. Quality costs can be categorized as prevention, appraisal, or failure costs.

A company incurs prevention costs to prevent the production of nonconforming products. Typical prevention costs include the expenses associated with in-process quality control methods, simulation tools, adaptive control systems, training in QC techniques, and the provision of technical support to vendors.

Appraisal costs are incurred to identify nonconformities before a product is delivered to the customer. The most common appraisal cost is the cost of product testing and inspection (e.g., costs associated with ATE, CMMs, machine vision systems). Appraisal costs are also associated with the inspection of incoming materials, depreciation of test equipment, and supervision of inspection activities.

If a product does not conform to its design specifications, the producer incurs failure costs, which are either internal or external. Internal failure costs result from the identification of poor conformance quality prior to product delivery — for example, cost of scrap, rework, repair, downtime as a result of quality problems, and retest of repaired parts. External failure costs (e.g., those associated with warranty repairs and replacements, product recalls, product liability, and responses to customer claims) are incurred when a nonconforming product is delivered to the customer. External failure costs also include such intangibles as the cost of lost sales that result when a company gets a reputation for producing low-quality products. Of the four types of quality costs, external ones are the costliest.

Exhibit 1. Relationship Between Quality Costs and the Time Nonconformity is Detected

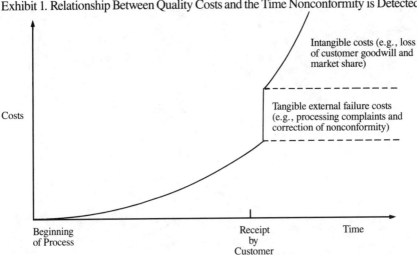

Internal failure costs are the product of an investment in appraisal costs. As a company boosts its appraisal activities, it can identify an increasing number of nonconforming products prior to their delivery and thus eliminate the possibility of losing customers as a result of poor quality. Moreover, repairs made in house cost less than repairs made in the field. But if all products were produced right the first time, scrapping, downgrading, and repairing nonconformities would be eliminated. For many companies, this would result in labor and material savings worth hundreds of thousands of dollars. Therefore, it is crucial to invest the appraisal dollar so that nonconformities can be identified as early in the production process as possible.

Exhibit 1 illustrates a hypothetical relationship between quality costs and the time at which a nonconformity is detected. Specifically, it shows that the later a nonconformity is identified in the production process, the greater the cost of correcting the nonconformity. Moreover, if the nonconforming product is identified by the customer rather than by the producer's appraisal activities, the costs rise dramatically. Thus, investing the quality dollar in prevention costs is the most desirable plan of action because preventing a nonconformity is often less expensive than identifying and correcting the quality problem. A key benefit of CIM and automation technology is the ability to detect or correct nonconformity as early as possible.

Exhibit 2. Relationship Between Quality Costs and Quality of Conformance

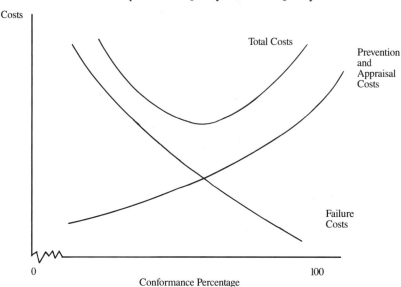

The Preferred Distribution

Exhibit 2 shows the theoretical relationship between quality costs and quality of conformance commonly used by quality professionals when helping managers allocate the quality dollar. The exhibit shows that total quality costs are high when quality of conformance is low. And when quality conformance is low, the bulk of the quality dollar is allocated toward failure costs.

Exhibit 2 also shows that total quality costs decrease as companies shift their quality expenditures from failure costs to prevention and appraisal costs. An improvement in outgoing quality also corresponds to this shift, and the decrease in costs continues until near-perfect quality is attained. At this point, management cannot realize further cost reductions through additional investment in prevention and appraisal. Rather, management must focus on achieving technological breakthroughs that will shift the prevention and appraisal curve down and to the right.

Exhibit 2 also shows that although prevention and appraisal costs are preferable to failure costs, quality is not free. There is no such thing as zero quality costs (although prevention and appraisal investments will pay for themselves), but such costs can be minimized. At the same time, Exhibit 2 can be misleading because the cost curves represent static relationships that are applicable to existing technology and knowledge. In reality, these relationships are dynamic. As technology and knowledge change, the curves shift. The search for quality improvements is therefore unending.

Exhibit 3. Quality Cost Pie Chart Suggesting a Maldistribution of Quality Costs

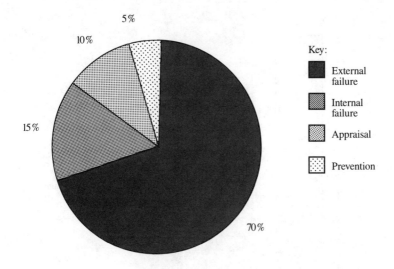

Uses of Quality Cost Data

Because quality costs and the initial commitment to a quality program are measured in the same way (i.e., dollars), a manager can more adequately assess the full results of quality accomplishments. For example, in addition to showing that rejection rates are lowered, a manager can show that the lower rejection rates have produced cost savings in excess of the initial investment.

The fact that quality costs are measured in dollars has other advantages as well. Money is a universal business language that all top-level executives understand. They may or may not understand the implications of attaining a state of process control or a reduction in rejection rates. Further, dollars are a common denominator that can be added across departmental and product-line boundaries to give an overall picture of a company's quality performance.

Quality cost information can also be used to highlight improper distributions of quality costs. An examination of a pie chart, like that shown in Exhibit 3, shows that a company's lack of prevention and appraisal efforts results in large external failure costs. A logical course of action would be to bolster appraisal efforts so that nonconformities can be identified internally. This action should reduce external failure costs while increasing internal failure and appraisal costs.

The quality cost trend analysis illustrated in Exhibit 4 indicates the effect of appraisal activities on various categories of quality costs. In Exhibit 4, the data in period 1 represents the maldistribution of quality costs depicted in Exhibit 3. The data in period 2 illustrates the results of an increase in appraisal efforts. The next step would be to

Exhibit 4. Quality Cost Trend Analysis Illustrating Desired Effect of Appraisal and Prevention Activities

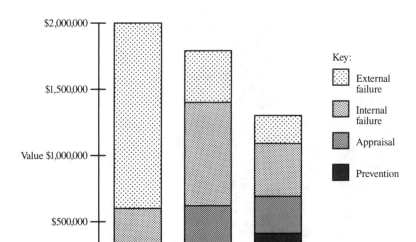

increase efforts to prevent the production of nonconforming products — a step that would, it is hoped, further reduce external failure costs, internal failure costs, and the need for appraisal, as exhibited in period 3.

Quality cost information also allows managers to assess the relative importance of quality problems so that they can devise a plan to resolve them. Without the benefit of quality cost information, managers are likely to rank quality problems according to such measures as rejection rates, even though the quality problem that creates the highest rejection rate may not be the most costly in terms of failure costs. By analyzing quality costs, managers can tackle quality problems that have the greatest financial impact before eliminating those of lesser financial significance.

Cost Data Limitations

Despite its benefits, quality cost information can be misused. Quality cost information must be viewed as a management tool; it does not solve quality problems or suggest specific ways to improve quality. The ingenuity, knowledge, and professional judgment of individuals throughout the organization are required to determine the best approaches to quality improvement.

Another potential pitfall in using quality cost information is short-term perspective. Some organizations have learned that an action that appears beneficial in the short run may be detrimental in the long run. One common example is the postponement of capital investment projects to preserve working capital. For example, although total quality costs can be reduced in the short term by reducing expenditures for prevention and appraisal, the long-term consequence is increased failure costs and potential loss of market share as a result of lost consumer confidence.

Management must also realize that quality improvement efforts and accomplishments rarely occur during the same period. Quality cost reports at the onset of a quality improvement program are likely to show an increase in total quality costs instead of the desired reduction in costs. This increase represents the initial investment in the improvement program — results will appear in time. Trends in quality costs can be monitored by using line graphs, bar charts, or pie charts to assess whether the desired change in the overall distribution of quality costs is occurring.

When reviewing quality cost reports, managers should remember that total quality costs fluctuate along with activity. Thus, a decrease in costs may not be due to elimination of a quality problem. Likewise, an increase in quality costs may not signify additional investments in prevention or new quality problems. Rather, these decreases and increases may simply reflect a change in activity. For this reason, it is often useful to compare quality costs from one period to the next on the basis of some activity measure (e.g., number of units produced, direct labor hours or dollars, or direct materials usage).

Quality cost reports may sometimes include inappropriate costs. Although waste is typically categorized as an internal failure cost, there are some instances in which waste is not a quality cost. One example is in textile manufacturing. If threads and yarns are attached to spools with glue, the last few inches or feet of the raw material may be unusable. That amount of waste cannot be controlled by the textile manufacturer and does not constitute a quality cost. Other items sometimes inappropriately included in quality cost reports are inventory write-downs because of obsolescence rather than quality problems, marketing concessions in which quality is not an issue, and the cost of redesign or rework requested by a customer for reasons other than quality.

Intangible Costs

A sincere attempt to estimate the intangible costs associated with external failures and to rationally allocate certain elements of overhead among the quality cost categories results in a report that is more accurate than one that omits such costs altogether. In making decisions based on the report, managers must simply be aware that some of the information is subjective and susceptible to measurement error.

Some companies unthinkingly omit important quality cost items from quality cost reports. One such cost is that associated with customer ill will and loss of competitive standing. This cost is often excluded from quality cost reports because it is difficult to measure the exact dollar value of the item. Still, a subjective estimate can be attempted. Otherwise, the omission of such external failure costs will distort the company's quality costs by implicitly assigning a value of zero to one of the most financially significant consequences of poor quality.

Overhead cost assignments are also frequently omitted from quality cost reports. This includes, for example, the depreciation of test and inspection equipment, the cost of utilities used in test and inspection areas of the factory, and allocations of supervisory time to prevention and appraisal activities. One reason for ignoring these costs is that they are not always measurable or easily assigned to cost categories, departments, and product lines. However, once again, the omission of these costs from quality cost reports is tantamount to assigning them a value of zero and can only distort reality.

Establishing a Quality Cost Reporting System

The support and commitment of top-level management are critical to establishing a workable quality cost system. Without such support and commitment, there is no guarantee that ongoing resources for this critical CIM subsystem will be available. One way to gain top management's attention is to develop preliminary estimates of the firm's total quality costs. If the company's quality problems can be phrased in financial terms, top management will be more motivated to take action to improve quality.

After top management supports the notion of a quality cost reporting system and has devoted resources to it, the next step is to establish a quality cost team that will be responsible for system installation. If possible, the team should include interested individuals from throughout the organization. Such diverse representation allows more viewpoints to be considered by the team and can make subsequent tasks much easier.

Next, the quality cost installation team must define a reporting unit that can be used as a prototype. It is usually easier if the team starts out small and then expands the system based on experiences with the prototype. Once the prototype is chosen, the team must identify who will be using the reports and who will be supplying the quality cost information so that their cooperation and support can be gained.

The next step is to define quality costs and to identify specific quality cost items. Quality cost concepts are new to most people. Therefore, the installation team must develop quality cost definitions that are easily understood and that clearly explain what types of costs are and are not included in the various categories. To identify specific quality cost items in each category, the team should ask for input from both the users and suppliers of quality cost information.

After identifying specific quality cost elements, the installation team should consider potential sources of data for each item. Some items to be included in the quality cost report may be available in the company's accounting system. Other elements may be more difficult to access and may require the collection of additional data, special allocation processes, or subjective estimates.

Finally, the quality cost installation team must devise a means of accumulating and disseminating the quality cost information. Often, the team may have to create special forms to accumulate raw data. The team must also design reporting formats, recognizing that one format may not be useful for all levels of management. Only through a careful analysis of management's needs from a quality cost reporting system can a useful management tool be devised — a tool that must become an integral element of the CIM architecture.

AVOIDANCE/FAILURE COSTS REVERSAL: AN ACTION PLAN

Andrew F. Grimm
Director, Quality Control
James G. Fox
Controller
C & F Stamping Company
Grand Rapids, Michigan

Traditional quality cost reporting models generally use the breakeven analysis approach to measure when the margin between the "dollar earned" equals the "dollar spent." This model works well when a firm institutes a quality cost reporting system to measure its historical quality cost base before implementing a comprehensive quality control program. In this scenario, companies usuallly find that the total failure cost area remains in an essentially level, high cost operational situation, and avoidance costs remain in a significantly level, low cost operational situation. Further, when observing the Appraisal Costs/Prevention Costs ratio within the avoidance costs area for a firm in this stage, Appraisal Costs are usually appreciably greater than the expenditures for prevention. This firm can logically use the breakeven analysis approach to managing quality through quality cost control. It can realize significant gains toward the breakeven point where one dollar spent on avoidance will equal one dollar of total failure cost reduction.

On the other hand, a dilemma arises for the firm that decides to install a comprehensive quality control program using all of the respected techniques such as SQC, SPC, Quality Function Deployment, Total Quality Control, and so on before determining the quality cost structure that results from such a program. At some point, this firm realizes that it must determine the economic effects and results of the program. Quality cost reporting is a method that can analyse the new program's impact on the business. The dilemma found in this scenario is that expenditures for avoidance far outweigh the corresponding expected reductions in Total Failure Costs. This dilemma was found in our company. We discovered that we were spending at a much higher level in the Avoidance Cost area than the level at which the failure costs were being generated. In fact, the total failure costs were very significantly lower than the avoidance costs which generated a significantly important negative margin.

The intent of this paper is to discuss the traditional quality cost reporting method, explore the negative margin dilemma as we discovered it functioning in our operations and to discuss a variety of strategies that may be employed to address and resolve this dilemma. These strategies should include the use and appreciation of the traditional quality cost reporting system as well as exploring new approaches to Quality Cost Accounting.

A Quality Cost System

A brief explanation of a Quality Cost Reporting System is necessary to prepare for the discussion about management strategy alternatives. There are two major cost areas and two subgroups for each of these major areas. Table 1 describes a typical however incomplete quality cost structure.

1.0 Avoidance Cost — Costs expended to achieve quality. Avoidance costs are composed of Prevention Costs and Appraisal Costs.
   A. *Prevention* — Costs expended for preventing poor quality from occurring in the first place.
      1. Quality System Management Team salaries, i.e., Q. C. Management.
      2. Training of Q. C. Management and in turn, training of the production and management personnel.
      3. SPC implementation.
   B. *Appraisal* — Expenditures for examining products and processes.
      1. Inspection and Inspectors wages.
      2. Testing costs.
      3. Calibration and Certification program.
      4. Quality maintenance of checking devices.
2.0 Failure Costs — Costs incurred for poor quality. Failure costs are composed of Internal and External Failure Costs.
   A. *Internal Costs* — Costs incurred for poor quality during production.
      1. Scrap.
      2. Rework or repair.
      3. Premium freight for replacement or behind schedule product.
      4. Downtime.
      5. Re-Test.
      6. Re-Inspect.
   B. *External Costs* — Costs incurred for poor quality after the customer accepts ownership of the product.
      1. Visits to customers (to investigate and explain product failures)
      2. Customer returned goods costs

Table 1

A table of this nature can be developed for each individual firm, or divisions within that firm, to describe the actual account elements found in the Quality Cost segment of the business. The source used to develop this table is your firm's Chart of Accounts.

To expand further on a traditional quality cost structure, a diagram can be constructed from Table 1 to explain how the quality cost categories are used to develop the breakeven analysis. The diagram (shown in Figure 1) demonstrates the classical structure of the quality cost model.

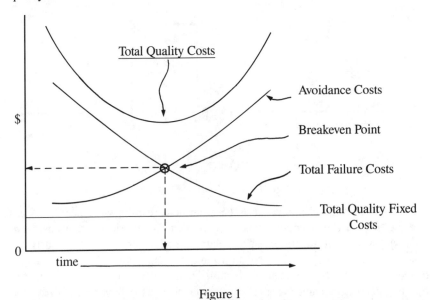

Figure 1

The dotted lines on the diagram indicate that at the rate of increase of expenditures in avoidance costs and the subsequent reduction in the total failure cost, a breakeven point will occur at the dollar level on the y-axis and at that time point on the x-axis. Theoretically, as the rate of increase or decrease of expenditures in the avoidance costs category is decided by management, the rate of total failure costs is affected in an indirectly proportional fashion. The breakeven point will change location on the diagram as the management decisions concerning expenditure level allocations are made. To illustrate this point, refer to Figure 2. Note that the location of the breakeven point shortened in time and lowered in dollars in figure 2(b) as compared to figure 2(a) when an apparent management decision was made to increase the rate of avoidance costs (AC). One word of caution concerning this change in the theoretical breakeven point location. As pointed out before, the Avoidance Cost area is composed of two categories, Prevention Costs and Appraisal Costs. If management decides to increase the Appraisal Cost and not the Prevention Cost expenditures, the usual result is to have little or no effect on the Total Failure Costs area rate of increase or decrease. Consider the situation where a company experiences high customer returned goods defect rates, an External Failure Cost Category account. If management decides to increase inspection on outgoing product (or for that matter also increase inspections at in process control points) there

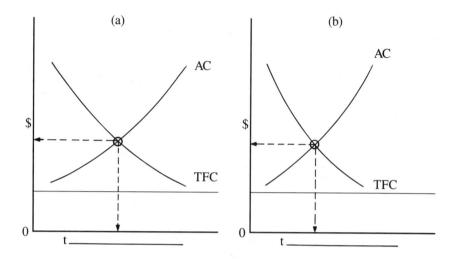

Figure 2

is an obvious increase in appraisal costs, hence an increase in the Avoidance Cost Area. The question must be asked, "Of what impact does an increase in appraisal costs have on solving a problem and preventing the recurrence of that problem?" Obviously, the answer is none, or hardly at all. What does transpire is that external failure costs might be reduced, but the cost of unquality is shifted to the Internal Failure Cost Category resulting in no appreciable overall effect on reductions in the Total Failure Cost Area other than some returned goods transportation costs savings that could show some slight reduction in the External Failure Cost Category.

The Cost Dilemma of a Strong QC System

Now to address the condition where a firm institutes a strong, comprehensive quality control program that is either required by its customers or by its own decision in order to protect its market share. In this scenario, it is common for these firms to initiate an all out control program prior to establishing a cost baseline, usually well in advance of preparing a quality cost reporting program. When the quality cost reporting program is finally addressed, more often than not, it is discovered that avoidance costs far exceed total failure costs. This situation presents a dilemma to management upon discovering that they are experiencing a negative marginal condition in their traditional quality cost breakeven analysis. Management must address the following questions:
- Should they continue at such a heavy quality control program cost rate?
- What should be done if an indicated quality control program cut back reduces the program below standards set by customer requirements?

Following is an example that actually occurred in our company that describes this scenario. Dollar amounts and timing have been changed, but the example remains relative. Table 2 shows the Quality Cost Report from which the Avoidance Cost/Failure Cost Reversal was discovered. Examination of the diagram in Table 2 shows the occurrence of the reversal shortly in the beginning of the 1st quarter, 1985. A "negative margin" condition occurred where it is costing our firm more money in terms of Avoidance Costs than a positive return on investment which would be the result of a reduction in failure costs. At the end of the first quarter, 1985, the negative margin is a return of $0.80 for every dollar spent.

QUALITY COST REPORT

	1984				1985
	1st Q	2nd Q	3rd Q	4th Q	1st Q
			(in '000$)		
Prevention					
QC Management	75	90	90	94	94
Training	5	10	10	10	10
	80	100	100	104	104
Appraisal					
Inspectors	24	48	52	52	57
Testing	13	13	13	13	13
Certification & Calibrations	9	11	11	11	11
	46	72	76	76	81
Total Avoidance Costs	126	172	176	180	185
Failure					
Scrap	70	63	57	50	40
Rework	90	92	83	77	65
Premium Frt.	30	28	20	15	5
Downtime	20	21	15	15	15
Retest/Reinspect	30	30	20	20	20
External					
Plant visits	5	1	10	5	3
Total Failure Costs	245	246	205	182	148

Table 2

QUALITY COST REPORT DIAGRAM

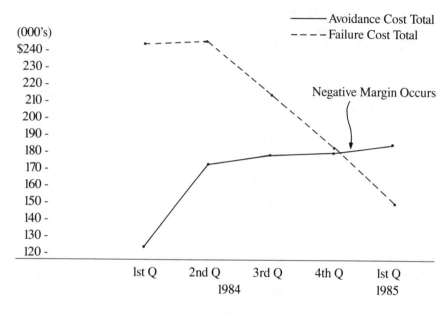

Figure 3

Management Strategies

There are a number of strategies that managements of firms faced with each scenario can consider for employment when faced with the negative margin condition in a traditional quality costs breakeven analysis. The management of the first scenario where the quality cost reporting program is installed prior to installing a comprehensive quality control program, faces the same dilemma as the firm in the second scenario when it discovers that the reduction in total failure costs is rapidly approaching the breakeven point. If nothing is done to apply the brakes to the program, these firms can easily slip into the negative margin situation. The following strategies suggest some approaches management may employ to resolve the potential or realized margin situation.

The "Quality Loss Function" Strategy

In recent years, and particularly in the automotive industry, interest has been gaining in the Taguchi, "Quality Loss Function." The "Loss Function" is defined by Dr. Genichi Taguchi as: the financial loss (of a part) with the functional specification through a quadratic relationship that comes from a Taylor Series expansion. A simpler definition might be: the degree of uniformity of "manufacturing" product quality around the design target rather than conformance to specification limits. In terms of the Loss Function, Clausing (Donald Clausing, PhD, Xerox Corporation) defines quality as:
"Quality Loss is the financial loss imparted
to society after a product is shipped."
With these concepts in mind, a powerful strategy that can be adopted by a company is to disregard the traditional quality cost approach and switch to the Loss Function approach.

The general structure of a Loss Function approach consists of analysing each product (or part) manufactured by the business using the Loss Function of each and then constructing a priority list of products. The products on the priority list may be analyzed using "Pareto Analysis" to determine the first group of projects where the variability about the part characteristic targets is reduced to achieve quality improvement. The return on individual project investment is determined by the summation of the elimination of losses at the next higher order assembly points in the production flow and eventually into the hands of the consumer. The Loss Function approach operates on the premise that "elimination of waste" more than offsets the cost of the initial project to reduce variability in the long run. This in turn contributes to productivity and quality improvements within the firm. Productivity and quality improvements lead to customer satisfaction and the development of a loyal customer base. These factors, when all taken in combination, lead to long term viability and profitability of the firm. The management focus is directed to improvements in quality of individual parts and products with financial rewards being realized after this attention to detail. This approach is reminiscent of the "old saying,". . ."Watch your pennies and the dollars will take care of themselves."

For those firms interested in a more detailed discussion of the "Loss Function" approach, the authors direct your attention to Sullivan's referenced paper.

Traditional Strategies for Firms Under Strong Market Pressures

The following strategies return to management's use of the traditional quality cost segment tool.

Another broad, but internalized approach to avert the dilemma condition is to immediately stop the increase in avoidance costs when it is noticed that the negative margin condition occurrence is imminent. This is an across the board hold on any further expenditure increases for both prevention and appraisal costs. A potential risk involved with this strategy is that the hold on expenditures could occur early in the quality system implementation plan. The result might be a partial plan completion status for some period of time until the quality cost picture can be sorted out by management. Management still faces the dilemma if the total system implementation must be reinstituted and the total failure cost picture remains at an improved level. If market pressures insist that management continue with their quality system implementation plan, a new strategy may need to be adopted. One of the other strategies discussed may have to be reviewed.

However, on the other hand, if the quality system implementation is almost or totally complete by the time the breakeven point is achieved, management will have found that the traditional approach to quality cost reporting was the most successful strategy for them to pursue. And, further, in light of the traditional approach, they may determine that the costs at the breakeven point are satisfactory in relation to business' expectations.

A similar strategy to the one just discussed is to stop any increases in the Appraisal Cost category of the Avoidance Costs area, but continue the increasing rate of expenditures in the Prevention Costs category segment of Avoidance Costs. This strategy is the usual practice recommended by the traditional quality cost model. The main reason is that prevention costs, at best, are found to be at a very low expenditure rate when first studies of a firm's quality cost structure are initiated. As an "investment," prevention costs will generate higher and quicker "rates of return on investment" when the resultant expectation of the business is to lower the Failure Cost Category's expenses. However, this will only be true if the firm has not exceeded the breakeven point and entered into the "negative margin" region of the quality cost diagram. Again, the "negative margin" dilemma appears if the breakeven point has been achieved.

The one, major difference between this strategy and the previous one discussed is the usually greater "return on investment" achieved when management accepts a "Prevention" mode of operation rather than retaining a strong "Appraisal" structure or the some what better, conservative "Prevention/Appraisal" balance of increasing expenditures. What is suggested in this strategy is not to keep a "Prevention/Appraisal" steady state ratio, but to abandon the ratio in favor of a rapidly increasing rate of expenditures for "Prevention" costs and holding or reducing "Appraisal" costs. The rapidly increasing "Prevention" costs strategy is to be retained for the short term until the quality system is fully implemented. Then, the "Prevention" cost category can be examined for reduction. An account in the "Prevention" cost category that would be effected by this strategy would be quality training. A rapid increase in quality training would be experienced for a short period of time as the work force is being trained. When the full training schedule has been completed, the quality training account can be greatly reduced to the point where training is given to new employees, and any further workforce quality training maintenance. Once these expenditures are completed and are shown to be effectively reducing failure costs, the breakeven point should be recalculated to see if the negative margin experience is turned around.

A Traditional Strategy for Firms Not Under Strong Market Pressures

A strategy to examine is one that must be addressed early in a quality system implementation program as suggested by first studies of the firm's quality cost structure. This strategy is very conservative and can be employed by those firms who are not yet under great pressures from world market conditions and competitors nor expect to be in the near future. However, these firms should also not be at product quality levels that seriously affect customer satisfaction. In other words, in terms of quality cost structuring, these firms should be enjoying low or no external failure costs, but whose internal failure cost problems are serious enough to lead to a strategy of improving quality and profitability.

The strategy suggested by this scenario is to plan slowly increasing expenditures in the Prevention Cost category, with any low level expenditures in the Appraisal Cost category necessitated by the Prevention Cost plan. An example of this type of situation is where a company might not have a laboratory for testing materials and products. The prevention cost in the Quality Engineering Planning account would call for test planning to develop an appropriate Inspection/Test Plan for laboratory testing of incoming materials and in process and/or final audits of products. Other Prevention Cost accounts that would be created and affected are the Laboratory Construction Expense and Laboratory Capital Funding accounts. Appraisal Cost accounts generated from a lab installation are: Laboratory Technician's Wages, Laboratory Supervision Salaries, Laboratory Materials Expense, Laboratory Perishable Tooling Expense and Laboratory Equipment Calibration and Maintenance Expense. Without market or competitive pressures, the timing of these expenditures in the Prevention Cost category can become very controlled and the phasing in of the Avoidance Cost category expenses can also be very controlled. Never the less, there would still be an overall increase in Avoidance Cost area expenditures.

A Strategy for Firms Under Strong Market Pressures

The final strategy parallels the first suggested strategy i.e., the "Quality Loss Function" concept, in that quality requirements are of the highest business priority and that, although short terms costs can be extremely high, long term results should more than offset the high initial program cost. Both strategies ignore the potentially high risk of occurrence of the negative margin generated by a traditional quality cost model using the breakeven analysis approach. A company placed in a position where it cannot afford to consider any other strategy than this one usually finds that its market or its customers within their markets impose quality system requirements that must be observed immediately in order to survive in those markets. This scenario is the one where the negative margin will rapidly surface in the quality cost breakeven analysis.

The only possible strategic controls that can be assumed by a firm placed in this critical situation are:
- If requirements are imposed by a customer:

  1. Determine through increased contacts with the customer the absolute parameters needed to meet the spirit of the requirements,
  2. Determine through a cooperative Advanced Quality Planning Program the timing expected by both parties for the final implementation of the program.

In this scenario, the customer is just as anxious for the success of its suppliers since its success and survival in its markets depends strongly on the success of its suppliers in meeting requirements generated from the demands of the markets!

- If requirements are imposed by direct market conditions:

In this scenario, markets may be populated by other firms, or by consumers directly. In this case the firm has more leeway in interpreting and addressing quality requirements generated by the markets. However, where direct customer requirements imposition is not evident, the firm must impose its own set of requirements upon itself and its suppliers in order to remain a viable presence in its markets. The leeway that it enjoys are in its interpretation of the market's product quality expectations. But, its timing for implementation may be as strong and restrictive as for company, described above, that has the requirements imposed by its customers.

Firms in this type of scenario must expect to go into a negative margin situation. This was the type of situation in which we found our company involved, as we have previously discussed. Our strategy is to completely install a quality control system that meets all of our customer's requirements as quickly as possible, regardless of being in the negative margin condition as indicated in figure 3. Once the system is in place, we will be establishing strategic business quality plans to maintain quality with the objective to return to a positive margin position on the breakeven analysis.

The Future of Quality System Costs

It is without a doubt that your company will be faced with serious decisions about the future of your firm due to worldwide economic market forces that exist today and appear to be getting more pervasive in the future. Your company will not be able to side step these issues and remain a viable business. Product offerings in U.S. markets today are from firms all over the world. Our U.S. markets are no longer passively protected because of a lack of industrialization in other nations. Some of our industries that no longer exist, or companies that are being currently pressured financially know well that many nations are positioned to make further incursions into U.S. markets. Further, American business is finding that competing in foreign markets is also becoming increasingly competitive. World economic factors and competitive posturing by more and more nations have made U.S. participation in these markets more difficult.

The purpose of this paper is to propose that our American companies seriously review the product and service quality and cost of quality aspects of their business as one of the prime defensive tools. Better yet, use these quality tools to create an aggressive marketing scheme to maintain, regain or make new inroads into U.S. markets and the World's markets.

Strategic Planning Recommendations

Regardless of which strategy discussed that best fits your company's needs, each should be preceded with an examination of the quality system financial data currently being generated by your company. A sound path to follow is to create the base quality cost reporting system from your company's Chart of Accounts. You can use the new ASQC publication, "Principles of Quality Costs" to help form the baseline Quality Cost Report. From this information, a Breakeven Analysis Diagram can be developed, singularly describing your company's quality cost experience and structure. When this diagram is studied and analyzed, you may understand the breakeven point for your firm in both the dollar levels and the achievement timing, the rates of increase (or decrease) in cost expenditures and savings, whether your firm is in a negative or positive margin situation, or how your company parabola (total cost curve from the Breakeven Diagram) stacks up in terms of the parabolic (quadratic) form of the Quality Loss Function. This analysis can then suggest the best strategy your firm believes best to meet your marketing objectives from a product quality and service standpoint.

A further recommendation is that with the firm's baseline Quality Cost Breakeven Diagram, new scenarios should be created from a series of "what if---" questions. Increasing rates of expenditure in the Prevention Cost category can be made to observe what rates of decreases can be experienced in total failure costs. The changes in breakeven point dollar levels and timing periods can lead to further refinements in the selected strategy.

Quality of your products and services is one of the most important tools your firm has to compete in today's and tomorrow's U.S. and World business environments. Strategic control of your firm's quality costs will help to control your product's and service's quality levels to assure your firm's profitability in competing with the product's and service's quality levels of the new worldwide competition.

Bibliography

Hagan, J., Editor In Chief, "Principles of Quality Costs," 1986 ASQC Quality Cost Technical Committee, Milwaukee, WI.

Sullivan, L. P., "Companywide Quality Control for Automotive Suppliers," June, 1985, Ford Motor Company, Body and Assembly Operations, Dearborn, MI.

DO CONTROLLER DEPARTMENTS MEASURE QUALITY COSTS?

Thomas N. Tyson
Assistant Professor Accounting
Clarkson University
Potsdam, New York

Abstract

Many organizations are now adopting quality improvement as their primary organizational objective. Effectively evaluating performance in quality improvement requires specific and regular measurement of activities underlying the quality area. The quality cost literature frequently and consistently calls for controller department participation in this measurement activity. Among other reasons, controller department participation lends validity, reliability, and credibility to quality cost numbers and generally increases the company-wide exposure of quality improvement programs.

This study investigated a number of issues surrounding controller department participation in quality cost measurement among the 1985 Fortune 500 industrial corporations. One hundred and twenty-five corporations were randomly selected from this population of firms. Corporate controller department personnel were contacted by telephone and asked a variety of questions relating to the quality area and the measurement of quality costs. This group of firms was selected for study primarily because many articles that describe successful quality cost programs are written by quality managers of major industrial corporations. Thus controller department personnel in these firms should be most knowledgeable about the purposes and benefits of quality cost measurement.

Telephone interviews were chosen as the mechanism for collecting data for a number of reasons. Since the study focuses on controller department participation among a large number of firms, the high costs of conducting face-to-face interviews necessarily limit the number of firms that can be contacted. Telephone interviews traditionally generate substantially higher response rates than mail questionnaires and lead to less biased results. This procedure also enabled the researcher to obtain insightful narrative comments from participants regarding the reasons measurement is or is not undertaken. The relatively high response rate attained in the present study and the request by over 90% of respondents to be furnished a summary of findings justifies the selection of this data collection technique.

Data was analyzed to identify statistically significant factors that may differentiate measuring from non-measuring corporate controller departments. Information was generated regarding the nature, extent, and scope of this measurement activity by corporate controller departments. Controller department personnel are also asked to express their beliefs about the relationship of quality to profitability. Results of this study should interest those individuals seeking greater controller department involvement in the quality area.

Introduction

The quality cost literature consistently calls for controller department participation in the measurement of quality costs. Controller department participation is sought for a variety of reasons. Alvin O. Gunneson writes that a quality cost measurement program must be perceived as a normal accounting function of the corporation.[1] John T. Hagan identifies three reasons for controller department participation: 1) the stamp of financial validity is provided, 2) the collection of costs remains within practical limits, and 3) the opportunity for teamwork between the controller and quality function is facilitated.[2] Daniel M. Lundvall discusses four other reasons why controller department participation is warranted: 1) accountants have extensive experience in reporting costs, 2) jurisdictional disputes between the controller and quality functions are minimized, 3) the credibility of quality cost numbers is increased, and 4) upper management looks to accounting for unbiased financial figures.[3]

Though the quality literature consistently encourages controller department participation, the accounting literature is generally silent on the matter. No standards exist regarding appropriate accounting treatment of quality costs and few references to this measurement activity appear in the popular accounting journals. This paper reports the results of a study that examines the nature and extent of quality cost measurement by the corporate controller's department of major industrial organizations.

Text

An extensive examination of the quality cost and controllership literature yielded a set of five factors that were expected to differentiate measuring from non-measuring controller departments. The five factors are: 1) the degree of top management's commitment to improving quality (COMMIT), 2) the level of available resources for conducting special studies like quality cost measurement (AVAIL), 3) the extent of participation in team projects that include quality department personnel (PARTIC), 4) the frequency of communications between controller department and quality department personnel (COMMUN), and 5) the degree of top management's reliance on formal cost reduction programs to improve profit (MGTREL).

Case studies that discuss successful quality cost programs involving the controller's department consistently mention the importance of strong COMMIT, frequent PARTIC and active COMMUN. The controllership literature also mentions the need for high AVAIL and strong MGTREL in order for nontraditional accounting activities, such as quality cost measurement, to be undertaken by the controller's department. The researcher believed that a regression model composed of these five factors could explain a controller department's participation in measuring quality costs. Survey participants were asked if quality costs are specifically and regularly measured by the corporate controller's department. Individuals responding that this measurement activity occurs were expected to report significantly higher levels for these five factors.

Advance letters were mailed to the corporate controllers of 125 randomly selected 1985 Fortune 500 industrial firms. Controller department personnel in 94 of these companies responded to a series of questions conducted in a telephone interview with the researcher. In 29 of these companies, corporate controller departments specifically and regularly measure quality costs, while in 65 they do not. Participants in this study were all members of corporate controller departments; therefore, the findings and conclusions presented in this paper relate only to this functional and hierarchical level in major industrial organizations.

The results of data analysis indicated that at the 0.05 level of significance, a model of these five factors can differentiate measuring from non-measuring controller departments. When each variable was tested individually for differentiating ability, responses to the levels of AVAIL, PARTIC, and COMMUN were each found to be statistically significant. In other words, corporate controller departments that specifically and regularly measure quality costs have higher AVAIL, greater PARTIC, and more frequent COMMUN that do non-measuring controller departments. Interestingly enough, the vast majority of all respondents describe their organization's top management as both strongly committed to improving quality and strongly reliant on formal cost reduction programs to improve profit. As a result, these two factors do not differentiate measuring from non-measuring departments.

Accountants have been widely accused of directing their measurements to matters relating to efficiency, such as budgets and schedules. Peter Drucker believes that accounting measurements should focus on activities more directly related to effectiveness, such as quality and productivity.[4] Specifically and regularly measuring quality costs appears to be an opportunity for the controller's department to become active in matters relating to effectiveness. However, if controller department personnel do not believe that quality improvement and profitability are highly correlated, this measurement activity may not be voluntarily undertaken.

In this study, respondents were requested to ordinally rank four items that may lead to higher profit. The four items are: 1) meeting cost budgets (MCB), 2) increasing employee productivity (IEP), 3) improving product quality (IPQ), and 4) meeting delivery schedules (MDS). Through a series of paired comparisons these four items were ranked in the following order of importance by both measuring and non-measuring groups: 1) IEP, 2) IPQ, 3) MDS, and 4) MCB. These results imply that controller department personnel in the surveyed firms are well aware of the relationship of quality and productivity improvement to profitability.

The quality cost literature identifies a number of factors that may lead controller departments to participate in quality cost measurement. No studies, however, report what controllers themselves believe are the most important factors. In the present study, controller department personnel were requested to ordinally rank four items on their ability to result in the specific measurement of quality costs. The four items are: 1) stronger leadership from quality management (LEAD), 2) greater quality commitment from top management (COM), 3) evidence of a stronger relationship between quality improvement and profit (REL), and 4) greater controller department resources available for quality cost measurement (RES). Respondents from both groups ranked these four items in the following order of importance: 1) COM, 2) REL, 3) LEAD, and 4) RES.

These rankings lead to several interesting conclusions. While the vast majority (82%) of all respondents view their organization's top management as strongly committed to improving quality, they also feel that quality cost measurement would most likely result from an even greater quality commitment. This apparent contradiction may imply that a strong commitment to quality improvement does not necessarily translate into corporate-wide quality cost measurement. Along this line, non-measuring respondents were asked to identify the reasons why these costs are not measured at the corporate level. The most frequently identified reasons are: 1) this measurement activity is an operating unit issue and/or violates the philosophy of decentralization, and 2) this measurement is not a top management priority and/or there is a lack of a strong top management commitment. In a number of cases, however, respondents from highly decentralized organizations acknowledge that these costs would be measured if so requested. In fact, in eight of 38 organizations described as having highly decentralized accounting functions, quality costs are specifically and regularly measured by the corporate controller's department.

That respondents ranked greater controller department resources for quality cost measurement the least important item is also surprising given that the level of available resources was found to be a significant differentiating factor in the regression model. A possible explanation is that a high level of resources available for special studies like quality cost measurement is a necessary but not sufficient condition for this activity to occur. That is, corporate controller departments must have the resources available to them for measurement to be undertaken; however, obtaining greater resources will not by itself result in quality cost measurement.

Survey participants were presented with four quality-related activities their organizations may have undertaken. The quality literature identifies these activities as demonstrating an organization's commitment to quality. The four activities are: 1) a formal policy statement pertaining to quality (POLSTMT), 2) a formal education program regarding quality cost concepts (EDUCPROG), 3) a formal company-wide program for quality improvement (QIPROG) and 4) a formal program of recognition for quality improvement (RECPROG). The researcher expected that each of these activities would be associated with a department's measuring status. Results indicated that organizations having EDUCPROG and QIPROG are statistically more likely to have corporate controller departments that specifically and regularly measure quality costs; however, firms with POLSTMT and RECPROG are not.

In a final question, respondents were asked if they feel that more time should be devoted to quality cost concepts in college and university curricula. When associating their responses with a department's measuring status, results show that the two variables are related. In other words, controller department personnel that represent measuring departments more frequently agree that more time should be devoted to quality cost concepts. In all, 62 of 94 respondents (66%) agreed.

Conclusion

This study has examined the extent and nature of corporate controller department participation in quality cost measurement. A number of issues were addressed and hopefully certain strategies are suggested for individuals seeking greater participation. Several future research projects are also suggested by this study's findings.

Perhaps the best line of related research is a study restricted to an industry group most likely to use quality cost information. A limitation of the present study is the confounding effect of industry and product type on this measurement activity. For example, in a number of commodity industries like chemicals and petroleum refining, quality cost measurement is essentially non-existent at the corporate controller department level. In the electronics, motor vehicles, and scientific equipment industries, however, this measurement activity is quite prevalent. An industry focused-study would be better able to isolate key variables and examine the relationship of quality cost measurement, quality improvement, and profitability.

Many controller department representatives were highly supportive of the need for quality cost measurements despite the fact that their departments were not yet participating in this activity. Several individuals mentioned that a corporate-wide measurement program was being evaluated or would shortly be instituted. Though the accounting community has been slow to accept the importance of quality improvement and its supportive measurements, results of this study indicate that changes in attitudes, if not actions, are occurring.

Footnotes

1. Alvin O. Gunneson, "How to Effectively Implement a Quality Cost System," *Effective Quality Cost Analysis System for Increased Profit and Productivity*, (1982) p. 2.
2. John T. Hagan, *Principles of Quality Costs*, Unpublished Manuscript, p. 16.
3. Daniel M. Lundvall, "Quality Costs," *Quality Control Handbook*, Third Edition, (1974) p. 5-21.
4. Peter Drucker, "Managing for Business Effectiveness," *Harvard Business Review*, (May-June 1963) p. 59.

Bibliography

Drucker, Peter, "Managing for Business Effectiveness," *Harvard Business Review*, Volume 41, May-June 1963 pp. 53-60.

Gunneson, Alvin O., "How to Effectively Implement a Quality Cost System," *Effective Quality Cost Analysis System for Increased Profit and Productivity*, Society of Automotive Engineers, Inc., Warrendale, PA, 1982 pp. 1-6.

Hagan, John T. *Principles of Quality Costs*, American Society for Quality Control, Unpublished Manuscript.

Lundvall, Daniel M. "Quality Costs," *Quality Control Handbook*, Third Edition, J. M. Juran editor, McGraw-Hill Book Company, New York, 1974, pp. 5-1 to 5-22.

A HIGH TECH SIGHT FOR COST TARGETS

John M. Ryan, PhD
Director of Quality
ACTON Computer Technology, Inc.
Goleta, California

Abstract

An inexpensive computerized quality cost analysis system which tracks and reports rework and scrap costs through out a repetitive manufacturing process is a definite asset to management. The system described is applicable to small companies (for under $25,000 including hardware and software) and for larger companies, may be easily duplicated for different parts of the process. Costs, defects, rework, scrap, etc. are displayed in different formats including SPC charts. The data base is easily established and is made additionally powerful through an enhanced statistical analysis package. The system helps to identify and analyze the payback potential of areas of the process which could benefit from tooling capital expenditures, improved input materials, enhanced training, or the addition or reduction of inspection and tests functions.

Text

When the corporate office wants to know where the profits are going, they are often asking what the costs of defects are, where they are located, and how they got there. They want to find the holes in the barrel of profit. We all know some of these holes are small enough to be ignored. Sometimes there are huge but well hidden holes, and some are very obvious yet so expensive to fix that the company has decided to live with them.

When it comes to sealing the fragile, yet essential, barrel of profit, there is no time to relax and kick back. The job is never done. As you read this article, new leaks are springing up.

Measuring this phenomenon is theoretically simple. Just constantly measure what goes in the barrel and what goes out, balance the numbers, make an adjustment or two to compensate for error and an acceptable cause or two, and you have yourself a number. Now maybe this number means a lot to you or me, but what about the corporate headquarters (or the guys upstairs) where your paycheck comes from. They're more interested in a different kind of number, one which is expressed as dollars.

Unfortunately, in reality, on the line, such numbers are difficult to obtain. There are too many work centers (barrels), too many defects (holes). Who is to find the holes in the barrels? Who will suggest the costly preventive measures required? Who is responsible to convince management of the need to invest in order to profit further, and in the future? That is your responsibility.

To do your job you need a tool that will provide you with the necessary and on-time data. That's where it starts.

One Alternative

At ACTON Computer Technology, Inc. we have stumbled our way onto something different. As a matter of fact, this "thing" may even help with the implementation of MIL-STD 9858A. We didn't design it for that. We designed it as a management tool. It works.

File Structure

If you look at your manufacturing line or fast food outlet as a series of operations and are able to estimate the percent of completion (percent of added value) added to the final product as it goes through the line, by the time the product gets to the end, it will be 100 percent complete. Suppose, however that a unit (or hamburger) needs some rework. If a standard good unit went through the line once and required 100 percent effort, a reworked unit might require 120 or 140 percent. The additional effort required by going back through once done operation is added to the total. More importantly, you would be 20 to 40 percent over the target cost established for manufacture of the burger or product. Now you know where the motto "Do it right the first time" came from. It's a lot cheaper.

If you took these "Percent of Completion" figures and put them into a computer file (See Figure 1). You would be on your way.

Many companies do not use percent of completion as means of measuring efficiencies, but use "Hour per Unit" (HPU's) instead. Suppose you know how many hours were expended per unit manufactured for each operation. If you were to take such a list and put it into an IBM AT or XT in place of the "Percent of Completion" file, using Lotus 1-2-3, you would have HPU instead.

Being a quality person, or at least one concerned about product quality, you probably already have a list of common defects which occur in each operation, and, you may even have used some method to "code" these defects for a computer. If no code is available, don't fret. You do not have to be a genius to set one up. Suppose such defects and their assigned codes become another computer file (Figure 2).

If you sat down for about one hour you could probably draw a flow chart of your entire manufacturing operation. By looking at this flow chart, you could take one defect at a time and determine which specific steps such a defect would have to go back through for rework or scrap purposes.

An example might help. Figure 3 shows a rework route due to a defect called "inductance failure" which was found at a final quality audit gate. In order to rework the part for this defect code, the part had to go back to the DM (Defective Material) unit, be torn down, go back through coil winding, through test, bonding, audit, route/crimp, tinning, pre-clean, batch clean, and into the final quality again. Quite a rework! The cost of the part was increased by almost double the labor cost involved.

After looking at each defect, and the route the defective would follow for rework or scrap, another file called a "Failure Routing File" could be set up for your computer. This might take some time because of the overall number of defect(ives) you might find in your plant. In order to keep complications to a minimum, only look at one defect code at a time, define its route and slowly but surely build your file (Figure 3).

There are a couple more bits of information your system needs. One is called the "burdened labor rate." The burdened labor rate is the average per hour cost of direct (sometimes it includes indirect) labor in your plant. Someone's got this figure, ask around. The other is the virgin cost of materials hitting your line. "Virgin" means that no one has messed with them yet.

Getting Data In

If you have in-line inspection gates, you are halfway home when it comes to getting data into the computer. What we call a "Lot Record Form" is developed which allows auditors or inspectors to record, in a simple format, the number of rejects per lot for each defect code audited at that particular inspection gate. There is a different form for each inspection gate; each forms calls out the main types of defects to be strangled at that point in the process. As inspectors do their job, they tally the number of defects of each type and record the tally on the form.

The form is collected and passed to data input people. If they put enough of this data in over a long enough period of time, a data base is established which gets mighty powerful.

Getting Reports Out

Now suppose you know enough about computer programming to get these files to work together to pump a report like Figure 4 out once a month, give you in-line process control charts, give you cost trends, establish a data base which can be researched with a good statistical package, provide your people with material traceability through lot control, and most importantly, convince those hard nosed (cheap appearing) managers above you that you can save money by doing things differently than their ancestors (the dinosaurs, and the company they worked for before for twenty-five thousand years) did? Would you like that?

If the prospect still excites you, looking at most summary reports (Figure 4) from this system you may notice that most problems are caught in the final quality check. This is where the costs are highest because rework routing requires that failures go all the way back to before the origin of the defect, the part must be torn down, and re-enter the build cycle. Now, it does not take a genius to figure that catching the defect close to the point of origin is cheaper. Maybe your manager is opposed to in-process inspection? Such inspection is indeed expensive, but is it more expensive than waiting until the part hits final inspection and fails (or do you ship it anyway due to delivery times your sales department promised?).

Based on the rate of occurrence of the defect and the cumulative costs across operations, a proposal can be written and cost justified which will show which approach is actually going to help the company become more profitable. "Business decisions" which turn out to be bull sh.. are exposed.

Let's take the example of the production or manufacturing manager who can see no value whatsoever in training new employees. But looking at our cost analysis, we can pick out those defect codes and costs which are attributable to workmanship error. Bingo! Why not write a little proposal projecting payback on training and learning curves. Hint: Many purchasing departments use learning curves to justify price reductions during negotiations. Maybe yours can help you.

Another favorite of mine is the purchasing department which buys only from the cheapest vendor (and uses prices to relentlessly beat vendors up). Why not look at the accumulated defect codes for a particular part number which are attributable to buy off's at receiving inspection. Track the costs through the entire line, add these costs to the purchase price, and you can end up with the "actual" price of a part. Note: The purchasing manager will no longer love you. Ho hum.

Let your creative juices flow. For those really creative, you can even move the engineering manager to take action on in-process problems which he could not previously justify.

Taking Preventive Action

Process control problems pop out like warts. They are unsightly, ugly, and not at all socially acceptable. The statistical charting capability of the system is exciting. The data base can be sorted by dollar amounts, part number, customer, process operation, defect code, lot number, or any of a variety of combination of variables. Control charts can be generated as you "swim upstream" from a summary report in your quest for trends which point directly to a cause.

At ACTON, we allow inspectors, operators, engineers, or anyone to "red flag" an operation. Performance teams from each area review their charts and may place a flag at the location of the probable cause of a defect.

Charts are also given to quality circles for analysis and action. Quality circles take only one priority at a time to analyze and attempt to solve. Financial rewards are given. The size of the reward is dependent on the cost savings calculated by the cost analysis computer. Some rewards are quite substantial, but the solutions are significantly more rewarding to the company than the rewards are a drain.

No one removes a red flag from a process until the problem is solved. Plant managers constantly direct their personnel to solve flagged problems in a manner designed to prevent reoccurrence rather than to give a quickie fix.

Summary

ACTON still has a long way to go. But we have a tool which is helping to put us out front of our competitors. Our ability to continue to reduce costs while simultaneously increasing our control over ever tightening industry quality requirements is a huge advantage.

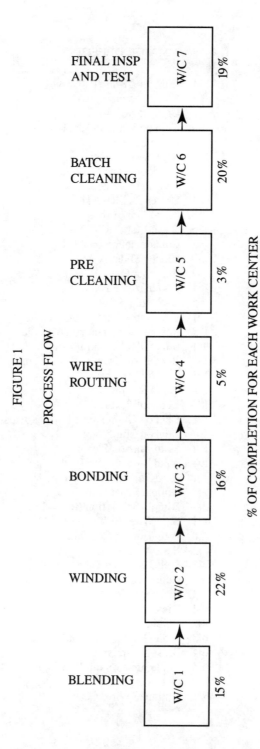

FIGURE 2

MAJOR DEFECTS

1	Chip on ABS L/E
4	Chip on Zone 1
8	Chip on Non-ABS
9	Chip on Winding Window
11	Pit on Zone 1
18	Breakaway on Zone 1
25	Crack on Zone 1
35	Scratch on Zone 1
41	Center Rail Recession
45	Metal Mark (Scrap)
46	Metal Mark (RWK)
52	Contamination on ABS
59	Broken Slider Nose
62	Improper Glass Bonding at W/W
65	Undersized ETW
66	Oversized ETW
151	Removable Contam on Flexure
153	Removable Contam on Mtg Blk
161	Burned Welds on Flexure
162	Burned Welds on L/Beam, Arm
168	Rust on Flexure
174	Bent Flexure
182	Dent on Flexure
186	Dent on Gram Angle Formed Area
187	Dent on Mtg Block
194	Solid Burr on Flexure
198	Solid Burr on Mtg Block
201	Burn Through on Flexure
218	Dimple Seperation (Scrap)
219	Dimple Seperation (Rework)
431	Bare Wire (Rewinding)
438	Poor Tinning Condition
442	Bare Wire (U-Coating)
546	Split/Damaged Tube
602	Collapsed Coil Layer
603	Pinched Wire
604	Kinked Wire
605	Lint in Coil Layer
613	White Particle in Wind'g Area
614	Broken Wire
615	Contamination on Wire
620	Cold Tin
621	Untinned Wire

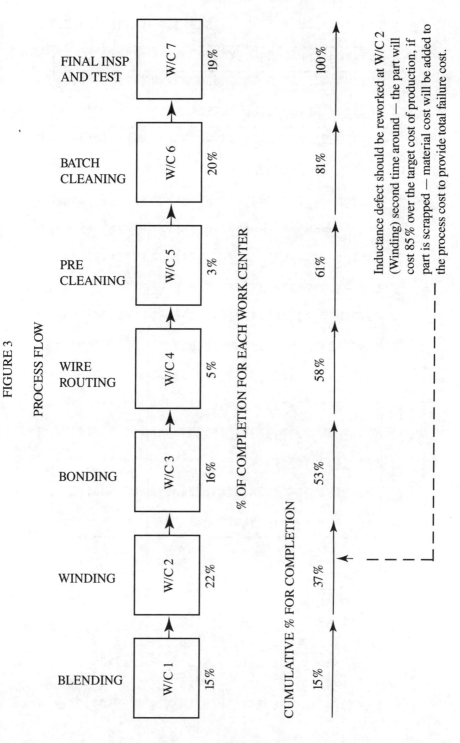

FIGURE 4

DATE: 9/10/86
SELECTION CRITERIA: MONTH-AUG.

******REJECT STATUS REPORT FOR ALL WORK CENTER******

	MAJOR DEFECTS	**BLENDING** Q'TY	**BLENDING** COST $	**WINDING** Q'TY	**WINDING** COST $	**BONDING** Q'TY	**BONDING** COST $	**R/CRIMPING** Q'TY	**R/CRIMPING** COST $	**P/CLEANING** Q'TY	**P/CLEANING** COST $	**B/CLEANING** Q'TY	**B/CLEANING** COST $	**FQC** Q'TY	**FQC** COST $	Q'TY	***TOTAL*** (%)	***TOTAL*** COST $	(%)
	MISCELLANY	1835	75.5	337	152.2	1599	239.9	1527	689.8	1431	518.5	25	2.6	935	94.3	7689	13.2	1772.8	12.2
901	HI-POT	0	0.0	6504	2406.5	0	0.0	0	0.0	0	0.0	0	0.0	96	34.2	6600	11.3	2440.7	16.8
52	CONTAMINATION ON ABS	0	0.0	0	0.0	0	0.0	0	0.0	64	1.9	121	10.9	5730	798.4	5915	10.2	811.2	5.6
46	METAL MARK (RWK)	0	0.0	0	0.0	0	0.0	0	0.0	1	0.0	0	0.0	5908	1063.4	5909	10.2	1063.4	7.3
605	LINT IN COIL LAYER	0	0.0	2848	89.2	132	75.2	1	0.0	333	196.0	2	0.2	147	32.3	3331	5.7	317.7	2.2
4	CHIP ON ZONE 1	23	3.0	869	451.9	172	4.7	67	43.6	646	471.6	9	0.1	1530	91.8	3276	5.6	1137.1	7.8
438	POOR TINNING CONDITION	0	0.0	0	0.0	0	0.0	2429	974.4	217	123.7	0	0.1	0	0.0	2818	4.8	1102.8	7.6
219	DIMPLE SEPERATION (REWORK)	0	0.0	0	0.0	178	66.7	49	3.4	785	47.1	10	1.6	1804	396.9	2648	4.6	449.0	3.1
431	BARE WIRE (REWINDING)	0	0.0	295	105.5	158	9.5	796	366.2	1080	626.4	1	0.6	197	122.1	2547	4.4	1287.5	8.9
218	DIMPLE SEPERATION (SCRAP)	1652	199.7	0	0.0	13	7.4	803	121.4	507	106.5	107	25.5	651	312.5	2226	3.8	575.4	4.0
59	BROKEN SLIDER NOSE	0	0.0	189	109.2	320	19.2	8	5.4	0	0.0	0	0.1	0	0.0	1863	3.2	312.7	2.2
174	BENT FLEXURE	0	0.0	0	0.0	1	0.0	975	138.1	263	55.2	3	0.7	104	49.9	1665	2.9	263.1	1.8
442	BARE WIRE (U-COATING)	0	0.0	1171	35.1	84	0.0	0	0.0	5	0.5	0	0.0	330	52.8	1508	2.6	88.5	0.6
602	COLLAPSED COIL LAYER	0	0.0	1144	34.3	2	2.5	117	5.5	82	7.7	0	0.0	41	7.7	1468	2.5	57.8	0.4
25	CRACK ON ZONE 1	70	9.1	108	56.2	145	1.2	0	0.0	0	0.7	0	0.0	1250	79.2	1431	2.5	146.3	1.0
603	PINCHED WIRE	0	0.0	26	9.5	82	65.3	301	138.5	243	140.2	0	0.6	489	291.0	1204	2.1	644.5	4.4
614	BROKEN WIRE	0	0.0	61	23.8	0	31.2	678	311.9	42	24.4	1	0.0	269	158.7	1133	1.9	550.5	3.8
62	IMPROPER GLASS BONDING AT W/W	803	104.4	0	0.0	0	0.0	0	0.0	0	0.0	0	0.0	0	0.0	803	1.4	104.4	0.7
8	CHIP ON NON-ABS	65	8.5	30	15.6	354	201.8	53	34.5	86	62.8	0	0.0	202	169.7	790	1.4	492.7	3.4
1	CHIP ON ABS L/E	89	11.6	127	66.0	22	12.5	17	10.5	355	259.2	7	0.1	81	4.9	698	1.2	364.8	2.5
707	TILTED SLIDER	0	0.0	0	0.0	367	25.7	8	3.7	0	0.0	0	0.0	0	0.0	375	0.6	29.4	0.2
35	SCRATCH ON ZONE 1	121	15.7	98	51.0	2	1.2	0	0.5	2	1.5	0	0.0	122	8.0	345	0.6	77.3	0.5
702	EPOXY ON FLEX	0	0.0	0	0.0	129	4.0	23	0.0	42	2.1	0	0.0	29	5.2	225	0.4	11.9	0.1
717	UNDERSIZED X-DIMENSION	0	0.0	0	0.0	188	13.2	0	0.0	0	0.0	0	0.0	0	0.4	188	0.3	13.2	0.1
709	INCOMPLETE TUBE CRIMPING	0	0.0	0	0.0	0	0.0	12	0.0	28	0.0	2	0.0	101	0.0	143	0.2	0.0	0.0
623	UNDERSIZED TOTAL WIRE LENGTH	0	0.0	3	1.1	3	0.0	128	58.9	3	1.7	0	0.0	2	0.8	139	0.2	62.4	0.4
620	COLD TIN	0	0.0	0	0.0	0	0.0	0	0.0	130	75.4	0	0.0	7	1.9	137	0.2	77.3	0.5
902	LOW INDUCTANCE	0	0.0	136	51.3	0	0.0	0	0.0	0	1.3	79	7.1	14	0.0	136	0.2	51.3	0.4
151	REMOVABLE CONTAM ON FLEXURE	0	0.0	5	1.8	2	0.1	8	0.5	43	5.5	0	0.0	12	2.2	136	0.2	10.6	0.1
604	KINKED WIRE	0	0.0	0	0.0	0	0.0	0	0.0	97	0.0	0	0.0	0	1.8	124	0.2	9.6	0.1
65	UNDERSIZED E.TW.	1	0.1	0	0.0	0	0.0	0	0.0	0	0.0	0	0.0	0	0.1	1	0.0	0.1	0.0
162	BURNED WELDS ON L/BEAM. ARM	0	0.0	0	0.0	0	0.0	0	0.0	0	0.0	0	0.0	1	0.5	1	0.0	0.5	0.0
41	CENTER RAIL RECESSION	0	0.0	0	0.0	1	0.0	0	0.0	0	0.0	0	0.0	0	0.0	0	0.0	0.0	0.0
727	LOW F/HEIGHT	0	0.0	0	0.0	0	0.0	0	0.0	0	0.0	0	0.0	0	0.0	0	0.0	0.0	0.0
194	SOLID BURR ON FLEXURE	0	0.0	0	0.0	0	0.0	0	0.0	0	0.0	0	0.0	0	0.0	0	0.0	0.0	0.0
726	HIGH F/HEIGHT	0	0.0	0	0.0	0	0.0	0	0.0	0	0.0	0	0.0	0	0.0	0	0.0	0.0	0.0
66	OVERSIZED E.TW	0	0.0	0	0.0	0	0.0	0	0.0	0	0.0	0	0.0	0	0.0	0	0.0	0.0	0.0
737	FLAKY BONDING EPOXY	0	0.0	0	0.0	0	0.0	0	0.0	0	0.0	0	0.0	0	0.0	0	0.0	0.0	0.0
714	UNDERSIZED LEAD WIRE LENGTH	0	0.0	0	0.0	0	0.0	0	0.0	0	0.0	0	0.0	0	0.0	0	0.0	0.0	0.0
198	SOLID BURR ON MTG BLOCK	0	0.0	0	0.0	0	0.0	0	0.0	0	0.0	0	0.0	0	0.0	0	0.0	0.0	0.0
186	DENT ON GRAM ANGLE FORMED AREA	0	0.0	0	0.0	0	0.0	0	0.0	0	0.0	0	0.0	0	0.0	0	0.0	0.0	0.0
187	DENT ON MTG BLOCK	0	0.0	0	0.0	0	0.0	0	0.0	0	0.0	0	0.0	0	0.0	0	0.0	0.0	0.0
	TOTAL Q'TY	4665		14034		4836		8035		6765		370		20269		58174	100		
	TOTAL COST ($)		428		3685		786		2912		2826		50		3838			14526	100

434

QUALITY COSTS IN STAFFS — MICRO, MACRO, OR BOTH?

William O. Winchell
Quality Consultant
General Motors Corporation
Warren, Michigan
Caroline J. Bolton
Cost of Quality Analyst
General Motors Corporation
Lansing, Michigan

Abstract

The application of quality cost has changed dramatically over the last few years. Traditionally, quality cost was used to track progress in making products — those items that were shipped from factories to customers. The costs could identify the activities that comprised the quality system for those products. Much prevention effort could be found in product engineering. An extensive investment in inspection could typically be found on the factory floor along with many dollars spent unwisely on rework and scrap. The service department was deeply involved in failure costs — fixing things causing customer inconvenience.

Now quality cost is being applied in staff departments. A staff department, or any activity within a department, can be assumed, in this new "micro" application for quality cost, to be producing a product or service for another department. The producing department or activity has its own internal failure cost for not "doing the job right the first time." If the department receiving the product or service does not find it suitable, then external failure costs for the providing department may result. In addition there are most likely inspection and prevention costs in the providing department.

The quality costs of many staff departments, using the traditional "macro" approach could be almost totally classified as prevention. But when using the "micro" approach, these same departments now have the full spectrum of quality costs — prevention, appraisal, internal failure and external failure. To truly reap the full benefits of quality cost, both approaches must be used.

Introduction

The focus of quality cost since its inception was on the costs involved in making a product or providing a service by a company to a customer external to that company. Within the last several years, an alternate focus for quality cost has developed. This new focus is tied to the awareness that each staff department or activity within a company also provides a product or service to a customer. In this case, however, the customer is internal to the company. For the purpose of this paper, the traditional approach will be called the "macro" approach. The alternate approach will be designated the "micro" approach.

What becomes apparent with the micro approach is that a formalized quality system for producing a product or service in a staff department is normally missing. This approach provides the framework for developing that quality system.

Macro Quality Cost

The traditional or macro approach to quality cost is detailed in the newly published "Principles of Quality Costs." In this publication the definitions of the elements of cost are intended to document the quality system in place for producing the product or service. The quality system starts in marketing which must identify the customer or user needs. Development or product engineering translates these needs into a tangible product or service and performs tests to validate that the needs of the customer or user will actually be met. The quality tasks of both marketing and development or product engineering are classified as prevention effort because they are intended to prevent problems from occurring in the hands of the customer.

With the switching of the inspection responsibility to manufacturing, the quality department is no longer a "traffic cop." It can devote its effort to support of the quality system — assist each department in performing appropriate quality tasks. In this new mode of operation, the efforts of the quality department could be classified as almost entirely prevention.

The manufacturing or operations department is responsible for conforming to the design of the product or service. It too has prevention effort. It also has appraisal costs to check whether the product or service is conforming to the design. If some products or services do not, they may have to be reworked or even scrapped before reaching their customer or user, resulting in internal failure costs.

The service department, for the most part, fixes the problems that the customer or user experiences. This is classified as external failure cost.

When the quality efforts of all the departments comprising the quality system are summed, we find that prevention is typically the smallest expenditure. Appraisal and failure costs are much larger. In many studies, the size of the external failure cost indicates that the needs of the customers or users are not being met very well and field fixes are all too common.

A common reaction to this situation is to increase prevention effort. An arbitrary, unfocused increase in prevention effort, however, may only add cost for the company since it may not address the true causes failures. It has been found, rather, that significant improvements have resulted by merely redirecting the existing prevention effort. The micro cost study approach provides insight regarding how to do this.

Micro Quality Cost

Plans for redirecting prevention efforts must be based on an analysis of the effectiveness and efficiency of current activities. Effectiveness means how well something is producing a desired effect. For a quality system, the desired effect is an improvement in the product or service. Efficiency means being productive without waste. For a quality system, this means eliminating the failure costs and minimizing the need for appraisal.

Conducting a micro quality cost study, is, in essence, the act of initially defining an appropriate quality system for the staff department involved, then comparing the effort of various elements in that quality system in order to make improvements.

Key to the usefulness of the micro study is how it is set up. First, it must be recognized that a viable staff department is essential to assure that the products or services meet the needs of the customers. Similar to the company quality system, the generic staff department quality system must include the following elements:

Prevention: effort within the department to determine the requirements and needs of the customer and to plan procedures which assure those requirements and needs are met.
Appraisal: effort within the department to check whether the requirements and needs are met before providing to the customer.
Internal Failure: rework within the department before providing to the customer after checking whether requirements and needs are met. This occurs because the job is not done right the first time.
External Failure: rework by the department after providing to the customer because requirements and needs were not met.

Here, then, is the first major difference between the macro and micro approaches to quality cost. Staff departments which may be classified as prevention under the macro approach, such as a modern quality department, are classified entirely differently under the micro approach. Under the micro approach, a staff department has the full spectrum of quality costs — prevention, appraisal, internal failure and external failure.

The second difference lies in the nature of the activities which make up the four elements above. Instead of traditional definitions, unique elements are defined by each activity in the micro approach. For example, if the activity being studied produces reports, a prevention effort may be to define requirements for the content of the report. If the activity is the repair garage, however, an appropriate prevention effort may be diagnosis of problems before processing.

Initial efforts for the micro quality cost study should involve departmental employees in defining the products or services of the staff department, who the customers of the department are and the activities which make up its quality cost elements. In this way the persons in the staff department "buy into" the study and its results up front, and also become more aware of the need for their own quality system. Suggested steps for accomplishing this initial part of the study are:

1. Define and list the products or services produced.
2. Define and list the customers to whom each product or service is provided.
3. Classify customers in the relative position they would be on organization chart as either:
   vertical — typically management
   horizontal (internal) — typically other activities within the department
   horizontal (external) — typically activities outside the department which are direct links to the company's end product
4. Identify typical external failure costs for each product or service that would be incurred dealing with particular customers.
5. Identify the types of prevention effort, like identification of customer requirements, that would avoid producing failures.
6. Identify the types of appraisal effort or checking that would avoid delivering products or services to customers that did not meet requirements or needs. Appraisal also indicates whether prevention effort is effective.
7. Identify the types of internal failures that would be found as a result of checking the products or services before delivery to the customers.

Completion of these first steps will result in a defined quality system for the staff department. The next step is to measure the amount of departmental resources spent on each of the elements above. This is achieved in employee interviews by collecting their estimates of:

1. their time spent on the quality tasks identified.
2. the distribution of the remaining departmental resources over these same quality tasks.

It should be noted that time and resources spent on quality tasks represent only part of the total effort expended on the product or service provided. But completion of the cost study and subsequent improvements may also result in a reduction in this area.

Effectiveness and Efficiency

As stated earlier, plans for redirecting prevention efforts within a staff department must be based on an analysis of the effectiveness and efficiency of current activities. The first part of the analysis of the micro quality cost study addresses effectiveness. A staff department that has more effort for vertical customers, which is management, than for the horizontal external customers, which are more likely a direct link to the company's end product, may very well not be as effective as desirable. Also a staff department that has more effort for internal horizontal customers who are within the same department, than the external horizontal customers may likewise not be as effective as desirable. The most desirable relationship for maximum effectiveness may be to have the greatest effort expended for those customers directly linked to the company's end product — the horizontal external customers.

The second part of the cost study analysis addresses efficiency. A staff department that has an efficient quality system has prevention effort to identify the requirements and needs of the customer. A small amount of appraisal to check whether the requirements and needs are met may also be necessary. Internal failure costs, however, are near zero, inferring that the jobs are done right the first time. In addition, external failure costs are near zero, indicating that the customer needs and requirements have been met.

Efficiency, then, can be evaluated by looking at the way quality costs for each product or service are distributed over the four elements. Let us look at some other typical cost distributions. The distributions presented here do not represent all of the possibilities.

A cost distribution where all four elements of quality cost are near zero indicates that no quality system really exists for the product or service. Rather than conclude that a quality system is really not needed, a subsequent check may very well confirm that the product or service does not mean a great deal to the customer and no complaints are made which generate external failure costs.

Another characteristic cost distribution for an activity is where all elements of quality cost are near zero except for external failure. This indicates the lack of a quality system in an activity furnishing customers with a desired product that does not meet their needs and requirements.

The lack of a fully developed quality system is indicated by a cost distribution for an activity having high prevention, appraisal and internal failure costs, but low external failure costs. This indicates a satisfied customer; however, the higher internal failure costs could indicate that more extensive needs and requirements are perceived by the activity than desired by the customer. It may also indicate that training of those within the activity is needed in order to do the job right the first time.

Opportunities for improvement should now be apparent. Ineffective and inefficient activities can be identified and improved by tailoring efforts to meet the needs of the customers — thus reducing failures. As in the macro approach, improvement efforts can be tracked and evaluated through periodic repetitions of the cost study.

Summary

The micro quality cost approach is valuable for determining and improving the effectiveness and efficiency of staff departments' activities. Use of the micro quality cost study approach also brings a greater awareness of the need to reduce waste to those in staff departments involved in the study. This approach demonstrates the need for a quality system tailored to each staff department, and conduct of the cost study in fact results in the definition of an appropriate quality system. The micro quality cost approach extends the never ending journey for quality improvement to the processes found in staff departments, supplementing the improvements tied to the end product produced by the company.

The traditional macro quality cost approach is still needed to document the entire company quality system and identify needed improvements that can be incorporated on a synergistic basis among departments. The micro quality cost approach for evaluating the quality process within staff departments is an essential supplement.

QUALITY COSTS — NEW CONCEPTS AND METHODS

Edgar W. Dawes
Quality Manager
Haydon Incorporated
Waterbury, Connecticut

Abstract

Optimized by application and proven over time, new quality cost concepts and methods are adding profits to a variety of American companies. This paper describes some of the concepts which have emerged and some of the methods which have been successful. First, the concepts:

1.) Planning for quality — The quality system and quality costs must reflect customer satisfaction issues and marketplace forces. An illustration of quality cost-market share relationships is developed in the paper.
2.) Improving Quality Control Operations — Effective quality control can exist only by treating suppliers-producers-customers as an entity. The paper compares quality costs for an entity vs. non-entity approach and details some cost considerations related to process capability and SPC techniques.
3.) Improving Quality Costs — The primary goal of the quality cost portion of the quality system must be quality cost improvement. This occurs through project by project identification of failures and corrective action. A review of this process is contained in the text.
4.) Perfection Is Possible — The organization must accept the concept that "Perfection is Possible." This attitude reflects not only a management philosophy, but an organization's style and skill in implementation. It results in a program of continuous improvement which does not necessarily increase costs as quality levels approach perfection. We will discuss the relationship of resource costs (prevention and appraisal) to failure costs and the achievement of improved quality at lower quality costs.

Planning for Quality

Quality can be defined as "conformance to customer expectations where the intended use is determined by the customer." In this context, we need to assure that 1) design quality fully achieves customer expectations (the planning issue), and 2) manufactured

FIGURE 1
CASE STUDY
PROFIT AND LOSS (P & L) STATEMENT
(ANNUAL $ IN 1,000s)

P & L Category	Included in the P & L category are the following quality costs	Customer Expectations Not Met	Customer Expectations Met
Sales		1000	1200
Selling Costs	External Failure	300	170
General & Administrative Costs		120	120
Fixed & Miscellaneous Expenses		160	160
Indirect Labor	Prevention, Appraisal, Internal Failure	130	78
Indirect Materials		10	12
Direct Labor	Internal Failure	110	126
Direct Materials		40	46
Income Tax		30	137
Net Income (After Taxes)		100	351
% Net Profit (Before Taxes)		13.0	40.7

product totally conforms to design quality needs (the control issue). Failure to do so results in the proverbial external failure quality costs which can include[1]:
- Complaint Investigations/Customer Service
- Returned Goods
- Retrofit Costs
- Recall Costs
- Warranty Costs
- Liability Costs
- Penalties
- Customer/User Goodwill
- Lost Sales

In today's world competitive quality race, failure to adequately plan for design quality and manufacturing conformance has dire consequences. An example of one company's experience follows.

The company marketed several consumer products with aggregate sales of $16 million per year. One product, introduced four years ago, was experiencing high customer complaints and loss of market share (see Figures 1 and 2). Management decided to have the marketing and quality functions test competitive offerings and compare them to the market guide, engineering specifications, and actual performance of the company's product. As might be expected, the study revealed a significant performance improve-

FIGURE 2
CASE STUDY
ANNUAL QUALITY COSTS ($1,000s)

P & L Category	Quality Cost		Customer Expectations	
	Category	Element	Not Met	Met
Selling Costs	External Failure	Complaints, Warranty, Lost Sales	230	100
Indirect Labor	Prevention	Quality Engineering	30	15
	Appraisal	Inspect, Test	39	13
	Internal Failure	Reinspect, Retest	13	2
Indirect Material	Internal Failure	Scrap	2	1
Direct Labor	Internal Failure	Rework	20	10
Direct Material	Internal Failure	Scrap	3	2
Total Quality Costs			337	143

ment for the competitive offerings. A redesign was undertaken and proved successful.

As a direct result of the study, the company came to two conclusions on how to improve it's competitive position:
- Future design quality planning would include a review of competitive offerings. In this case, customer expectations had been conditioned by actual superior performance by its competitors. (It should be noted that a growing number of companies will not market a new product unless they can assure superior performance over competitors.)
- Manufacturing conformance to design requirements must be assured. In this case, not only was there a substantial number of customer complaints but the product was difficult to produce and manufacturing operations were incurring an unacceptable amount of internal failure costs.

For this company, the quality cost and profit improvement associated with meeting customer expectations was significant (Figures 1-2-3):
- Sales increased 20%
- Total quality costs were reduced by 57% ($194,000)
- Net income increased 251% ($251,000)
- Net profit increased 26.7%

FIGURE 3
CUSTOMER QUALITY EXPECTATIONS

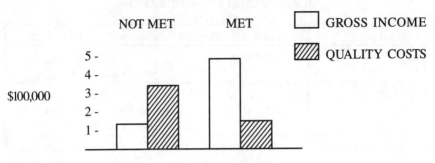

Finally, the structuring of most profit and loss (P & L) statements where quality costs are an integral part of P & L categories (such as selling costs and direct-indirect labor and materials) provides a powerful tool for the quality professional. Constant emphasis should be placed on the profit contribution to be made through quality cost improvement. Discussions with management should concentrate on "net profit increase," "equivalency per share," and other meaningful profit measures. In this case, a substantial portion of the increased net income and profit improvement was the result of quality cost improvement.

Improving Quality Control Operations

The need for higher efficiency at lower costs is a compelling force for all companies. Computer integrated manufacturing, flexible manufacturing systems, just-in-time, and ship-to-stock systems are just a few of the responses. In turn, these responses require the attainment and maintenance of high quality levels. For companies using these techniques, their supplier quality levels are an important key to success.

This common need for high quality (both suppliers-users)[2] invariably results in discussions concerning quality costs. Figure 4 details two possible outcomes.

One consultant has made the observation that, ten years from now, most of the suppliers and users that follow the traditional approach will be out of business.

In the course of satisfying the actual requirements, today's quality systems are employing increasingly effective statistical tools, including the coupling of economics to the statistical results. For the quality cost practitioner, this is vital.

Taguchi's quadratic loss equation (Figure 4) and its relationship to both specification and process capability positioning are extremely important. We now have a tool to evaluate incremental dollar loss probabilities in relation to departure from target mean for a quality characteristic (Figure 5).[3]

FIGURE 4
SUPPLIER-USER
QUALITY COST CONSIDERATIONS

Factor	Traditional Approach	Actual Requirement
Cost/Piece	Lowest Bidder	Cost/Piece including the user's quality costs.
Quality Level	What is the user's A.Q.L.? (Acceptable Quality Level)	The supplier's products must totally conform to the user's requirements.
Supplier Selection	The user employs less than complete evaluations of the supplier's quality capability.	The supplier's process must be totally qualified and capable of producing to the user's requirements.

Improving Quality Costs

The primary goal of the quality cost portion of the quality system must be quality cost improvement. There are certain conditions that must exist for continuous improvement to become a reality. These are:
- A belief that improvement is needed.
- An understanding of the company's goals, objectives and plans.
- A measurement system which shows the extent and causes of problems. In the case of quality failure costs, this equates to measuring items such as scrap, rework and complaint losses and ranking them in the well known Pareto maldistribution.
- Rewards to the individual based on performance as measured against goals.
- More education and training.
- Close and continuous interface with the customer or user.
- A program with suppliers that ensures their participation in the process of continuous improvement.
- Finally, a strong upper management mandate that continuous improvement must be employed.

These factors must be "orchestrated" so that a "fine tune" results. To improve company performance, management must perceive there is a need to do so and then assure support for the process. In the process, all individuals in the enterprise should become involved and feel "ownership" through pride and recognition. The process requires communication, education and training and both suppliers and customers benefit through participation, understanding and active support.

FIGURE 4
THE QUADRATIC LOSS FUNCTION

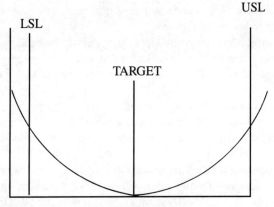

$L=k(Y-T)^2$ L = Loss In Dollars
k = Cost Coefficient
Y = Value of Quality Characteristic
T = Target Value

FIGURE 5
GAUSSIAN NORMAL CURVE

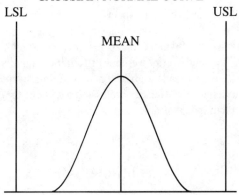

Perfection Is Possible

An increasing number of companies, both in the United States and overseas, are reporting success in improving "Bottom Line Profits" through the use of effective Quality Cost Systems. Spurred on by the need to achieve success in the marketplace, and realizing the strategic advantage of Quality as a weapon in that fight, smart managers are using cost improvement and cost avoidance tools afforded by Quality Cost Systems to improve financial performance.

As companies report their success stories, a pattern of quality cost improvement is emerging which indicates the following:
- Project by project improvement opportunities must first be identified.
- Trained teams must be assigned and must feel "ownership" for these cost improvement projects. They must then be supported in their progress toward solutions.
- In successful companies there appears to be no end to the identification and solution of quality cost problems. In other words, attainment of ever increasing quality at lower quality costs appears to be not only desirable but feasible (Figure 6).

These successes have resulted in revisions to the classic model of optimum quality costs. Previously, prevention and appraisal costs were portrayed as rising asymptotically as defect free levels were achieved (Figure 7). Now, it is realized that the processes of improvement and new loss prevention are, in themselves, subject to increasing cost effectiveness. We can, indeed, achieve total conformance without a disproportionately high expenditure of resources.

Summary

To achieve fulfillment of customer expectations, a company must first assure that design quality and manufacturing conformance respond to that expectation. Quality cost techniques offer a powerful tool to measure that response and, when the response is found inadequate, to identify improvements. In addition, quality cost improvement is now recognized as a significant contributor to profit improvement.

In the pursuit of excellence, the quality professional will find the quality cost portion of the quality system an indispensable help.

Footnotes

[1] Note the presence of loss of customer goodwill and loss of sales. These losses should be estimated in establishing a company's marketplace strategy and in calculating quality costs.

[2] In this context, the user is the manufacturer. In the previously cited case study, the user was the end customer of the product. In a broader sense, we are all consumers of a wide variety of products and services with our own expectations of perfection.

[3] Performance departures from a target mean must be individually evaluated. The loss function may not be quadratic and Taguchi's experimental methods determine that.

FIGURE 6
VARIOUS COMPANIES EXPERIENCE
QUALITY COST IMPROVEMENT

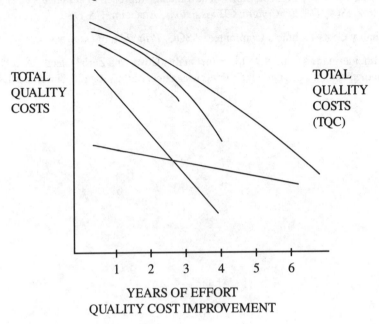

YEARS OF EFFORT
QUALITY COST IMPROVEMENT

FIGURE 7
OLD AND NEW CONCEPTS OF TOTAL QUALITY COSTS

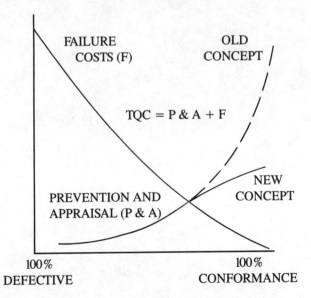

Bibliography

1. Bryne, Diane M; Taguchi, Shin, "The Taguchi Approach to Parameter Design," Page 168, *ASQC Quality Congress Transactions,* Anaheim, CA 1986.

2. Quality Costs Technical Committee, ASQC, *Principles of Quality Costs,* 1986.

3. Schneiderman, Arthur N. "Optimum Quality Costs And Zero Defects" Are They Contradictory Concepts. *Quality Progress* Magazine, ASQC, Page 28, November 1986.

QUALITY COSTS: PREVENTION THE TOOL FOR REDUCTION

Frank J. Corcoran
Singer Kearfott Guidance and Navigation Division

Prevention is the heart of a Quality Cost System. Our purpose is to analyze this category of Quality Costs stressing its importance, describe prevention costs and relate these costs to Quality Improvement and cost reduction.

If Quality Improvement is the Goal of a Quality Cost System, what is the strategy for achieving this? From the ASQC publication "Principles of Quality Costs" we find this strategy is quite simple:
1. Take direct attack on failure costs to try to drive them to zero.
2. Invest in the "right" prevention activities to bring about improvement.
3. Reduce appraisal costs according to results achieved.
4. Continuously evaluate and redirect prevention effort to gain further improvement.

Immediately it is obvious that prevention is a major thrust of this strategy. We know that the premise for this strategy is:
1. For each failure there is a root cause.
2. Causes are preventable.
3. Prevention is always cheaper.

However, determining the root cause can be extremely difficult. One of our major faults is not completing the analysis and/or stopping short at what might be considered the apparent cause. A definition and explanation of root cause follows:

> The root cause can be defined as the real cause of a problem. This is quite different than the apparent cause which appears after a superficial investigation. A frequently asked question is how can a person know when the root cause of the problem has been found. The root cause has been found when the problem can be turned on or off by adding or removing the cause.
>
> Once the root cause has been found, an irreversible corrective action must be implemented so that there is no foreseeable situation in which the root cause can return. Adhering to this practice ensures permanent improvement.

We must therefore include the "right" prevention activities in our Quality Cost System and evaluate how much we are expending on them. Trend analysis will measure our Quality Improvement and analysis of the Quality Cost Categories can direct us to adjustments to obtain the optimum situation. Standard Operating Procedure for a Quality Cost Program with emphasis on prevention is given in the following paragraphs.

The Quality Cost Categories are:

Prevention Costs: Are the costs of all activities specifically designed to prevent non-conformance in deliverable products or service. Details will follow later in this paper.

Appraisal Costs: Are the costs associated with measuring, evaluating or auditing products or services to assure conformance with quality standards and performance requirements. These include the inspection, test or audit of purchased materials, manufacturing or process operations, operations support documentation and materials, and installation or field trials. Includes labor and fringe benefit costs, as well as expenses and depreciation.

Failure Costs: Are the costs required to evaluate and either correct or replace products or services not conforming to requirements or customer/user needs. This includes purchased materials, and associated product or service design and support materials because they failed to meet requirements or customer/user needs. Includes both material and labor costs, with fringe benefits.

- Internal failure costs occur prior to completion or shipment of the product, or furnishing of a service.
- External failures costs occur after shipment of the product, and during or after furnishing of a service.

A quality cost program consists of the following steps:
- establish a quality cost measurement system.
- develop a suitable long range trend analysis.
- establish annual improvement goals for total quality costs.
- develop short range trend analyses with individual targets that collectively add up to the incremental demands of the annual improvement goal.
- monitor progress against each short range target and take appropriate corrective action when targets are not being achieved.

To determine in what categories the cost elements are assigned, the following questions are asked:
1. Prevention — Is this cost related to the prevention of non-conformances within an activity?
2. Appraisal — Is this cost related to evaluating the conformance of products or services to quality standards and performance requirements?
3. Failure — Is this cost related to non-conforming product or service?
   a. Internal — Found before shipment of product or providing service.
   b. External— Found after shipment or providing service.

Many of the elements in the Prevention Category are the pre-determined corrective action of the Quality System. The thing to remember about corrective action is that you have to pay for it once, whereas failure to take corrective action may be paid for over and over again. Therefore, the Quality Assurance function is very important in the Improvement Program. The efficiency of performing these functions (market analysis, contract review, design review, planning and education) has a direct bearing on the Appraisal and Failure Categories.

The following excerpted from Appendix B of "The Principles of Quality Costs" gives elements considered Prevention Costs and detailed descriptions of those pertaining to Quality Planning and Quality Training.

Prevention Costs

The experience gained from the identification and elimination of specific causes of failure cost are utilized to prevent the recurrence of the same or similar failures in other products or services. This is achieved by examining the total of such experience and developing specific activities for incorporation into the basic management system that will make it difficult or impossible for the same errors or failures to occur again.

The prevention costs of quality have been defined to include the cost of all activities specifically designed for this purpose. Each activity may involve personnel from one or many departments. No attempt will be made here to define appropriate departments since each company is organized differently.

1. Marketing/Customer User.
2. Marketing Research.
3. Customer/User Perception Surveys/Clinics.
4. Contract/Document Review.
5. Product/Service/Design Development.
6. Design Quality Progress Reviews.
7. Design Support Activities.
8. Product Design Qualification Test.
9. Service Design-Qualification.
10. Field Trials.
11. Purchasing.
12. Supplier Reviews.
13. Supplier Rating.
14. Purchase Order Tech Data Reviews.
15. Supplier Quality Planning
    The total cost of planning for the incoming and source inspections and tests necessary to determine acceptance of supplier products. Includes the preparation of necessary documents and development costs for newly required inspection and test equipment.
16. Operations (Manufacturing or Service)
    Costs incurred in assuring the capability and readiness of operations to meet quality standards and requirements; quality control planning for all production activities; and the quality education of operating personnel.
17. Operations Process Validation.

18. Operations Quality Planning

    The total cost for development of necessary product or service inspection, test, and audit procedures; appraisal documentation system; and workmanship or appearance standards to assure the continued achievement of acceptable quality results. Also includes total design and development costs for new or special measurement and control techniques, gages, and equipment.

19. Design and Development of Quality Measurement and Control Equipment

20. Operations Support Quality Planning

    The total cost of quality control planning for all activities required to provide tangible quality support to the production process. As applicable, these production support activities include, but are not limited to, preparation of specifications and the construction or purchase of new production equipment, preparation of operator instructions; scheduling and control plans for production supplier; laboratory analysis support; data processing support; and clerical support.

21. Operator Quality Education

    Cost incurred in the development and conduct of formal operator training programs for the express purpose of preventing errors — programs that emphasize the value of quality and the role that each operator plays in its achievement. This includes operator certification programs in subjects like, statistical quality control, process control, quality circles, problem solving techniques etc.

    This item is not intended to include any portion of basic apprentice or skill training necessary to be qualified for an individual assignment within a company.

22. Operator SPC/Process Control

    Costs incurred for education to implement program.

23. Quality Administration.

24. Administrative Salaries.

25. Administrative Expenses.

26. Quality Program Planning

    The cost of quality (procedure) manual development and maintenance, inputs to proposals, quality record keeping, strategic planning and budget control.

27. Quality Performance Reporting

    Costs incurred in quality performance data collection, compilation, analysis, and issuance in report forms designed to promote the continued improvement of quality performance. Quality cost reporting would be under this category.

28. Quality Education

    Costs incurred in the initial (new employee indoctrination) and continued quality education of all company functions that can affect the quality of product or service as delivered to customers. Quality education programs emphasize the value of quality performance and the role that each function plays in its achievement.

29. Quality Improvement.

30. Quality Audits.

31. Other Prevention Costs.

Obviously it is much more difficult to capture prevention costs than appraisal and failure costs as they are considered normal business costs and are expected as operating expenses. These costs are hidden and uncontrolled. Segregation may be expensive and the pay back must be weighed. Prevention Costs are dispersed across the various departments of the company which affects collection and presentation. Proper use and control will result in continual improvement.

Summary

Prevention costs are most important in a quality cost system as Quality Improvement is obtained from Corrective Action. Corrective Action depends on the determination of root cause and implementing an irreversible action. In addition, there are "right" prevention activities that are difficult to segregate and considered normal costs of doing business. These activities in effect are pre-determined corrective actions which improve the performance and efficiency of the business.

SPENDING PREVENTION DOLLARS EFFECTIVELY

Martin W. Wirt
Director of Product Development
RAYOVAC Corporation
Madison, Wisconsin

Across the country, management teams of manufacturing and service companies are focusing on quality improvement as the means to fend off increasing competition from abroad. This has been associated with an increased interest in quality costs, both as a measure of "How are we doing today vs. our competition?" and "What return will be gained by a quality improvement decision, activity, or project?"

After viewing, often with shock, a first tabulation of their quality costs, management asks themselves "How do we reduce our quality costs? Where do we invest? How do we know we are working on the right problem area? How do we know our problem solving activity will pay off?"

Developing the Framework

The first step, of course, is to develop the framework for measuring the gravity of the problem and assessing the results of corrective action. This framework is a well thought-out, thorough and consistent quality cost reporting system. This first fundamental step provides management with a wealth of information, including:
1. The competitive advantage/disadvantage of our operation vs. the competition, given industry or sector quality cost statistics.
2. A view of the quality "maturity" of the organization by looking at ratios of failure costs to appraisal costs and prevention costs.
3. A guide to action — where the most promising areas are for quality cost reduction.

One use not commonly made of the quality cost framework is its use as an investment analysis tool for evaluation of quality improvement projects and also an audit tool for determining whether projects deliver the promised results.

After the quality cost report is structured, first results reluctantly swallowed and digested, and a couple of months' reports show that things aren't getting any better, management is faced with the challenge of reducing the total quality cost bill.

One of the first observations will likely be that quality costs are ranked in order of value by (1) failure costs, (2) appraisal costs, and (3) prevention costs. A typical manufacturing or service industry will have 3-5 percent of its total quality cost bill represented by the prevention category. Yet, most purists would assert that an "excellent" organization's cost categories would be ranked in the reverse order, whereby comprehensive and effective prevention activities are resulting in negligible failure costs and, therefore, a reduced "need" for inspection (appraisal costs).

Assessing Prevention Costs

Subsequently, most management teams will soon discover that prevention costs are a difficult category to clearly identify. In a manufacturing organization, failure costs can be tracked through most accounting system records of scrap, yield losses, and customer returns. In service organizations, this obviously becomes more complex because liability and warranty costs and judgments of lost business and reputation play key roles in failure costs.

Appraisal costs can likewise be tabulated by adding up the inspection payroll and related overhead accounts. This again gets more difficult as organizations increasingly devolve the inspection function to the production operators and allocations of payroll, etc. costs are required to calculate inspection costs.

In contrast to failure and appraisal costs, prevention costs nearly always require more judgment in identification and allocation. To review, prevention costs include any costs associated with assuring that defects do not arise in the manufacture of the product or provision of the service. This includes:

1. Development of quality, inspection and reliability plans.
2. Preparation of QC Manuals and procedures.
3. Design reviews and test plan development for new products.
4. Preparation of training programs for improving quality performance.
5. Process control development where the end goal is improving conformance to requirements or fitness for use.
6. Development of the quality data acquisition system and analysis of the data.
7. Development of quality reporting systems for upper management review and decision-making.
8. Consulting fees or other charges for outside assistance (e.g., statistical software, QA motivational programs) where the objective is the reduction of failure or appraisal costs.
9. Diagnostic experimentation and analysis in the process of isolating the main causes of defects.
10. Assistance to vendors in developing quality systems.
11. Any other Q/C Engineering costs.

The Classical Model

The classical relationship of prevention costs to failure and appraisal costs in a manufacturing setting is shown in Figure 1, where costs are plotted against percent defective. Here the basic "no loss" manufacturing cost is shown as a flat line regardless of the percent defective. To get to the total product cost, losses due to defectives (failure cost) and costs to minimize or eliminate defectives (prevention and appraisal costs) are added to develop a total cost curve. As shown, if no action is taken to prevent or inspect for defects, a 100% defective product will almost assuredly result. Then, as effort is expended and money spent to reduce percent defective, total costs are reduced. In the

FIGURE 1

RELATION BETWEEN QUALITY AND QUALITY COST

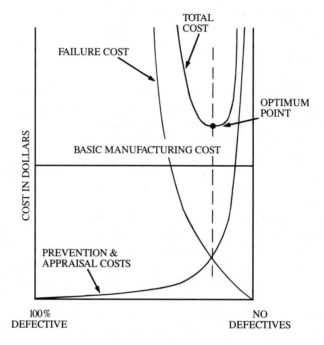

classical presentation the optimum point, where minimum cost is achieved, occurs approximately at the point of intersection of failure cost with the total of prevention and appraisal costs. Beyond this point, further expenditure of appraisal and preventions dollars yields a reduced improvement in failure costs; as a result total cost starts to increase.

Whether the optimum cost point is truly at the equivalence of failure costs with the total of appraisal and prevention costs will be left for others to argue. It will be accepted, however, that for most organizations wise use of prevention dollars will result in substantial reduction of failure costs and, as a result, the total cost of the product or service.

Targeting the Right Prevention Projects

The key word here is "wise." Many organizations have proven that "throwing money at a problem" is an ineffective and wasteful use of the organization's resources. How do we know that we are investing preventive dollars in solving the right problem in the most productive manner?

FIGURE 2

PARETO ANALYSIS — PLANT SCRAP BY PRODUCT/PROCESS

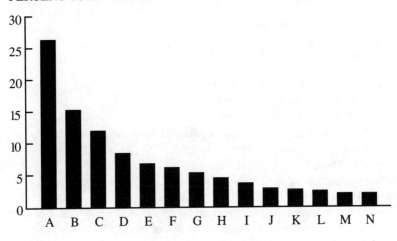

My experience has shown that use of the widely accepted Pareto analysis points the way to the right problem area(s) in a company's quality costs. In Figure 2 we see an actual Pareto distribution of my company's failure costs in one of its major manufacturing facilities. The code letters A through N represent most of the individual product model numbers manufactured in the plant. Obviously, a management which is serious about reducing quality costs will attack the top causes for scrap in order to get the largest and most immediate payback. In the plant represented by Figure 2, product A represented over twenty-six percent of the total scrap losses in the plant, and it became the first target in the plant's quality improvement program.

The next step is to develop options whereby these costs can be reduced and evaluate these options to determine which one(s) to pursue.

It is proposed that this step can be treated like any other investment option. As with other investment options, dollars are spent for expected future returns.

Evaluating the Options

Several methods are used for analysis of investments: these include payback, return on investment (ROI), discounted cash flow (DCF) analysis, net present value (NPV), and individual, specially tailored investment indices are created by companies for their own use based upon special company or industry factors.

Of these options, I subscribe that net present value (NPV) offers a widely applicable index for evaluating a prevention investment option, in that it values future cash flows in terms of today's dollars and will rank investment options in terms of total dollar impact.

The formula for NPV is:

$$NPV = C_0 + \sum_{t=1}^{N} \frac{C_t}{(1+r)^t}$$

Where: C_0 is the initial cash flow.
C_t is the cash flow at time t.
N is the number of periods.
r is the opportunity cost of capital (the interest rate that applies).

Using the above formula, an investment yielding a positive NPV is a good one. Judgments among competing investments can be made by selecting the ones with the largest NPV.

The critical inputs for the evaluation are the estimates of future cash flows and the determination of the opportunity cost of capital. For this, the following guidelines are offered:

1. Future cash flows could include avoidance of internal/external failure costs, changes in appraisal costs, or changes in other operational costs. In evaluating these cash flows, use only incremental, or "out-of-pocket" costs. Inclusion of burden or fixed overhead allocations can distort results, when in truth these fixed overhead costs are not affected in the near term in absolute dollars.

2. For simplicity's sake, use a current interest rate which would apply if money had to be borrowed to the latest statistical designs, the costs were minimized for experimentation to assess the major contributions to defects. The results of the experimentation led to design improvements which have virtually eliminated the major cause for scrap on product A.

Effective Expenditure of Prevention Dollars = Recipe for Success

In summary, to spend prevention dollars effectively a company management needs to proceed through several logical steps which span the disciplines of QA and Finance:

1. Develop the framework for measurement and analysis of its quality problems in financial terms — a thorough quality cost reporting system.
2. Fully understand the elements of prevention costs and their role in reducing total quality costs.
3. Target the main causes of quality costs and identify options to eliminate them, using latest statistical techniques to minimize the cost of effective analysis.

4. Evaluate prevention costs like any other investment using, for example, net present value (NPV) as a technique for determining which projects to pursue.

My company is continuing to search the fertile fields of quality cost reduction through wise prevention expenditures. At some time in the future, analysis will show that the NPV of these opportunities are either very small in relation to other investment opportunities or do not yield a positive NPV. At that time, we will truly have achieved our optimum point of expenditure of prevention dollars.

Bibliography

Berger, Roger W. *Quality Control: Meeting the New Competition*, American Management Associations, 1981

Bouillon, Lincoln III. "Quality and Finance," *Quality*, June, 1987

Feigenbaum, A. V. *Total Quality Control Engineering and Management*, McGraw-Hill Book Company, 1961

Juran, Joseph M. (ed.) *Quality Control Handbook*, 3rd Edition, McGraw-Hill Book Company, 1974

Juran, J. M. and Gryna, Frank M. Jr. *Quality Planning and Analysis*, 2nd Edition, McGraw-Hill Book Company, 1980

ROCKY MOUNTAIN CONFERENCE PROCEEDINGS

QUALITY COSTS: PRINCIPLES AND IMPLEMENTATION

Jack Campanella
Eaton Corporation AIL Division
Deer Park, New York

Quality Is Free[1] is the title of a very popular book by Phil Crosby...the very same Phil Crosby who originated the idea of zero defects programs more than 20 years ago. Phil wrote the book while corporate vice president for quality at ITT but has since left ITT, to establish a highly successful quality college in Florida. His ideas have a habit of catching on... not only because of his catchy phrases but because they are based on sound principles, and because they work!

But this one, *Quality Is Free*, how can he say that? Everyone knows quality costs money. Wrong! Quality is free, it's the unquality that costs...the rework, the scrap, the reinspection, the engineering changes. Why if there was no unquality, that is if quality were perfect, there would not even be a need for a quality department.

Because of product defects, and employee error, there is the need for product inspection and test with resulting rejections causing scrap and rework costs as well as inspection costs. Efforts to reduce these costs require expenditures for prevention. The result is that the company's actual costs are much higher than they would be if there were no errors. The actual cost beyond the theoretical cost of the perfect company is the cost of quality. And guess what? The cost of quality can run as high as 20 percent or more of sales. What were your sales figures last year? The cost of quality in a 100 million dollar company, for example, could run as high as 20 million dollars or more. Think of the impact in profit a mere 10 percent improvement would make. It is a fact, too often unrecognized, that every dollar saved in the total cost of quality, is directly translatable into a dollar of pretax earnings. Quality pays!

How can you find out what the cost of quality is in your company? First you need to know a little about what quality costs are.

History

The concept of quality costs can be traced back to the early 1950's. Chapter one of Dr. Juran's first *Quality Control Handbook*[2] published in 1951 was titled the "Economics of Quality" and contained discussions of the "Cost of Quality" and his famous analogy of "gold in the mine."

Among the earliest articles on quality cost systems as we know them today were Harold Freeman's 1960 paper on "How To Put Quality Costs To Use"[3] and chapter 5, "Quality Costs" of Dr. A. V. Feigenbaum's famous text *Total Quality Control*[4] in 1961. They are among the earliest writings categorizing quality costs into the costs of prevention, appraisal, internal and external failure.

In December 1963, the Department of Defense issued MIL-Q-9858A, "Quality Program Requirements"[5] making "costs related to quality" a requirement on many Government contractors and subcontractors. From that point on, the subject became extremely popular. In 1964, the Industrial Engineering Department at Stanford University did a research study for the Air Force Systems Command and published the Q.U.I.C.O. (Quality Improvement Through Cost Optimization) system[6] ultimately leading to the May 1967 release by the Government of Quality and Reliability Assurance Technical Report TR8, "A Guide To Quality Cost Analysis."[7]

The ASQC Quality Costs Technical Committee was formed in 1961 with a primary objective of promoting broader application and use of quality cost principles. In 1967, the committee published *Quality Costs — What and How*,[8] the most popular document on the subject and the largest seller of any ASQC publication. After serving well for almost 20 years, *Quality Costs — What and How* was replaced as ASQC's standard work on the subject, by *Principles of Quality Costs*[9] published in 1986. Other popular publications of the Quality Costs Committee are the 1977 *Guide for Reducing Quality Costs*,[10] and the *Guide for Managing Supplier Quality Costs*,[11] published in 1987.

The more than 300 excellent articles and papers listed in the current "Bibliography of Articles Related to Quality Cost Concepts and Improvement,"[12] maintained by the Quality Costs Committee further attest to the importance of the subject and the recognition given to it throughout industry. A compilation of the best of these papers has been published and can be purchased through ASQC. It is entitled *Quality Costs — Ideas and Applications*.[13]

Philosophy

The real value of a quality program is ultimately determined by its ability to contribute to improved customer satisfaction and profits. This is the environment in which quality management exists and is the principal reason why quality costs should be an integral part of an effective quality management system.

To develop the concept of quality costs, it is necessary to establish a clear picture of the difference between quality costs and the cost of the quality department. It is important that we do not view quality costs as the expenses of the quality function. Fundamentally, each time work must be redone, we are adding to the cost of quality. The most obvious examples are the reworking of a manufactured item, the retesting of an assembly, or the rebuilding of a tool because it was originally unacceptable. Other examples may be less obvious, such as the repurchasing of parts or the response to a customer complaint.

In short, any cost that would not have been expended if quality was perfect, contributes to the cost of quality. Unfortunately, many such costs are overlooked or unrecognized simply because most accounting systems are not designed to identify them. It is for this reason that the technology or system of quality costs was created. It was designed to recognize the costs of "doing things over" as a significant addition to the cost of quality management, and to show them collectively as an otherwise hidden opportunity for profit improvement.

As seen in Figure 1, the most costly condition exists when a customer finds defects. Had the manufacturer, through much inspection and testing, found the defects himself, a less costly condition would have resulted. However had the manufacturer's quality program been geared toward prevention of defects, defects and their resulting costs would have been minimized — obviously the most desirable condition.

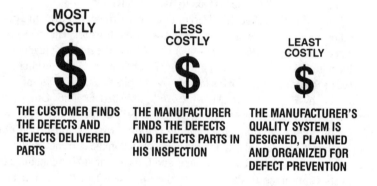

Figure 1. Comparative Cost of Quality

Quality Cost Categories

Quality costs must be categorized in order to be managed. The three major categories commonly used are prevention costs, appraisal costs, and failure costs.
- *Prevention Costs* are those costs expended in an effort to prevent poor quality, such as the costs of quality planning, training programs, and supplier quality rating.
- *Appraisal Costs* are those costs associated with the measurement of quality characteristics to assure conformance to requirements, such as the costs of inspection, test, and calibration.
- *Failure Costs* are those costs expended as a result of products not meeting requirements. They are usually divided into two types:
  - *Internal Failure Costs* result from discrepancies found prior to delivery of the product to the customer such as the cost of scrap, rework, and material review.
  - *External Failure Costs* result from discrepancies found after delivery of the product to the customer, such as the costs associated with the processing of customer returns, customer complaints, warranties, and general loss of reputation and good will.

Total Quality Cost is the sum of these costs: prevention plus appraisal plus failure. It represents the difference between the actual cost of a product, and what the reduced cost would be if there were no possibility of failure of the products nor defects in their manufacture. It is as described by Juran,[14] "gold in the mine." To repeat, quality costs can run as high as 20 percent or more of sales in some companies. That is a lot of gold.

The objective is to bring the total quality cost to a minimum while improving quality. The basic concept is that an increase in the cost of prevention should result in a larger decrease in the cost of failures, thereby reducing the total cost of quality. The old saying "An ounce of prevention is worth a pound of cure" was never more appropriate than when applied to quality costs.

While an increase in prevention should result in a decrease in the cost of failures, it may also cause appraisal costs to decrease somewhat. The reduction in failure costs may justify an increase in sampling inspection due to the confidence gained in improved quality, thereby reducing the amount of inspection performed. A double benefit.

How to Start

Well, now that we are all familiar with the quality cost concept, how do we start? How do we implement a quality cost program?

The first thing to do is to determine the need for the program. This should be presented to management in a way which will justify the effort, and interest them in participating. To interest management, justify the need.

One way to do this is by establishing a trial program. It can be simple. For this purpose, only major costs need to be gathered, and only readily available data need be included. Hopefully, you may find that much of the data you require is presently available, and if you have to, you may even estimate some of these costs.

When setting up the trial program, do not try to do the whole place... there's plenty of time for that later. Select a program, division, facility, or area of particular interest to management. The results should be sufficient to sell them on the need for the program.

Most trial runs will show eye-opening results... spectacular enough to make management sit up and take notice. They will see quality costs running as much as 20 percent or more of sales dollars, and opportunities for significant savings will be obvious.

And, now that you have sold top management, getting the much needed cooperation of the accounting people should be easy.

Defining Quality Cost Elements

With management sold, and with accounting "ready to go," you must now determine the specific quality costs to be collected. Tasks must be classified as to prevention, appraisal, or failure, and listed together with the departments responsible for them. Remember that quality costs are not only incurred by the quality department.

To determine the prevention costs to be collected, the tasks performed in your company, in an effort to prevent discrepancies, should be listed together with responsible departments.

In a like manner, appraisal cost elements are determined by listing those tasks associated with the inspection and test of products for the detection of discrepancies.

For failure costs, you need to determine those costs which would not have been expended if quality were perfect. If quality were perfect, you would not have any rework, nor would you have to respond to customer complaints, or take corrective action, and do not forget to divide failure costs into internal and external categories.

Quality cost elements or tasks may be different from company to company, especially so from industry to industry; however, the overall categories of prevention, appraisal, and failure are always the same.

Cost Collection

Now that you have decided on the specific costs to be collected, you need to develop a method to collect them. Collection of quality costs should be the responsibility of the controller. The finance and accounting department is the cost collection agency of the company and what we are doing here is collecting costs. Besides . . . having the controller collect the costs adds credibility to the data.

If top mangement was properly sold on the program, the controller would have been charged with the task of heading this effort. With the help of the quality manager, the controller should review the list of costs to be collected, determine which of these are already available under the existing accounting system, and decide where additions to the existing system are needed. Sometimes, the simple addition of new cost element codes to the present charging system is sufficient. However, if necessary, the present system could be supplemented by separate forms designed especially for this purpose.

Ideally, a complete system of cost element codes could be generated. They could be coded in such a way that the costs of prevention, appraisal, and internal and external failures could be easily distinguished and sorted. Then these codes could be entered on a labor or time card, together with the hours expended against the cost element or task represented by the code. The labor hours would later be converted to dollars by data processing.

An exception to this system, of collecting quality costs as they are incurred, is scrap. No one knows they are incurring scrap costs while they are actually incurring them. The work needs to be inspected, rejected, and dispositioned first. In many companies, the existing scrap reporting documents are forwarded to estimating, where the costs of labor and material expended, to the stage of completion of the scrapped items, are estimated.

The accounting department should provide all collected quality costs to the quality department in a format suitable for analysis and reporting. Of course, training programs will be necessary to assure that all personnel are informed as to how to report their quality cost expenditures. The training must be repeated periodically and the collection system should be audited on a regular basis.

Summary and Analysis

There are many ways quality costs can be summarized. They may be summarized by company, by division, by facility, by department, or by shop. They may be summarized by program, by type of program, or by all programs combined. What is the best way? It is the way that is best for you! The decision must be based on the individual needs of your organization.

Analysis can include comparison of the total quality cost to an appropriate measurement base. Some commonly used bases are sales, cost input, and direct labor. Again, the base selected will depend on what is appropriate for your needs. Comparing quality cost to a measurement base will relate the cost of quality to the amount of work performed. An increase in quality cost with a proportionate increase in the base is normal. It is the nonproportionate change that should be of interest. The index, total quality cost over the measurement base, is the factor to be analyzed. The goal is to bring this index to a minimum through continuous quality improvement. The index may be plotted so that trends representing present status in relation to past performance and future goals may be analyzed.

Other methods of analysis include study of the effect that changes in one category have on the other categories, and on the total quality cost. For example, was the increase in prevention cost effective in reducing failure cost? And was this reduction in failure cost sufficient to cause a reduction in total quality costs? This technique can provide insight into where the quality dollar can most wisely be spent. Increases in failure costs must be investigated to determine where prevention costs must be expended to reverse the trend and reduce the total quality cost. Losses must be defined, their causes identified, and preventive action taken to preclude recurrence.

To identify significant problems, other existing quality systems, such as your defect reporting system, can be used in conjunction with the quality cost program. The defect reporting system can help you define the causes of your scrap, rework, and other failure costs.

While the losses are distributed among many causes, they are not uniformly distributed. A small percentage of the causes will account for a high percentage of the losses. By Pareto's Principle,[15] these causes are the "vital few," as opposed to the "trivial many." Concentration of prevention of the "vital few" causes will achieve maximum improvement at a minimum of cost. This will have the effect of improving quality while reducing costs.

Reporting

There are almost as many ways to report quality costs as there are companies reporting them. That is because how they are reported depends on who they are reported to and what the report is trying to say. The amount of detail included in the quality cost report generally depends upon the level of management the report is geared to.

To top management, the report might be a score card, depicting through a few carefully selected trend charts, the status of the quality program...where it has been, and the direction it is heading. Savings over the report period and opportunities for future savings might be identified.

To middle management, the report might provide quality cost trends by department or shop to enable identification of areas in need of improvement.

Reports to line management might provide detailed cost information, perhaps the results of a Pareto analysis, identifying those specific areas where corrective action would afford the greatest improvement. Scrap and rework costs by shop are also effective charts, when included in reports to line management.

The examples provided as Figures 2 through 6 are samples of reporting techniques that have worked in companies with established quality cost programs.

Use by Management

Once the quality cost program is implemented, it should be used by management to justify and support improvement in each major area of product activity. Quality costs should be reviewed for each major product line, manufacturing area, or cost center. The improvement potential that exists in each individual area can then be looked at and meaningful goals can be established. The quality cost system then becomes an integral part of quality measurement. The proper balance is to establish improvement efforts at the level necessary to effectively reduce the total cost of quality; and then, as progress is achieved, adjust it to where total quality costs are at the lowest attainable level. This prevents unheeded growth in quality costs and creates improved overall quality performance, reputation, and profits.

A quality cost program, based on the concept and methods of implementation broadly presented here, can be used by management as an aid towards achieving its goal of an optimum quality program at a minimum quality cost. The program will measure the value of the quality effort, identify the strong and weak points of the quality program, indicate how the quality dollar can be spent most effectively, and provide quality improvement while reducing costs.

QUALITY COSTS — BY PROGRAM

MONTH AUGUST 1978

CODE	ELEMENT DESCRIPTION	PROGRAM (IN THOUSANDS)					TOTAL ALL PROGRAMS
		280Z	200X	101Q	RX7	MISC	
K	MATERIAL REVIEW ACTIVITY	71.8	—	0.7	7.7	—	80.2
L	CORRECTIVE ACTION	249.8	—	0.9	2.7	—	253.4
X	TROUBLESHOOTING/FAILURE ANALYSIS	47.2	—	—	0.8	—	48.0
R	REWORK/REPAIR	128.6	0.4	6.3	26.1	—	161.4
P	SCRAP	19.2	—	—	0.5	—	19.7
V	REWORK/REPAIR/SCRAP — VENDOR RESP	27.1	—	0.1	2.6	—	29.8
U	PROCESSING OF CUSTOMER COMPLAINTS	5.4	—	—	1.4	—	6.8
I	PROCESSING OF CUSTOMER RETURNED MATL	23.3	—	—	—	—	23.3
J	FIELD SERVICES	—	—	—	—	—	—
Y	WARRANTY COSTS	0.1	—	—	—	—	0.1
	TOTAL "UNQUALITY" COSTS	572.5	0.4	8.0	41.8	—	622.7
	QUALITY PREVENTION AND APPRAISAL COSTS	718.2	0.9	8.3	69.8	38.6	835.8
	TOTAL QUALITY COSTS	1290.7	1.3	16.3	111.6	38.6	1458.5
	MFG DIRECT LABOR COSTS	6973.8	7.5	116.5	882.8	396.5	8377.1
	SCRAP/REWORK/REPAIR AS % OF MFG D/L	2.1	5.3	5.4	3.0	—	2.2
	COST INPUT	20943.4	66.1	978.4	1946.6	104.1	24038.6
	TOTAL QUALITY COSTS AS % OF COST INPUT	6.2	2.0	1.7	5.7	37.1	6.1

Figure 2. Quality Costs by Program

QUALITY COST REPORT FOR THE MONTH ENDING _____ (In Thousands of U.S. Dollars)						
DESCRIPTION	CURRENT MONTH			YEAR TO DATE		
	QUALITY COSTS	AS A PERCENT OF		QUALITY COSTS	AS A PERCENT OF	
		SALES	OTHER		SALES	OTHER

1. PREVENTION COSTS
  1.1 Product Design
  1.2 Purchasing
  1.3 Quality Planning
  1.4 Quality Administration
  1.5 Quality Training
  1.6 Quality Audits
    TOTAL PREVENTION COSTS
    PREVENTION TARGETS						

2. APPRAISAL COSTS
  2.1 Product Qualification Tests
  2.2 Supplier Product Inspection and Test
  2.3 In Process and Final Inspection and Test
  2.4 Maintenance and Calibration
    TOTAL APPRAISAL COSTS
    APPRAISAL TARGETS						

3. FAILURE COSTS
  3.1 Design Failure Costs
  3.2 Supplier Product Rejects
  3.3 Material Review and Corrective Action
  3.4 Rework
  3.5 Scrap
  3.6 External Failure Costs
    TOTAL FAILURE COSTS
    FAILURE TARGETS						

TOTAL QUALITY COSTS						
TOTAL QUALITY TARGETS						

MEMO DATA	CURRENT MONTH		YEAR TO DATE		FULL YEAR	
	BUDGET	ACTUAL	BUDGET	ACTUAL	BUDGET	ACTUAL
Net Sales						
Other Base (Specify)						

Figure 3. Quality Costs by Element

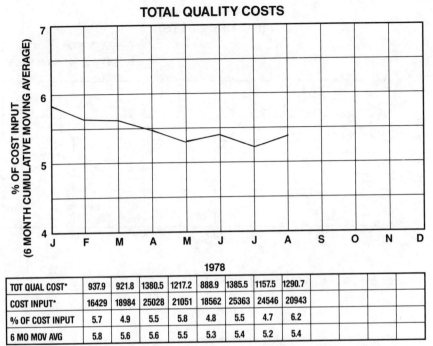

Figure 4. Total Costs by Month

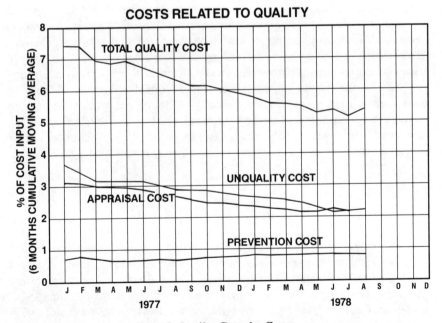

Figure 5. Quality Costs by Category

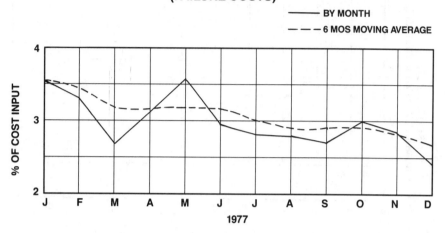

Figure 6. Failure Costs by Month with a 6 Month Moving Average for Trending

Quality Costs as a Budgeting Tool

Still another benefit to be gained from a quality cost program is its ability to be used as a budgeting tool. As costs are collected against quality cost elements, a history is generated. This history can then be used to determine the average cost per element. In other words depending on how detailed you have established your elements, you will know how much you have been spending for various functions or tasks. This information can be used as the basis for future quotes and estimates. Budgets can be established for each element. Then, going full circle, the actuals collected against these elements can be bounced against the budget amounts to determine budget variances. Action can then be initiated to bring over or under running elements into line. Figure 7 illustrates a report providing actual-versus-budget figures for each quality cost element.

Reducing Quality Costs

Quality improvement results in cost improvement. Designing and building a product right the first time always costs less. Solving problems, by finding their causes and eliminating them, results in measurable savings. To cash in on these savings, past quality performance must be improved. Ways and methods to do this are described in the ASQC publication *Guide for Reducing Quality Costs*.[16] It describes techniques for using quality cost data in programs to reduce costs and improve profits.

QUALITY COST REPORT

DATE _____
MONTH _____

PREVENTION		MONTH		YEAR TO DATE	
		ACTUAL	BUDGET	ACTUAL	BUDGET
P1.1	ENGINEERING EFFORT	20.0	21.0	200.0	200.0
P1.2	QUALITY ADM, PLNG AND SERV	5.0	5.0	50.0	50.0
	P1 TOTAL	25.0	26.0	250.0	250.0
APPRAISAL					
A1.1	INCOMING INSPECTION	20.0	25.0	200.0	250.0
A1.2	IN-PROCESS AND FINAL INSPEC	50.0	55.0	600.0	700.0
A1.3	TESTING	50.0	50.0	600.0	600.0
A1.4	PRODUCT AUDIT	10.0	12.0	100.0	150.0
A1.5	OTHER TESTS — RELIAB ETC.	10.0	11.0	100.0	100.0
A1.6	EQUIP CALIBRATION	10.0	12.0	50.0	50.0
	A1 TOTAL	150.0	165.0	1650.0	1850.00
INTERNAL FAILURES					
N1.1	DEFECTIVE WORK/JUNK	30.0	35.0	350.0	400.0
N1.2	REMAKES, REWORK, AND RETEST	30.0	20.0	400.0	300.0
N1.3	INVESTIGATION	10.0	5.0	100.0	150.0
	N1 TOTAL	70.0	60.0	850.0	850.0
EXTERNAL FAILURES					
E1.1	WARRANTY — COMPLAINTS	10.0	20.0	100.0	200.0
E1.2	FIELD REPAIR (IN WARRANTY)	2.0	5.0	50.0	50.0
E1.3	FIELD REP (OUT OF WARRANTY)	1.0	5.0	50.0	75.0
E1.4	OTHER	2.0	5.0	50.0	25.0
	E1 TOTAL	15.0	35.0	250.0	350.0
Q1	TOTAL	260.0	286.0	3000.0	3300.0
	RATIO TO TOTAL SALES	3.2%	5.0%	3.0%	3.2%
	RATIO TO TOTAL STANDARD COST OF OUTPUT	5.1%	4.6%	4.6%	4.7%

Figure 7. Quality Cost Actuals vs. Budget

Corrective Action

A key factor in the reduction of quality costs is corrective action. Quality costs do not reduce themselves, they are merely the scorecard. They can tell you where you are at and where your corrective action dollar will afford the greatest return. Quality costs do identify targets for corrective action.

Once a target for corrective action is identified, through Pareto or other methods of analysis, the action necessary must be carefully determined. It must be individually justified on the basis of an equitable cost trade-off. You do not want to resolve a $500 problem with a $5000 solution. At this point, experience in measuring quality costs will be invaluable for estimating the payback for individual corrective action investments. Cost benefit justification of corrective action should be a continuing part of the program.

Some problems have fairly obvious solutions, such as the replacement of a worn bearing or a worn tool. They can usually be fixed immediately. Others are not so obvious, such as a marginal condition in design or processing, and are almost never discovered and corrected without the benefit of a well organized and formal approach. Marginal conditions usually result in problems that can easily become lost in the accepted cost of doing business. Having an organized corrective action system, justified by quality costs, will surface such problems for management's visibility and action. The true value of corrective action is that you only have to pay for it once; whereas failure to take corrective action may be paid for over and over again.[17]

Auditing Quality Costs

Implementation of a quality cost program will be effective only as long as it continues to accurately measure true quality costs within the organization. The financial establishment has long recognized that adequate initial implementation of sound and reasonable procedures is, in itself, insufficient to maintain an accurate reporting system. Periodic audits are required to determine if the system is functioning as it was designed and if it is still conceptually adequate. Some companies use their financial auditors to review the cost collection system, and their quality auditors to review the balance of the program. An annual "conceptual" review of the total quality cost system with more frequent "functional audits" of major portions of the system is usually sufficient.

Conclusion

Quality costs have expanded to where they have become a principal management tool. Definitions and standards have been developed and refined along with techniques and methods for implementation. The quality cost program is the bridge between line and executive management. It provides a common language, measurement, and evaluation system which proves that quality pays in increased profits, productivity, and customer acceptance.

References

1. Crosby, Philip B., *Quality Is Free,* McGraw-Hill Book Company, 1979.
2. Juran, J. M., *Quality Control Handbook,* First Edition, McGraw-Hill Inc., 1951.
3. Freeman. H. L., "How to Put Quality Costs to Use," Transactions of the Metropolitan Conference, ASQC, 1960.
4. Feigenbaum, A. B., *Total Quality Control,* McGraw-Hill, Inc., 1961.
5. Department of Defense, MIL-Q-9858A, "Quality Program Requirements," 16 December 1963.
6. Morgan, E. D. and Ireson, W. G., *Quality Cost Analysis Implementation Handbook,* Stanford University, Department of Industrial Engineering, 27 March 1964.
7. Office of the Assistant Secretary of Defense (Installations and Logistics), Quality and Reliability Assurance Technical Report TR8, "A Guide to Quality Cost Analysis," 31 May 1967.
8. Quality Cost Effectiveness Technical Committee, ASQC, *Quality Costs — What and How,* American Society for Quality Control, 1967.
9. Quality Costs Committee, ASQC, *Principles of Quality Costs,* American Society for Quality Control, 1986.
10. Quality Costs Technical Committee, ASQC, *Guide for Reducing Quality Costs,* American Society for Quality Control, 1977.
11. Quality Costs Committee, ASQC, *Guide for Managing Supplier Quality Costs,* American Society for Quality Control, 1987.
12. Quality Costs Committee, ASQC, "Bibliography of Articles Relating to Quality Cost Concepts and Improvement."
13. Quality Costs Technical Committee, ASQC, *Quality Costs — Ideas and Applications,* American Society for Quality Control, 1984.
14. Juran, J. M., *Quality Control Handbook,* Third Edition, McGraw-Hill, Inc., 1974.
15. Juran. J. M., op. cit., (Third Edition).
16. Quality Costs Technical Committee, ASQC, *Guide for Reducing Quality Costs,* American Society for Quality Control, 1977.
17. Hagan, J. T. "Quality Costs," ITT, N. Y., 1981.

QUALITY COST ANALYSIS: EXTEND THE BENEFITS

(September Issue)

William O. Winchell
Associate Professor
Industrial Engineering Division
Alfred University

Caroline J. Bolton
Quality Cost Analyst
Product Assurance
General Motors
Lansing, Michigan

The focus of the quality cost concept has traditionally been on the costs involved in making a product or providing a service to an external customer. Within the last several years, an alternative focus for quality has been developed. This new focus is tied to the awareness that each staff department or activity within a company also provides a product or service to a customer. In this case, however, the customer is internal to the company. The traditional approach can be called the "macro" approach, while the alternate approach can be called the "micro" approach. Most staff departments do not have a formalized quality system for producing a product or service. The micro approach to quality cost provides a framework for developing that quality system.

Macro Quality Cost

The traditional or macro approach to quality cost is detailed in *Principles of Quality Costs*.[1] This book defines the elements of quality cost to help a company document its quality system. The quality system starts in marketing, which must identify the customer or user needs. Development or product engineering translates these needs into a product or service and performs tests to validate that customer needs will be met. The quality tasks of both marketing and development/product engineering are classified as prevention effort because they are intended to prevent problems from occurring in the hands of the customer.

With the inspection responsibility shifting to manufacturing, the quality department is no longer a "traffic cop." It can devote its effort to supporting the quality system by helping each department perform appropriate quality tasks. In this new mode of operation, the quality department's efforts could be classified as almost entirely prevention. The manufacturing or operations department, which is responsible for conforming to the design of the product or service, incurs prevention and appraisal costs. If some products or services do not conform, they may have to be reworked or even scrapped before reaching the customer, resulting in internal failure costs. The service department mainly fixes the problems that the customer or user experiences. This is classified as external failure cost.

The cost of prevention is typically the smallest cost incurred by all departments in the quality system; appraisal and failure costs are much larger. In many studies the size of the external failure cost indicates that customers' needs are not being met very well — and field fixes are all too common. A common reaction to this situation is to increase prevention effort. An arbitrary, unfocused increase in prevention effort, however, may only add cost for the company since it may not address the true causes of the failures. Instead, merely redirecting the existing prevention effort can yield significant improvements. The micro cost study approach provides insight on how to do this.

Micro Quality Cost

Any change in prevention efforts must be based on an analysis of the effectiveness and efficiency of current efforts. Effectiveness means how well something is producing a desired effect. The desired effect of a quality system is an improvement in the product or service. Efficiency means being productive without waste. For a quality system, this means eliminating the failure costs and minimizing the need for appraisal.

Conducting a micro quality cost study is, in essence, the act of first defining an appropriate quality system for the staff department involved, then evaluating current efforts to make improvements. How the micro study is set up determines its usefulness. First, remember that a viable staff department has customers. Second, a quality system within the staff department is essential to assure that the products or services meet customer needs. Like the company quality system, the generic staff department quality system must include the following elements:

• Prevention: effort within the department to determine customer requirements and needs and to plan procedures that assure those requirements and needs are met.

• Appraisal: effort within the department to check whether the requirements and needs are met before the product or service reaches the customer.

• Internal failure: rework within the department before the product or service is provided to the customer. This occurs because the job is not done right the first time.

• External failure: rework by the department after the product or service reaches the customer because requirements and needs were not met.

Here, then, is the first major difference between the macro and micro approaches to quality cost. Staff departments, which may be classified under the macro approach (such as a modern quality department), are classified entirely differently under the micro approach. Under the micro approach, a staff department has the full spectrum of quality costs — prevention, appraisal, internal failure, and external failure. The second difference lies in the nature of the activities that make up the four elements above. Instead of traditional definitions, unique elements are defined by each activity in the micro approach. For example, if the activity being studied produces reports, a prevention effort may be to define requirements for the content of the report. If the activity is the repair garage, however, an appropriate prevention effort may be to diagnose the problems before processing.

In the initial efforts of the micro quality cost study, department employees should help define the products or services of the staff department, who the customers of the department are, and the activities that make up its quality cost elements. Thus, the department employees "buy into" the study and its results up front, and also become more aware of the need for their own quality system. Suggested steps for accomplishing this initial part of the study are:

1. Define and list the products or services produced.

2. Define and list the customers to whom each product or service is provided.

3. Classify customers in the relative position they would fill in the organization chart. Classify as either:
- vertical — typically management.
- horizontal (internal) — typically other activities within the department.
- horizontal (external) — typically activities outside the department that are direct links to the company's end product.

4. Identify typical external failure costs for each product or service that would be incurred in dealings with particular customers.

5. Identify appropriate types of prevention effort that would avoid producing failures, such as identification of customer requirements.

6. Identify the types of appraisal effort that will help avoid the delivery of products or services that do not meet requirements or needs. Appraisal also indicates whether prevention effort is effective.

7. Identify the types of internal failure that would be found through checks of products or services before delivery to the customers.

Once these first steps are completed, the staff department will have a defined quality system. The next step is to measure the amount of departmental resources spent on each of the elements above. This is done by interviewing employees to collect estimates of the time they spent on the quality tasks identified and on the distribution of the remaining departmental resources over these same quality tasks. Time and resources spent on quality tasks represent only part of the total effort expended on the product or service provided. But completion of the cost study and subsequent improvements may also reduce that total effort.

Effectiveness and Efficiency

As stated earlier, plans for redirecting prevention efforts within a staff department must be based on an analysis of the current activities. The first part of the analysis of the micro quality cost study addresses effectivenesss. A staff department that has more effort for vertical customers — management — than for the horizontal external customers — more likely a direct link to the company's end product — may not be as effective as possible. Also, a staff department that has more effort for internal horizontal customers who are within the same department than for external horizontal customers may likewise not be as effective as possible. The most effective situation may be to have the greatest effort expended for those customers directly linked to the company's end product — the horizontal external customers.

The second part of the cost study analysis addresses efficiency. A staff department that has an efficient quality system has prevention effort to identify customer requirements and needs. A small amount of appraisal to check whether the requirements and needs are met may also be necessary. Internal failure costs, however, are near zero, implying that the jobs are done right the first time. In addition, external failure costs are near zero, indicating that customer needs and requirements have been met. Efficiency, then, can be evaluated by looking at how quality costs for each product or service are distributed over the four elements. Some typical cost distributions include:

- a distribution where all four elements of quality cost are near zero. This indicates that no quality system really exists for the product or service. Rather than conclude that a quality system is really not needed, a subsequent check may confirm that the product or service does not mean a great deal to the customer and no complaints that generate external failure costs are made.

- a distribution where all elements of quality cost are near zero except for external failure. This indicates the lack of a quality system and that customers are furnished with a desired product that does not meet their needs and requirements.

- a distribution that has high prevention, appraisal, and internal failure costs, but low external failure costs — characteristics of an underdeveloped quality system. This indicates a satisfied customer; however, the high internal failure costs could mean that more extensive needs and requirements are perceived by the activity than desired by the customer. It may also indicate that training of those within the activity is needed to help them do the job right the first time.

Opportunities for improvement should now be apparent. Ineffective and inefficient activities can be identified and improved by tailoring efforts to meet customer needs — thus reducing failures. As in the macro approach, improvement efforts can be tracked and evaluated through periodic repetitions of the cost study.

An Essential Supplement

The micro quality cost approach is valuable for determining and improving the effectiveness and efficiency of staff departments' activities; employees also become aware of the need to reduce waste in their department. This approach demonstrates that the quality system needs to be tailored to each staff department. Conducting the cost study, in fact, helps define an appropriate quality system. The micro quality cost approach extends the never-ending journey for quality improvement to the processes found in staff departments, supplementing the improvements tied to the end product. The traditional macro quality cost approach is still needed to document the entire company quality system and identify needed improvements that can be incorporated synergistically among departments. But the micro quality cost approach is an essential supplement.

This paper was delivered at the 41st Annual Quality Congress.

Reference

1. ASQC Quality Costs Committee, John T. Hagan, editor, *Principles of Quality Costs* (Milwaukee, WI: ASQC, 1986).

THE QUALITY REVIEW

THE COST OF POOR QUALITY

(Spring Issue)

(Part A)
Walter F. Raab
Chairman and CEO
AMP Incorporated

(Part B)
Edward P. Czapor
Group Vice President
Quality and Reliability Parts Operations
General Motors

A recent Gallup Survey* asked executives, "How much does poor quality cost your company?" Two commentators analyze the findings and see a serious gap between perception and reality.

Part A

It is startling to see that nearly 50 per cent of the interviewed executives believed (guessed?) that the cost of poor quality was less than 5 per cent of the gross sales in their companies. This statistic suggests two rather disturbing observations.

First, it implies that *most* executives have not really evaluated their firm's actual cost of poor quality. (An additional 16 per cent *admitted* they "didn't know.") Not knowing the true cost, they apparently would naturally guess a low percentage rather than admit to a higher unverified figure.

Second, the low guess shows an ignorance of the extensive publicity over what American industry has tacitly admitted. Based on studies to date, the cost of poor quality, when properly evaluated, has consistently been in the 5 to 15 per cent range; and when the cost of conformance is added, the total costs range from *20 to 25 per cent of the total sales dollar.*

When AMP began its comprehensive emphasis on quality improvement in the early 1980s, one of the initial decisions was to determine accurately our *total* quality costs, including the cost of poor quality as well as the cost of conformance. We knew we could lower our nonconformance costs by spending more on conformance activities such as testing, inspection, and review. But to enjoy a valid return on any investment we might make, we knew we had to reduce *all* costs associated with quality.

Our corporate quality improvement process is dedicated to this end. When we implemented the process four years ago, our first "guess" was 4.7 per cent for the cost of poor quality (i.e., nonconformance). We, too, would have been counted in the Gallup poll as part of the current 50 per cent of underestimators. Our estimate for conformance costs was 5.8 per cent for a total estimate of 10.5 per cent — in retrospect, much lower than actual.

We immediately set out to capture an extremely vital aspect of any quality cost measurement — the amount of time our employees, at all levels, spend in both conformance and nonconformance activities. Since 1984 we have been using a time-reporting system that provides a departmental, divisional, and corporate management tool to identify and reduce all costs of poor quality. The system is supplemented by inputs from other natural accounts such as scrap, lost discounts, etc., that are real costs of poor quality not reflected in time reporting.

By the end of 1984 our figures were confirmed at 8 per cent for conformance and 7.7 per cent for nonconformance. Our newfound awareness hinted that we could probably dig even deeper and find another per cent or two we missed — but we stuck with a practical, workable system. Of paramount importance was the fact that now all departments knew their actual quality costs, particularly of nonconformance, and were challenged to manage them downward.

We can now look at essentially three years of quality cost performance (as a percentage of sales). In 1984, conformance costs were 8 per cent and nonconformance 7.7 per cent for a total of 15.7; in 1985, the figures were 7.6 and 7.1 per cent, for a total of 14.7. The 1986 numbers were 6.5 and 6.1 per cent, for a total of 12.6.

Notice that our first measured figures in 1984 were almost 60 per cent higher than our 1983 "guess" of 10.5 per cent. We were not surprised to see this rise, because we now had a proper measurement system. The encouraging reductions we saw in 1985 and 1986 represent a return of several million dollars per quarter for each percentage point of reduction in these costs. Our investment is paying off.

Part B

It is not surprising that many American executives either don't know or underestimate the cost of quality in their operations. Experts like Philip Crosby tell us the price of nonconformance can run between 20 and 30 per cent of a manufacturing company's sales and can go even higher in a service firm. Despite the obvious magnitude of these figures, cost of quality isn't an easy concept to grasp.

Part of the problem is that an exact figure is hard to fix. Much depends on the way it is calculated and what costs are included or excluded. At GM, we have found that exact amounts are really not all that important. Cost of quality is not a detailed accounting system. Instead, it is a tool that helps focus decisions of quality improvement and cost reduction. It is a system for helping managers understand that the path to low cost is through achieving the highest quality.

At GM, we define cost of quality (COQ) as "the total cost of doing things wrong plus the cost of preventing mistakes." It is the cost of producing, finding, correcting, and preventing product and service defects.

COQ puts quality into dollars-and-cents terms that managers can grasp. In consumer-oriented businesses like automobile manufacturing, we're well aware that the sticker price of an automobile is one of the important factors in a customer's decision to buy. For a business, whether it's involved in manufacturing or services, cost of quality acts as a "sticker price" for quality, allowing comparison between various courses of action and clarifying the decisions to be made by managers.

A cost of quality system in itself will not solve quality problems. Just keeping track of the various expenses related to quality will not reduce the cost of prevention and detection. Instead, the collection and analysis of COQ data was only a starting point. Based on these figures, improvement opportunities could be found, resources assigned, and corrective action achieved.

One of the areas where COQ has helped focus decisions for GM is that of problem resolution, the early detection and resolution of discrepancies that could affect customer satisfaction. Attaching a cost to a problem such as a poor paint job, inadequate engine performance, or a malfunctioning window lift system makes it much clearer to managers that such discrepancies must be found and corrected early in the design and testing process. It is important to note that the cost figure must include more than just warranty. For example, the average auto industry cost to replace a power window regulator assembly is about $90. But hidden costs, such as administration, inventory adjustment, shipping, repair of the problem at the manufacturing plant, and loss of future sales due to customer unhappiness, can multiply the costs by a factor of three or four.

Another area where cost of quality has helped clarify issues is process control and capability. For a company operation like GM, it's absolutely essential to reduce the variability of our manufacturing processes. First, we must be sure that we have control over all processes and that we have capable manufacturing operations that produce parts within established tolerances. Then as we improve consistency, tolerances can be tightened for a significant cost reduction. Here again, a COQ analysis can make clear how desirable such improved process control is.

For example, if the variability of piston sizes and engine bores is reduced, it is no longer necessary to try to match the diameter of a piston to a cylinder bore. All the pistons for a specific engine will fit the bores. The expensive process of size matching is longer required. Considering that the average cost of an hour of labor in the automobile industry is approaching $25, plus the cost of support services, it is clear that quality is not only free but reduces piece cost.

GM's experience in its joint venture with Toyota, New United Motor Manufacturing, Inc. (NUMMI) in California, has been a striking illustration of the importance of suppliers in reducing cost of quality. When suppliers provide high-quality parts on time, in-plant repairs and warranty costs can be greatly reduced. For example, NUMMI receives engines shipped from Japan almost fully dressed. Certain of the high quality of these engines, the plant does not test them or even start them until the car is fully assembled.

No manager can afford to ignore or underestimate cost of quality and its use for improving the profit performance of his operation. As American business wrestles with the issue of growing international competition, all business people must get a better hold on this important contributor to our quality improvement goals.

*The Gallup Survey, "Executives' Perceptions Concerning the Quality of American Products and Services," was based on telephone interviews with a representative cross-section of 698 executives in Fortune 500 companies and a sample of smaller companies in the manufacturing and service industries, conducted during the summer of 1986. For results based on samples of this size there is a 95 per cent confidence that error attributed to sampling and other random effects could be ± 4 per cent.

MANAGEMENT ACCOUNTING

QUALITY & PROFITABILITY: HAVE CONTROLLERS MADE THE CONNECTION?

(November Issue)

Thomas N. Tyson
Internal Revenue Service

More and more organizations are adopting quality improvement as a primary corporate objective. Effectively evaluating this objective requires accurate and reliable measurement of the costs underlying quality-related activities. The quality literature suggests that these costs be segregated into the following categories:
- Costs of preventing product defects (job training, quality circles),
- Costs of ensuring that products conform to specifications (inspection, testing), and
- Cost of failure, whether they are discovered internally prior to shipment (rework, yield loss) or externally by the customer (warranty charges, complaint adjustments).

Corporations that fully commit to quality often measure their progress as the reduction in the total amount of quality costs. Controllers are asked to measure these costs because accounting measurements generally are perceived as objective, reliable, and credible by other organization members. Accounting information also receives much greater company-wide exposure than does information generated from the quality function.

Controllers Surveyed

To find out the nature and scope of quality cost measurement in major industrial corporations, I contacted 125 randomly selected corporate controllers of the 1985 Fortune 500. Telephone interviews were completed with personnel in 94 of these firms, resulting in a 75.2% rate of response (see Table 1). Referrals to other members of the corporate controller's department were accepted, but in no case were interviews conducted with personnel from lower organizational levels. Therefore, information reported here relates exclusively to corporate controller department involvement in quality cost measurement.

I asked survey participants: "At your corporation, does the corporate controller's department specifically measure quality costs on a regular basis?" Their responses indicated that these costs are measured in 29 (31%) of the 94 sampled firms. Table 2 illustrates that measurement is conducted far more often among certain industries than among others. For example, in the industrial and farm equipment and motor vehicles and parts industries, quality costs are measured by the majority of surveyed firms. In the chemicals, metals, and forest products industries, however, quality cost measurement

TABLE 1/JOB TITLES OF RESPONDENTS

Corporate Controller	37
Vice President & Controller	20
Assistant Corporate Controller	15
Manager of Accounting	4
Director of Financial Control & Planning	2
Manager of Financial Reporting	2
Director of Corporate Accounting	2
Manager of Corporate Business Analysis	2
Vice President Accounting	1
Financial Vice President	1
Assistant Corporate Treasurer	1
Manufacturing Controller	1
OEM Operations Controller	1
Director of Manufacturing Costs	1
Deputy Controller	1
Director & Controller of Operations	1
Manager of Internal Audit	1
Cost & Investment Analysis Manager	1

TABLE 2/RESPONDENTS BY INDUSTRY

Code #	Description	Measurers	Non-measurers	Total
40	Motor Vehicles & Parts	4	0	4
45	Industrial & Farm Equipment	4	3	7
36	Electronics	4	5	9
38	Scientific & Photo. Equipment	2	1	3
44	Computers & Office Equipment	2	1	3
32	Building Materials	2	3	5
20	Food	2	5	7
29	Petroleum Refining	2	8	10
22	Textiles	1	0	1
30	Rubber Products	1	1	2
42	Pharmaceuticals	1	2	3
10	Mining & Crude Oil Production	1	3	4
27	Publishing & Printing	1	4	5
26	Forest Products	1	6	7
28	Chemicals	1	10	11
23	Apparel	0	1	1
25	Furniture	0	1	1
41	Aerospace	0	2	2
34	Metal Products	0	4	4
33	Metals	0	5	5
	TOTAL	29	65	94

rarely or never takes place. These results indicated that measurement may occur more frequently among manufacturing industries forced to confront strong foreign competition.

Why Measure Quality Costs?

What factors contribute to a controller department's participation in quality cost measurement? According to the quality cost and controllership literature five factors should explain measurement.
- The level of top management's commitment to improving quality,
- The level of available resources for conducting special studies,
- The extent of controller department participation in team projects that include quality personnel,
- The frequency of communication with quality function personnel, and
- The level of top management's reliance on formal cost reduction programs to improve profit.

A regression analysis of these five factors established that the levels of available resources, participation in team projects, and communication with quality function personnel all encourage the use of quality measures. For example, controller departments that have resources frequently available for special studies activities and that frequently interact with quality department personnel are much more likely to undertake measurement. In the 1950s, Herbert Simon and others similarly reported that resources for special studies and effective two-way communication were essential elements for conducting nontraditional accounting activities.[1] It appears that their observations are still valid and apply to quality cost measurement.

Because the overwhelming majority of respondents indicated their top management was strongly committed to improving quality (82%) or strongly reliant on formal cost reduction programs to improve profit (67%), these two factors did not seem to encourage this measurement activity. For example, simply having top management strongly committed to improving quality does not mean that the controller's department will specifically measure quality costs. This fact is understandable because many controllers may be largely unfamiliar with quality cost measurement procedures or may not have the resources necessary to do the job.

Perceptions of Quality and Profitability

Accountants often are accused of being preoccupied with matters of efficiency, such as cost budgets and delivery schedules, rather than with matters of effectiveness, such as employee productivity and product quality. If this preoccupation were so, measurements underlying these latter areas might not be inititated by the accounting function unless they were perceived as directly relating to higher profit.

FIGURE 1/RELATIVE IMPORTANCE AMONG PROFIT IMPROVING FACTORS

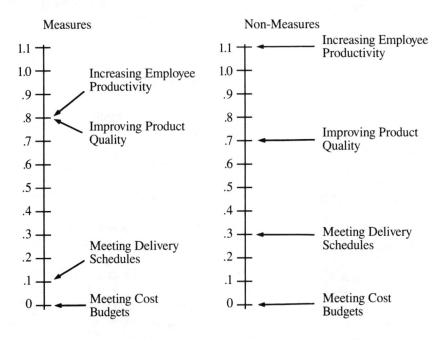

To discover their perceptions regarding quality and profitability, respondents were asked to rank four factors that might lead to higher profit. The four factors were ranked in the following order of importance by both the measurer and nonmeasurer subgroups: 1) increasing employee productivity, 2) improving product quality, 3) meeting delivery schedules, and 4) meeting cost budgets. By ranking productivity and quality as the two more important factors, controller department personnel appear to be well aware of the direct relationship of these two factors to profitability.

Respondents' overall preference rankings were converted to interval scale measures by Thurstone's Case V procedure.[2] The two scales shown in Figure 1 illustrate the relative importance among these four factors by the measurer and nonmeasurer subgroups. Though both subgroups generated identical overall rankings, measurers appear to be cognizant of the equal importance of quality and productivity improvement to higher profit and the equal relative unimportance of meeting delivery schedules and cost budgets. It seems, therefore, that a strong corporate awareness of and commitment to quality may occur when there is greater controller department participation. It also suggests that measurers may more clearly differentiate matters of effectiveness and efficiency.

To assess the level of corporate commitment to quality, respondents were asked if four quality-related activities had been undertaken by their organizations. The four activities are: 1) a formal policy statement relating to quality, 2) a formal education program regarding quality cost concepts, 3) a formal company-wide program for quality improvement, and 4) a formal program of recognition for quality improvement.

The data revealed that when the corporate controller's department specifically measures quality costs, a formal education program, a quality improvement program, and a quality recognition program all occur with far greater frequency. The presence of a formal policy statement pertaining to quality does not appear as a relevant factor, primarily because a majority of both measuring and nonmeasuring organizations have established such a statement. These results indicated that controller department participation occurs far more frequently in firms that go beyond stating the importance of quality by allocating resources to a variety of quality-related programs.

Quality Cost Reports

Measuring department representatives were asked to identify the levels of management receiving specific quality cost reports, the frequency of these reports, and the purposes for which these reports are used. Reporting parameters are shown in Table 3. This table reveals that there is a tendency for the frequency of reporting to increase as one moves closer to the source of cost incurrence. This reporting pattern is consistent with traditional responsibility accounting concepts.

The quality literature regularly identifies three reasons for measuring quality costs: 1) for budgeting activities, 2) in performance evaluation, and 3) simply to identify quality costs. Measuring department respondents were asked if quality costs are used for these three purposes.

Of the 29 measuring department representatives questioned, 10 indicated they measure quality costs for budgetary purposes, 23 said they use quality costs in performance evaluation, and 24 indicated they use quality costs to identify the nature and magnitude of these costs.

Measurers also were asked to identify any other reasons that quality costs are specifically measured. Their responses are quite varied, as shown in Table 4, but do reveal that in some organizations, a strong and clear correlation between quality and profitability has been made. In addition, these measurements are shown to support marketing and production activities in more than one of these companies.

TABLE 3/HOW FREQUENTLY REPORTS ARE SENT TO:

CORPORATE MANAGEMENT

Time Interval	Frequency	Percent
Weekly	3	10.3
Monthly	15	51.7
Quarterly	5	17.2
Not Reported	6	20.7
TOTAL	29	100.0

DIVISION MANAGEMENT

Time Interval	Frequency	Percent
Daily	1	3.4
Weekly	10	34.5
Monthly	14	48.3
Quarterly	2	6.9
Unknown	2	6.9
TOTAL	29	100.0

LINE MANAGEMENT

Time Interval	Frequency	Percent
Daily	6	20.7
Weekly	7	24.1
Monthly	10	34.5
Unknown	6	20.7
TOTAL	29	100.0

TABLE 4/WHY MEASURE QUALITY COSTS?

Reason for Measurement	Frequency
Critical in profitability assessment	5
A good marketing tool	4
Identifies quality problems	3
Evaluates overall production costs	1
A component of our value improvement program	1
Used in long-term forecasting	1
An element of productivity improvement	1
An integral part of the planning process	1
Quality is our number one corporate priority	1
Used in setting prices	1
Used in monitoring production & capacity	1
Helps regarding engineering issues	1
Increased government regulation	1
Pressure in industry to lower prices	1
Used in establishing standard costs	1

Top Managements' Commitment

Respondents were presented with four factors that might lead a controller department to specifically measure quality costs. These four factors were ranked in the following order of importance by both the measurer and nonmeasurer subgroups: 1) greater quality commitment from top management, 2) evidence of a stronger relationship between quality improvement and profit, 3) stronger leadership from quality management, and 4) greater controller department resources available for measurement.

In conjunction with the earlier finding that the level of available resources differentiates measuring from nonmeasuring departments, these latter results seem to imply that having adequate resources may be a necessary condition, but it is certainly not a sufficient condition for measurement to occur. In other words, simply increasing the level of available resources will not lead a controller department to measure quality costs. Furthermore, at least at the corporate level, obtaining greater resources is of relatively minor concern once conditions prompting measurement are in place. The minimum conditions appear to be a strong top management commitment to quality along with a belief that quality and profitability go hand-in-hand.

On the whole, management accountants appear to be well aware of the direct relationship between quality and profitability. Assuming that sufficient resources are available for conducting these measurements, that top management is sincerely committed to quality, and that an opportunity exists for interaction with quality function personnel, then the level of controller department participation should increase.

WHY QUALITY COSTS ARE IMPORTANT

(November Issue)

Wayne J. Morse
Professor of Accounting
Chairman of the Faculty of Accounting and Law
Clarkson University

Harold P. Roth
Associate Professor of Accounting
University of Tennessee
Knoxville, Tennessee

Many of the concepts management accountants use on a daily basis were originally developed for the purpose of product costing in a labor-intensive environment. The following trends, however, are having a significant impact on these traditional cost accounting concepts: recent economic growth has been in nonmanufacturing services, the competition for manufactured goods has become global, consumers have become more interested in quality and less interested in brand or national loyalty, and advances in technology have produced a new manufacturing environment. Comparing today's manufacturing costs with those of yesteryear, indirect costs have increased, fixed costs have increased, labor costs have become more fixed, and labor as a percent of total manufacturing costs has decreased.

As a result of these changes, this is an exciting time to be a management accountant. Old ways are being questioned, and a search is under way for new cost models. While the task of improving management accounting theory and practice may be daunting, it does offer us an opportunity to make an important contribution to our profession and society. The problem, however, is where to begin. For many, a starting point is the study of quality costs.

As a part of the search for a new cost model, new terms are appearing in the literature. The development of new terms helps to comprehend new ideas and overcome the connotations of older terms. Among the avant garde, the new emphasis is on "cost management," rather than "cost accounting," and "cost drivers," rather than "allocation bases."

The terms "cost management" and "cost driver" are destined to become an important part of new management accounting cost models. The perceived deficiency of "cost accounting" is its emphasis on determining product costs for external reporting. The new emphasis on "cost management" is intended to imply the development of cost data for the primary purpose of managing costs, rather than for the primary purpose of managing costs. What's more, cost management is concerned with all costs incurred within an organization rather than primarily product costs. It is likely that future managers and accountants will take courses in cost management, with cost accounting being regarded as an important subdiscipline.

An important part of any cost management system will be the search for "cost drivers" — those activities that cause the incurrence of costs. Once identified and their relationship to costs understood, cost management will focus on the management of cost drivers. Within the subdiscipline of cost accounting, cost drivers will serve as a basis for cost allocation. Of course, external reporting requirements may, at times, require the use of something other than a cost driver for cost allocation, or the cost allocations be made even when a cost driver cannot be identified.

During the past few years, management's interest in quality cost information has increased. Although the accountants' attention to quality costs has lagged, quality cost measurement appears to be an idea whose time has come. This is partly due to the significance and potential usefulness of quality cost information. Yet, there seems to be a more fundamental reason for this interest; namely, the development and use of quality cost information serves as a prototype cost management system because of the following:

- Quality cost systems incorporate measures of significant organizational success factors (e.g., defect rates, number of warranty claims, cost of rework).
- Quality cost systems emphasize the development and use of quality cost data for internal rather than external purposes.
- Management is interested in managing quality costs and to do this management needs to understand the drivers of quality costs.
- Quality cost systems must be specifically designed to each organization's operating environment and needs.
- Quality cost systems cut across organizational boundaries (design, purchasing, manufacturing, marketing, administration, etc.) to accumulate costs for a specific purpose.
- Nonaccountants often take the initiative in the development of quality cost measurement.
- Quality cost information should be used in isolation, but only in conjunction with other measures of quality and productivity.
- Measuring quality costs involves adjusting data currently available in an organization's accounting system and developing new, sometimes subjective, data.

Quality costs usually are defined as costs incurred because poor quality may or does exist. Using this definition, quality costs are classified into four categories: prevention, appraisal, internal failure, and external failure. Prevention costs are incurred to ensure that poor quality products are not produced. Appraisal costs are incurred in identifying poor quality products and ensuring that they are not processed further or shipped to customers. Internal failure costs are costs of poor quality materials and products that are identified before they are delivered to customers. External failure costs are costs incurred because poor quality products are delivered to customers.

Many quality cost systems use these broad categories for identifying and classifying costs because they are relevant in many situations. The objective of a quality cost system, however, should be to help improve quality rather than simply measuring and reporting quality cost data. To accomplish this objective, the system should not be limited to four categories if other classifications are more useful. Because a quality cost system is a management tool, it should be designed to provide information that will help management in planning and controlling quality.

Using a quality cost system as a management tool also means that the system should not be constrained by financial accounting concepts. For example, a quality cost system based on the financial accounting principle of objectivity may not provide useful information to management because many quality cost items require estimates and accuracy may not be as important as timeliness. However, one principle that should be considered is consistency because comparisons of costs in different time periods will not be meaningful unless the costs are measured in the same manner.

If a quality cost system is designed as a management tool, it can provide many benefits to an organization including improved quality, higher productivity, and better cost management.

The potential uses of quality cost measurements were considered in a recently published NAA research study we co-authored with Kay M. Poston titled *Measuring, Planning, and Controlling Quality Costs*. One way a quality cost system can assist in cost management is by helping to identify cost drivers. A knowledge of cost drivers can then be used for such purposes as helping to reduce costs and analyzing investments in new technology.

As we noted, cost drivers cause the incurrence of costs. For quality costs, the cost drivers are the activities that result in prevention, appraisal, and failure costs. An identification of these drivers can show which activities should be emphasized to obtain maximum benefits from quality improvement efforts. For example, an analysis of internal failure costs should lead to the identification of materials, parts, or components causing the failures. Once the cause is identified, management can take steps to correct the problem. This should lead to reductions in future failure costs.

Identifying the drivers causing quality costs also might help managers evaluate new manufacturing technology. A criticism of current capital budgeting methods for evaluating new technology is that all benefits (i.e., cost savings) often are not included in the analysis. A better understanding of the costs that will be affected by installing new technology should help improve these decisions. If improved quality results from using new machines, a system identifying current quality costs and their drivers can provide valuable information for evaluating capital investment proposals.

Although quality cost concepts have been known for decades, only recently have accountants been introduced to these concepts. Current quality cost systems often are developed and operated by nonaccounting personnel. The use of these systems by managers for planning and controlling operations provides validity to the criticism that current accounting systems are not providing relevant information for operating in the new manufacturing environment. If management accountants want to overcome this criticism and provide information that is useful for cost management, they need an understanding of quality costs and the relationships between quality, productivity, and the new manufacturing environment.

AUTHOR INDEX

Andre, Yves A. 292
Aubrey, Charles A., II 46

Bhatty, Robert F. 176
Bolton, Caroline J. 435, 474
Bradshaw, Charles W., Jr. 142

Campanella, Jack 102, 460
Caplan, Frank 152
Chauvel, Alain M. 292
Corcoran, Frank J. 102, 449
Crosby, Philip B. 95
Czapor, Edward P. 479

Daisley, Peter A. 216
Dale, B.G. 216
Dawes, Edgar W. 440
Demetriou, James 1

Fox, James G. 409

Gilmore, Harold L. 87
Goeller, William D. 235
Golomski, William S. 279
Grimm, Andrew F. 30, 409
Groocock, J.M. 201
Gryna, Frank M. 268

Hagan, John T. 245
Hollocker, Charles P. 355

Kano, Noriaki 331
Karnebjer, M. 208
Keats, John B. 57
Kildahl, David D. 326
Krishnamoorthi, K.S. 315
Kroehling, Hugh E. 299
Kume, Hitoshi 282

Lenane, Dean-Michael 374
Luega, Miguel Arenas 67

Morse, Wayne J. 125, 134, 400, 490

Mundel, August B. 10
Murthy, Vyasaraj V. 5

Noz, William C., Jr. 38

Ortwein, William J. 225
Oyrzanowski, Bronislaw 81

Petersek, Lawrence R. 299
Plunkett, J.J. 216
Poston, Kay 400

Quinn, Michael P. 176

Raab, Walter F. 479
Reames, J.P. "Jerry" 167
Redding, Bradley R. 38
Roth, Harold P. 134, 490
Ryan, John M. 426

Scanlon, Frank 18
Schneiderman, Arthur M. 394
Schrader, Lawrence J. 384
Siff, Walter 25
Sink, D. Scott 57
Stephenson, Anthony R. 346
Sullivan, Edward 116, 122
Szymanski, Earl T. 257

Tsiakals, Joseph J. 98
Tyson, Thomas N. 420, 483

Ullberg, K. 208

Vocht, Rudiger K. 191

Ware, Paul A. 38
Wayne, James J. 30
Wawak, Tadeusz 81
Winchell, William O. 54, 309, 368, 435, 474
Wirt, Martin W. 454

Zimbler, Debra A. 46

TITLE INDEX

Accounting for Quality Costs — A Critical Component of CIM 400
Accounting for the Real Cost of Quality 374
Application of Economic Principles to Quality 142
Avoidance/Failure Costs Reversal: An Action Plan 409

Business Management and Quality Cost: The Japanese View 282

Consumer Product Quality Control Cost Revisited 87
Costing Quality for Sensitivity Analysis 299
The Cost of Poor Quality 479
Cost of Quality and Productivity Improvement 176
Cost of Quality System — A Management Tool 1
Costs Related to Quality Present Situation in the Federal Republic of Germany (FRG) and Future Aspects 191

Do Controller Departments Measure Quality Costs? 420
Don't Be Defensive About the Cost of Quality 95

The Effects of Regulation on Quality Costs 279
An Engineering Organization's Cost of Quality Program 384

Focusing Quality Costs Using the Basics 309
Finding the Cost of Software Quality 355
Follow-Up and Analysis of the Costs of Rejections 67

Guide for Managing Vendor Quality Costs 368

A High Tech Sight for Cost Targets 426
How to Institutionalize Quality Cost Improvement 201
How to Succeed in Action to Reduce Quality Costs. Theory and Reality. 208

Increased Profits Through Company-Wide Commitment 225

Let's Help Measure and Report Quality Costs 134

Management Team Seeks Quality Improvement from Quality Costs 98
Managing Cost of Quality 5
Managing for Success Through the Quality System 152
Measuring Quality Costs 125

Optimum Quality Costs and Zero Defects: Are They Contradictory Concepts? 394
On the Road to Quality Savings 235

Predicting Quality Cost Changes Using Regression 315
Principles of Quality Costs 102
Profit Improvement Through Scrap Reduction 167

Quality and Economy More Emphasize the Role of Quality on Sales Rather than on Cost 331
Quality and Profitability: Have Controllers Made the Connection? 483
Quality Cost Analysis: Extend the Benefits 474
Quality Costs; Better Prevent than Cure 292
Quality Costs Breakthroughs in U.S. Production 10
Quality Costing in the UK 216
Quality Costs and Profits — Myth or Reality 346
Quality Costs: Current Applications 116
Quality Costs: Current Ideas 122
Quality Costs — Failures and Potentials 268
Quality Costs in a Non-Manufacturing Environment 18
Quality Costs in Staffs — Micro, Macro, or Both? 435
Quality Costs in the Process Industries 25
Quality Costs — New Concepts and Methods 440
Quality Costs: Prevention the Tool for Reduction 449
Quality Costs: Principles and Implementation 460
Quality Costs II: The Economics of Quality Improvement 245
Quality Costs: We Know Where We're Going! Do You? 30
The Quality Manager's Job: Optimize Costs 38
Quality + or − Quality Costs Equals Productivity 46

Reducing Failure Cost and Measuring Improvement 54
Relationship of Financial Information and Quality 257

The Significance of User-Consumer Quality Costs in the World of Unlimited Resources 81
Solving the Quality Cost Equation 326
Spending Prevention Dollars Effectively 454

Using Quality Costs in Productivity Measurement 57

Why Quality Costs Are Important 490

JACK CAMPANELLA is manager of quality engineering at the AIL Division of Eaton Corporation. His 30 years of experience in the quality field have included positions with Fairchild Republic, Hazeltine, General Instruments, and Sperry Gyroscope. A graduate of Brooklyn College, Campanella has presented and published many papers and articles on quality costs and other quality-related subjects. He is a Fellow of the American Society for Quality Control.

Quality Press offers the most complete information available on quality costs:

Quality Costs: Ideas and Applications, Volume 1
ASQC Quality Costs Committee; Andrew F. Grimm, editor
1987. 588 pages. ISBN 0-87389-046-9. Hardcover.
Order H0565

Principles of Quality Costs
ASQC Quality Costs Committee; John T. Hagan, editor
1986. 81 pages. ISBN 0-87389-019-1. Softcover.
Order T166

Guide for Reducing Quality Costs
ASQC Quality Costs Committee
1987. 79 pages. ISBN 0-87389-029-9. Softcover.
Order T106

Poor-Quality Cost
H. James Harrington
1987. 198 pages. ISBN 0-8247-7743-3. Hardcover.
Order H0534

Available Spring 1990:

Principles of Quality Costs, Second Edition
ASQC Quality Costs Committee; Jack Campanella, editor
ISBN 0-87389-084-1
Order H0593

**For more information call 1-800-952-6587 or write
Quality Press, 310 W. Wisconsin Avenue, Milwaukee, WI 53203.**